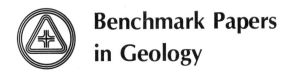

Benchmark Papers
in Geology

Series Editor: Rhodes W. Fairbridge
Columbia University

PUBLISHED VOLUMES

RIVER MORPHOLOGY / *Stanley A. Schumm*
SLOPE MORPHOLOGY / *Stanley A. Schumm and M. Paul Mosley*
SPITS AND BARS / *Maurice L. Schwartz*
BARRIER ISLANDS / *Maurice L. Schwartz*
ENVIRONMENTAL GEOMORPHOLOGY AND LANDSCAPE
    CONSERVATION, VOLUME I: Prior to 1900, VOLUME II: Urban
    Areas, and VOLUME III: Non-Urban Regions / *Donald R. Coates*
TEKTITES / *Virgil E. Barnes and Mildred A. Barnes*
GEOCHRONOLOGY: Radiometric Dating of Rocks and Minerals /
    C. T. Harper
MARINE EVAPORITES: Origin, Diagenesis, and Geochemistry / *Douglas*
    W. Kirkland and Robert Evans
GLACIAL ISOSTASY / *John T. Andrews*
PHILOSOPHY OF GEOHISTORY: 1785–1970 / *Claude C. Albritton, Jr.*
GEOCHEMISTRY OF GERMANIUM / *Jon N. Weber*
GEOCHEMISTRY AND THE ORIGIN OF LIFE / *Keith A. Kvenvolden*
GEOCHEMISTRY OF WATER / *Yasushi Kitano*
GEOCHEMISTRY OF IRON / *Henry Lepp*
SEDIMENTARY ROCKS: Concepts and History / *Albert V. Carozzi*
METAMORPHISM AND PLATE TECTONIC REGIMES / *W. G. Ernst*
SUBDUCTION ZONE METAMORPHISM / *W. G. Ernst*
PLAYAS AND DRIED LAKES: Occurrence and Development / *James T.*
    Neal
GEOCHEMISTRY OF BORON / *C. T. Walker*
GLACIAL DEPOSITS / *Richard P. Goldthwait*
PLANATION SURFACES: Peneplains, Pediplains, and Etchplains / *George*
    F. Adams

Additional volumes in preparation

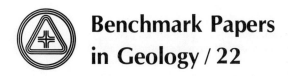

Benchmark Papers
in Geology / 22

A BENCHMARK® Books Series

PLANATION SURFACES
Peneplains, Pediplains, and
Etchplains

Edited by
GEORGE F. ADAMS

Dowden, Hutchinson
& Ross, Inc.
Stroudsburg, Pennsylvania

Distributed by

HALSTED PRESS *A Division of John Wiley & Sons, Inc.*

Copyright © 1975 by **Dowden, Hutchinson & Ross, Inc.**
Benchmark Papers in Geology, Volume 22
Library of Congress Catalog Card Number: 75–12942
ISBN: 0–470–00786–9

77 76 75   1 2 3 4 5
Manufactured in the United States of America.

LIBRARY OF CONGRESS CATALOGING IN PUBLICATION DATA
Main entry under title:

Planation surfaces : peneplains, pediplains, and
  etchplains.

  (Benchmark papers in geology ; 22)
  Bibliography
  Includes indexes.
    1. Peneplains—Addresses, essays, lectures.
2. Plains—Addresses, essays, lectures. I. Adams,
George Finiel, 1907–
GB571.P55   551.4'5   75–12942
ISBN 0–470–00786–9

Exclusive Distributor: **Halsted Press**
A Division of John Wiley & Sons, Inc.

ACKNOWLEDGMENTS AND PERMISSIONS

ACKNOWLEDGMENTS

GEOLOGICAL SOCIETY OF AMERICA—*Bulletin of the Geological Society of America*
Canons of Landscape Evolution
Ep-Archean and Ep-Algonkian Erosion Surfaces, Grand Canyon, Arizona
Geomorphology of the Ruby-East Humboldt Range, Nevada
Pre-Fountain and Recent Weathering on Flagstaff Mountain near Boulder, Colorado

UNIVERSITY OF CALIFORNIA PRESS—*University of California Publications, Geology*
The Epigene Profiles of the Desert

PERMISSIONS

The following papers have been reprinted with the permission of the authors and copyright holders.

AMERICAN JOURNAL OF SCIENCE (YALE UNIVERSITY PRESS)—*American Journal of Science*
Dynamic Equilibrium and the Ozark Land Forms
Geomorphic Development in Humid and Arid Regions: A Synthesis
Interpretation of Erosional Topography in Humid Temperate Regions
Time, Space, and Causality in Geomorphology

AUSTRALIAN AND NEW ZEALAND ASSOCIATION FOR THE ADVANCEMENT OF SCIENCE—*Search*
How Old is Old?
The Nature of Old Landscapes

THE BRITISH CROWN—*Journal of the Geological Society of Australia*
The Age and Geomorphic Correlations of Deep-Weathering Profiles, Silcrete, and Basalt in the Roma-Amby Region, Queensland

CAMBRIDGE UNIVERSITY PRESS—*The Geological Magazine*
The Influence of Climate and Topography in the Formation and Distribution of Products of Weathering

COLUMBIA UNIVERSITY PRESS—*Stream Sculpture on the Atlantic Slope*
Excerpts

GEBRÜDER BORNTRAEGER, BERLIN-STUTTGART—*Zeitschrift für Geomorphologie*
Double Surfaces of Leveling in the Humid Tropics (*English summary*)
The Inselbergs of Uganda
The Weathered Land Surface in Central Australia

Acknowledgments and Permissions

GEOGRAPHISCHES INSTITUT DER UNIVERSITÄT, BONN—*Colloquium Geographicum*
The Relief Types of the Sheetwash Zone of Southern India on the Eastern Slope of the Deccan Highlands Toward Madras (*English summary*)

GEOLOGICAL SOCIETY OF AMERICA
Bulletin of the Geological Society of America
Erosion on Miniature Pediments in Badlands National Monument, South Dakota
The Geological Society of America, Inc. Special Paper 84
Contribution to the Study of the Brazilian Quaternary

GEOLOGICAL SOCIETY OF AUSTRALIA—*Journal of the Geological Society of Australia*
Deep Weathering and Erosion Surfaces in the Daly River Basin, Northern Territory

INSTITUTE OF BRITISH GEOGRAPHERS—*Institute of British Geographers Transactions*
Some Geomorphical Implications of Deep Weathering in the Crystalline
Rocks of Nigeria

UNIVERSITY OF CHICAGO PRESS—*The Journal of Geology*
Rock Floors in Arid and in Humid Climates

SERIES EDITOR'S PREFACE

The philosophy behind the "Benchmark Papers in Geology" is one of collection, sifting, and rediffusion. Scientific literature today is so vast, so dispersed, and, in the case of old papers, so inaccessible for readers not in the immediate neighborhood of major libraries that much valuable information has been ignored by default. It has become just so difficult, or so time consuming, to search out the key papers in any basic area of research that one can hardly blame a busy man for skimping on some of his "homework."

This series of volumes has been devised, therefore, to make a practical contribution to this critical problem. The geologist, perhaps even more than any other scientist, often suffers from twin difficulties—isolation from central library resources and immensely diffused sources of material. New colleges and industrial libraries simply cannot afford to purchase complete runs of all the world's earth science literature. Specialists simply cannot locate reprints or copies of all their principal reference materials. So it is that we are now making a concerted effort to gather into single volumes the critical material needed to reconstruct the background of any and every major topic of our discipline.

We are interpreting "geology" in its broadest sense: the fundamental science of the planet Earth, its materials, its history, and its dynamics. Because of training and experience in "earthy" materials, we also take in astrogeology, the corresponding aspect of the planetary sciences. Besides the classical core disciplines such as mineralogy, petrology, structure, geomorphology, paleontology, and stratigraphy, we embrace the newer fields of geophysics and geochemistry, applied also to oceanography, geochronology, and paleoecology. We recognize the work of the mining geologists, the petroleum geologists, the hydrologists, the engineering and environmental geologists. Each specialist needs his working library. We are endeavoring to make his task a little easier.

Each volume in the series contains an Introduction prepared by a specialist (the volume editor)—a "state of the art" opening or a summary of the object and content of the volume. The articles, usually some thirty to fifty reproduced either in their entirety or in significant extracts, are selected in an attempt to cover the field, from the key papers of the last century to fairly recent work. Where the original works are in foreign languages, we have endeavored to locate or commission translations. Geologists, because of their global subject, are often acutely aware of the oneness of our world. The selections cannot, therefore, be restricted to any one country, and whenever possible an attempt is made to scan the world literature.

To each article, or group of kindred articles, some sort of "highlight commentary" is usually supplied by the volume editor. This commentary should serve to bring that article into historical perspective and to emphasize its particular role in the growth of the field. References, or citations, wherever possible, will be reproduced in their entirety—for by this means the observant reader can assess the background material available to that particular author, or, if he wishes, he, too, can double check the earlier sources.

A "benchmark," in surveyor's terminology, is an established point on the ground, recorded on our maps. It is usually anything that is a vantage point, from a modest hill to a mountain peak. From the historical viewpoint, these benchmarks are the bricks of our scientific edifice.

RHODES W. FAIRBRIDGE

PREFACE

Some forty years ago, the writer was privileged to hear William Morris Davis give a short series of lectures at Columbia University. Davis may properly be called the "Father of Geomorphology" because of his introduction of the trinity of structure, process, stage. This trinity has been canonized over the years, with varying emphasis on the three components. One group seeks to glorify "process" and denigrate "stage" and its penultimate result, the broad planation surface. Other workers have chosen to consider "structure" to be the matter of overriding importance.

Recently there has been a revival of the emphasis on "stage," as attested by various symposia on planation surfaces. At a meeting of the Eastern Section of the Geological Society of America in 1974, in an afternoon devoted to topics related to planation, the importance of the saprolite cover was emphasized. Many important papers are coming from Australia and much work has been done recently in Africa, India, and Europe. The scope of these studies has expanded beyond the original peneplain of Davis to include two other important types, the pediplain and the etchplain.

Finally, we may seek new insights into planation surfaces from concepts of plate tectonics and sea-floor spreading, with their inferred paleolatitudes and paleoconfigurations of land and sea. Specific considerations of such factors are only now beginning to appear in the literature.

GEORGE F. ADAMS

CONTENTS

Contents

PART II: PEDIPLAINS

PART III: ETCHPLAINS

Contents

CONTENTS BY AUTHOR

INTRODUCTION

A benchmark is a point that does not, in itself, give direction to any other point but merely supplies data to aid in its location. Relative locations are derivative and usually show a random pattern. There are, however, broad trends among benchmarks which elongate in preferred directions and with which is associated a certain history of development. I have tried to select those benchmarks which might guide the reader along paths that lead to active areas of research and, thus, may have overlooked some significant works obscured by time or language. It has also been necessary to do some excerpting, but this has been kept to a minimum.

The benchmark articles included herein lead to our present understanding of *planation surfaces*—land surfaces modeled by surface or near-surface wear on a rock mass, where the result of the wear is reasonably plane (planate). Many planate surfaces exist. Among these are smooth valley walls, some retreating cliffs, and other steep and smooth elements of tectonic background. These are more properly considered under the heading "slope morphology" (see Schumm and Mosley, 1973).

Planation surfaces, as used in this book, include the broader, flatter, and more time-dependent surfaces, such as peneplains, pediplains, and etchplains. A perfectly plane and horizontal erosion surface would reveal little of its manner of formation and it is, admittedly, doubtful if such a surface exists. But far from lamenting this probability, we should welcome it, for it is the departure from flatness that may yield clues to origin and permit us to add generic meanings to descriptive terms.

Ideally, imperfect planation surfaces could all be included under one heading—"almost-plains"—used in a purely descriptive sense. Unfortunately, the translation "peneplain" has been virtually preempted by Davis, and the term has come to be associated with a particular origin, not agreed to by all. The German equivalent, "Fastebene," could be used if it were limited to describing the surface; or the English "endplain" could be used to include the idea of having been formed or abandoned near the end of a cycle, without specifying the nature of that cycle. This problem of nomenclature has not yet been solved to everyone's satisfaction. Some of the complications will appear in the subsequent discussion.

With regard to the surfaces themselves, how much departure from a perfect plane can be allowed in "almost a plane" is a common question. In broad regional terms, the curvature of the earth would enter the picture, but this is seldom considered. In a narrower, although still general framework, we may expect a perfect plane to have the slight concavities of stream gradients or that such concave or even flat slopes may be facets of even broader surfaces.

More immediately, we must ask how close to a plane a planation surface should be to be so defined. A rigorous answer would involve the analysis of stated numbers of points, stated distances apart, on profiles spaced at stated intervals considered to represent the surface in question. By regression analysis, lines (straight, concave, or convex) that best fit the points could be determined. At the same time, maximum departures of the original profile from these lines could be agreed upon. This is not common practice, perhaps because of the inherent difficulties and because the allowable departures would be subjective.

In a gross way, it is often stated or implied that a surface cannot be considered to be a planation surface when there is excessive relief. But a mere difference in elevation need not be significant. It is really a question of slope values. A relief of 50 m/km is quite different from 50 m/50 km. In other words, it is more significant to agree on slope limits than simple relief limits. One approach to such an agreement would be through functional limits, associated with the angles of repose of materials undergoing soil creep. Slopes of coarse talus (35° or less) would presumably be well above a planation limit, being common in nonplanated areas. Lower angles of repose are associated with finer or mixed material. If a figure could be established for stable slopes of soil accumulation, an acceptable upper limit for planate slopes might be formed. Progress in assigning slope limits has been made (Büdel, Paper 25) but application to date has been rather limited.

Relief limits of interfluvial elements expressed as slope may be different from those along watercourses. Channel (thalweg) gradients should ordinarily be less along senile streams.

The idea of slope limits is introduced at this point not to settle the question, but to alert the reader to the problem when relief is mentioned in connection with the surfaces being studied in the ensuing papers. For the same reason we note that a planated rock mass may move tectonically so as to alter the original regional slopes and shapes of a surface.

Apart from the factor of maximum allowable slope, there is the *slope-pattern* factor. This matter is more often discussed in terms of "what kind of planation surfaces" rather than "is a planation surface present?" Such matters as concave or convex interfluves figure largely in the discussion.

Slope patterns are presently being considered in connection with the general erosional results not only of planation but also of the overlying (genetically associated) *sedimentary cover* (e.g., weathering and alluvial products associated with the development of the surface). Such cover may be quite distinct from or else merge into overlying sediments of a later date and origin, formed after the planation surface has ceased to be functional (e.g., where the sea advances over the planation surface).

Associated with the weathered products overlying planation surfaces is the problem of the climate within which they were generated: Is it the same or has it changed from time to time? This is but a part of the general problem of time dependency of planation surfaces. We recognize that it may take a long time to produce the surface, and that the functional life of the surface must be fitted into the geologic time scale. In addition to problems associated with time and climate, there are those related to the timing, rates, and amounts of crustal plate motion and seafloor spreading. These may well be key elements to our understanding of planation surfaces.

There is a basic question to be answered in all cases: Is the surface still being shaped by the agents that contributed to its production; that is, is the cycle still in progress, resulting in the forming of an ever-smoother surface, or is the surface now isolated from its original producing regime (or region)? In the latter case we are critically interested in the terminal date, the time when the surface ceased to develop. The sculpturing of an Upper Cretaceous pediplain, for example, must almost certainly have been interrupted by that time. The accuracy of the dating usually depends on the nature of the geologic evidence at hand. The statement of this upper limit, however closely it may be established to the cessation of formation, should be accompanied by an estimate of a lower starting limit, which must postdate the youngest rock eroded. The resulting time bracket is useful for correlation purposes, keeping in mind that planation surfaces, like lithologies, need not be everywhere synchronous. Furthermore, this time span provides a basis for estimating tectonic stability, erosion–denudation rates, sedimentation rates in adjacent basins, and related sedimentary problems.

Time considerations for planation surfaces have generated the need for a status classification that will indicate in a general way the vicissitudes to which the surface has been subjected subsequent to its terminal date. It is hoped that the following expressions and their definitions will fulfill this need:

1. *Active surface:* one still being shaped with respect to its base level (i.e., it is still functional). All such surfaces are Holocene.

2. *Dormant surface:* one whose active shaping has ceased temporarily (perhaps because of climatic change) and is expected to function again in the near geologic future. Primarily Pleistocene to Holocene.

3. *Exotic surface:* one formed under climatic conditions that no longer exist at the site and are not firmly expected to return. No age limit.

4. *Defunct surface:* one that has been removed from the active zone of planation by uplift, depression, or climatic change. Probably Tertiary to Holocene.

5. *Buried surface:* one covered (and preserved) by sediments not related to its shaping or sealed by lava flows. Points on this paleogeomorphic surface may be discovered by drilling. Lines may be exposed in valleys, caves, mines, and so on. Geophysical methods may indicate their existence. Range from Precambrian to Tertiary.

6. *Exhumed or fossil surface:* buried surface partly exposed by removal of nongenetic cover such as later sediments or flows. These surfaces may be subject to dissection from the time of exhumation onward. Range from Precambrian to Tertiary.

IDENTIFICATION OF DISSECTED SURFACES

In dealing with planation surfaces, we must first agree to the existence of a particular example, with its dimensional characteristics, and then proceed to the problem of its origin. Whatever the origin, surfaces that are no longer active and which have become defunct by uplift are subject to dissection, which may be active or subdued. The question then arises: How much of the original surface must remain before identity is lost? Where headward erosion in the present cycle has started in a belt near the sea and left inland portions almost intact, there is not much problem (Paper 31), but it may be a question of sets of valleys pervading the whole area of the surface in question. If these pervasive streams have cut narrow valleys and left broad interstream areas of accordant heights and erosional origin, the former existence of a broad planation surface may be reasonably inferred. Difficulties arise when divide areas are either rounded or sharp-crested, because there may be introduced in a variety of ways not involving a planation interval. Such evidence is equivocal (Tarr, Paper 2).

In areas of complex structure and rock type, the factor of *differential rock resistance* should be considered in more detail. For instance, in a series of folds involving the gamut of sedimentary types, long-continued erosion may produce a high ridge-and-valley relief in which some rock types are typically found in ridges, others in valleys. Proponents of former peneplains would say that the diversity of rock types means that rocks of different resistance to erosion had been worn down to a common level, followed by uplift and dissection. Opponents would say that on a priori grounds (lacking other means) the fact that rocks of different types maintain similar ridge height is purely an indication of equal resistance to erosion; therefore, they may have simply worn down from a higher and possibly more uneven surface to the present ridge and valley stage (Tarr, Paper 2; Hack, Paper 8).

To resolve this dilemma, let us assume, for example, that this ridge-and-valley relief becomes worn down to a fairly even planation surface graded to a contemporary base level. If we now judge the resistances of the same rock masses, we find that since they are all on the same level, again on a priori grounds, they must have the same resistance. Stated somewhat differently, we may say that the equally resistant formations that formerly underlay the ridges now show the same resistance as the more or less equally resistant valley rock types. Very simply, at this stage, we can postulate that rock weathering resistance has become a time function, overriding all others.

If we could now demonstrate the inheritance of a quasi-uniform rock resistance from a planation surface as shown in the present landscape, we might have a powerful tool of recognition. For example, a shale formation responsible for the existence of a broad lowland might also be found in a monoclinal ridge alongside a resistant sandstone, which is found by itself elsewhere in other ridges. We may accept the frequent occurrence of sandstone in present ridges as indicating high resistance in the present cycle, just as we accept the generally low altitude of the shale area as indicating lower resistance in the present cycle. Where the two formations retain the same level, however, we may explain it as a remnant condition of a former cycle.

To cite a case in point, the Shawangunk Mountain ridge of New York State (in one area) is underlain by the Silurian Shawangunk Conglomerate and the Ordovician Hudson River Shale, which stand shoulder to shoulder, forming a broad ridge summit where westerly-dipping beds (10° to 15°) are truncated. Areas immediately north or south have only the Shawangunk Conglomerate at the ridge summit. To the southeast a broad lowland is developed on the Hudson River Shale some 200 m lower. (The special circumstances that permitted the shale to remain at the ridge crest cannot be discussed here.)

This illustration is introduced to lend substance to the idea that rock resistance may be a function of time and to the possible bearing that this may have on the discussion of erosional remnants (rather than weathered products) as evidence for former planation.

The most recent work on planation surfaces has leaned more toward the study of ancient pedogenic products (paleosols) of the sediments produced by weathering in conjunction with the planation cycle itself, of the later history of these sediments, and discussion of what would be revealed if all such sediments were eroded away.

Before proceeding to this aspect, we must consider a residual erosional matter dealing with the quantitative aspects of the denudation process. On theoretical and empirical grounds, some students have rejected the Davisian peneplain (Davis, Paper 3). They say that, in effect, Davis deduced the peneplain without knowing much about the behavior of streams nor the general operation of open systems. Nor are they happy about the lack of numerical restraints in the description of peneplains. These may be valid criticisms, but they imply that time has little importance in the scheme of things and that in an open system a dynamic equilibrium is rather quickly reached after the initiating disturbance. In erosional systems, the steady state of this equilibrium is associated with high-relief ridge and ravine topography, or what Davis called *mature dissection*. This is cited as the present normal topographic condition and that peneplains, lacking modern examples, are rather improbable.

DEVELOPMENT OF CONCEPTS

A. C. Ramsey, writing on the "Denudation of South Wales" in 1846, is credited with the early recognition of a planation surface attributed to marine agencies. In describing the action of the sea on the land he wrote:

> The line of greatest waste on any coast is the average level of the breakers. The effect of such waste is obviously to wear back the coast, the line of denudation being a level corresponding to the average height of the sea. Taking *unlimited* time into account, we can conceive that any extent of land might be so destroyed, for though shingle beaches and other coast formations will apparently for almost any ordinary length of time protect the country from the further encroachments of the sea, yet the protections to such beaches being at last themselves worn away, the beaches are in the course of time destroyed, and so, unless checked by elevation, the waste being carried on forever, a whole country might gradually disappear.
>
> If to this be added an *exceedingly slow depression* of the land and sea bottom, the wasting process would be materially assisted by this depression, bringing the land more uniformly within the reach of the sea, and enabling the latter more rapidly to overcome obstacles to further encroachments, created by itself in the shape of beaches. By further gradually increasing the depth of the surrounding water, ample space would also be afforded for the outspreading of the denuded matter. To such combined forces, namely, the *shaving away* of coasts by the sea, and the spreading abroad of the material thus obtained, the great *plain* of shallow soundings which generally surrounds our islands is in all probability attributable.
>
> The power of running water has also considerably modified the surface, but the part it has played is trifling compared with the effects that have sometimes been attributed to its agency. . . . In the larger valleys, where the streams are sluggish, instead of assisting in further excavations, the general tendency is often rather to fill up the hollow with alluvial accumulations, and so help to smooth the original irregularities of the surface.

A useful review of the work of early observers, which traces the change of emphasis from marine to continental agencies, is given by Davis (1896).

By 1889, Davis had accepted the efficacy of fluvial erosion and coined the word *peneplain* in the following statement: "I believe that time enough has already been allowed and that strong Jurassic topography was really worn out somewhere in Cretaceous time, when all this part of the country was reduced by erosion to a featureless plain at a low level, a 'peneplain' as I would call it." No mention is made in this definition of cyclic stages leading to the development of the surface.

Influenced by the work of early explorers in the western United States (Gilbert, 1880a, b; Dutton, 1882), Davis deduced a series of changes through which the landscape would pass, eventually developing a planation surface, the peneplain. This was synthesized in his "Geographical Cycle" (Paper 1). Since the time for perfect planation would be practically infinite, Davis agreed that in finite time only the penultimate stage in the cycle could ever be reached. This would yield a rolling surface of low relief and low slopes, the peneplain, with the double implication that it was almost a plain and almost at the end of a cycle.

By the turn of the century, Davis had transferred his attention to the New England–Appalachian region. Here he believed he saw remnant peneplains in the dissected uplands as shown by the accordant and sometimes broad hilltops where diverse rock types had been truncated. But where large areas of the supposed peneplain had been removed by dissection (in a later cycle, according to Davis), the concept was difficult to sustain. Application of the peneplain idea to New England was quickly challenged (Tarr, Paper 2; and others) on the grounds that it was possible to produce accordant divide areas by headward stream erosion and differential weathering without passing through the peneplain stage. The Davis reply (Paper 3) was apparently successful, as judged by the reporting of peneplains in various regions in North America, both east and west, and in other parts of the world. Davis himself had to advise caution in the naming of new surfaces. During this period (1900 to 1940) many workers investigated the correlative ideas of intersecting peneplains with their unburied, buried, and exhumed elements. Papers by Henry Sharp (Paper 5), Robert Sharp (Paper 6), and Douglas Johnson (Paper 7) are representative. Other workers, such as Reginald Daly (1905), while conceding peneplanation, would limit its application to lower altitudes. In Alpine-type mountains, for example, accordance of summits was thought to be either inherent in mountain building or achieved shortly thereafter by a minimum of denudation. Elsewhere, on lower terrains within relatively easy reach of the sea, the peneplain was considered to have been present but obliterated by a flight of marine terraces (Barrell, Paper 4).

Early in this period, Davis (1922) made a strong plea for constructive criticism of his cycle concept rather than the use of piecemeal evidence to denigrate it. In 1924, Walther Penck responded by pointing out that an erosion cycle could be initiated not only by rapid uplift followed by stillstand, as pictured by Davis; but also by a slow accelerating uplift with peak velocities variously distributed in time and space. According to Penck, this alternative condition would produce a set of landscape features quite different from those visualized by Davis.

Indeed, Penck appeared to claim that all planation features that he had studied conformed to this second model. The question is discussed at length in Baulig (1939) and Simons (1962), in the latter's critique of the English translation of Penck's book (1924).

PEDIMENTS AND PEDIPLAINS

While the peneplain idea was being exploited in the western United States, workers in that region rediscovered a different type of planation surface, the *pediment,* already defined in its modern sense in 1897. As the name suggests, it is a feature usually found at the foot of a mountain. It has been likened to the pediment of a Greek temple, but with somewhat confusing results. An architectural pediment is the part of the facade of the building that rises above the entablature. It consists of three cornices arranged in a low isoceles triangle filled with a vertical wall, the tympanum. A horizontal cornice tops the entablature while twin cornices are simply gabled roof ends extending beyond the tympanum. The word "pediment" presumably derives from the horizontal cornice, which may serve as a common pedestal for freestanding statuary.

The term "pediment" had been previously applied by Gilbert (1880b) to *alluvial fans* flanking the mountains near Lake Bonneville's shorelines as an expression of the work of streams. in a semiarid region. Since the fans had considerable slope, were constructional in nature, and had no protruding hills, when seen in profile at a cross-cutting stream, they would indeed resemble a pediment. While working in the Henry Mountains, Gilbert (1880a) observed "hills of planation" whose planate summits extended from points close to the mountain base toward adjacent basins and were now left high and dry by later stream dissection. Although these hills seen in profile also resembled an architectural pediment, Gilbert did not apply the name in this case, possibly because he had already used the term in a depositional sense.

Aside from the fundamental difficulty that an architectural pediment is made by construction rather than destruction, we are left in confusion as to which part of the pediment should be applied in the geomorphic sense. The closest architectural analogy (but not the accepted geomorphic usage) was applied by Clarence Dutton in 1882. Dutton, in descending the Toroweap Valley toward the Grand Canyon of the Colorado, described as a pediment the *facade* (i.e., the valley wall) in terms of the flat cornice-like surface of a resistant horizontal bed with its statue-like remnants of overlying beds. He completed the picture with the gable-like shape of the more massive background remnants of higher topographic and stratigraphic levels. Although these features are architecturally analogous, they are not today considered to be pediments geomorphically, but only stripped rock benches.

Much more could be said about the difficulties of the architectural analogy from which it would appear that it should not be rigorously applied and were better abandoned. It is recommended here that the analogy be dropped in favor of the simple idea that a pediment is a smooth plano-concave-upward erosion surface, typically sloping down from the foot of a highland area and graded to either a local or more general base level. It is an element of a piedmont belt, which may include depositional elements such as fans and playas. The pediment, as defined, excludes such depositional components, although an alluvial cover is frequently present.

An alternative name for a pediment, introduced by French writers, is *glacis*. Architecturally, the glacis is the slope graded to a low angle that forms an apron around a castle wall; this produces a surface sloping away from the castle and makes it difficult for an attacking force to use cover or concealment. The term has a certain functional as well as geometric similarity to the natural surface that makes it more acceptable in these terms than is the word "pediment." At present "glacis" is largely restricted to certain European literature.

Unlike the case of peneplains, almost everyone agrees that pediments exist, that they have recognizable features, and that they are still being formed in many places. Unlike peneplains, they are, or can be, very small and local features that may coalesce over broad regions to form *pediplains*. The relation between peneplains and pediplains has been the subject of much discussion (Davis, Paper 14; King, Paper 18; Holmes, Paper 19).

ETCHPLAINS

Until recently, the study of planation surfaces has emphasized the discovery and altimetric description of the eroded rock surface itself. A variety of survey and statistical procedures has been established in the last decade or so for this purpose. More recently, there has been a growing tendency to deal with planation surfaces in terms of the materials that may overlie the rock surface and are associated with the planation process. On the one hand, we may investigate the alluvial cover of a contemporaneously eroded surface, thereby seeking clues to its origin. On the other hand, we can study the weathered products (*saprolite*) that have accumulated as the rock surface beneath has been lowered. Especially where the saprolite attains considerable thickness, it may afford data for determining the origin of the rock surface.

Etching effects on the weathering front as well as in the saprolite have produced a third important class of planation surfaces, *etchplains*. The name carries implications that the etching process or *etchplanation* may be related to the art of etching. Having seen the difficulty of applying architectural terms to a geomorphic surface, we may be prepared for similar difficulties in the geomorphic applications of the art of etching. In hard-ground etching, for example, a metal sheet is covered with an inert substance such as wax. The artist removes part of this, exposing the metal to acid attack. Lines and areas are registered as hollows, which later hold ink or tints. After removal of the inert substance, the metal, with its grooves and hollows, becomes the *etch plate*.

In geomorphology we picture an entire rock surface (the metal sheet) exposed to soil acids, which produce their own residual coating as they "etch" the underlying rock. Occasional washing away of the weathered products to expose the unweathered rock would produce the *etchplain* analogous to the etch plate.

This fairly straightforward analogy becomes complicated as the weathering becomes deeper. Unless the saprolite is periodically washed away, the bedrock surface (the *weathering front*) becomes lower and lower as weathering products continue to be developed above it. Eventually, an equilibrium may be achieved between rate of weathering and rate of soil wash. Then the surface is related to an etching process but the bedrock is not exposed. Should this soil surface or the one developed within the soil be called an etchplain? The short pioneer paper by Wayland (Paper 23) should be read carefully with this question in mind. The term "etchplain," in its present connotation, is now used to include both possibilities. Two additional matters should be noted: (1) that etchplains may be expected to develop from other planation surfaces and not from dissected uplands of high relief or from high youthful mountains, and (2) deposits from deep weathering often develop hard layers near the surface of the saprolite, called *duricrusts* by Woolnough (Paper 21), which may be subject to later etching or dissection effects themselves. However, once formed on a surface of low relief, the duricrust tends to be a semipermanent "armor-plating," and further erosion becomes inhibited (i.e., the rock surface becomes dormant).

CONCLUSION

Although peneplains, pediplains, and etchplains are distinct surfaces with different modes of origin, they are subject to many types of overlap, at least in terms of recognition. For example, a peneplain may be transformed into an etchplain by removal of all or part of the contemporary cover. Thus, perhaps some surfaces identified as peneplains should more properly be called etchplains. Also, during removal of the deep weathering cover, pediments may be cut into it which could become merged into a recognizable pediplain. The rather abrupt slope changes of such a pediplain may be softened by subsequent weathering or creep (including periglacial solifluction), developing extensive colluvium to form an apparent peneplain. Recognition features of all three types of planation surfaces are needed to distinguish one from the other.

A specialized type of planation restricted to Quaternary periglacial landscapes of moderate relief is known as *altiplanation*. Above the permanent snowline, cryoclastic processes are dominant and tend to operate more rapidly than the humid processes of lower altitudes. A more or less planar surface or terrace is likely to develop around such elevated regions. They are often in multiples and represent no more than relics of multiple snowlines, but may be confused, especially with pediments, by the unwary.

The term *panplanation* was proposed by Crickmay (1933) for a process that emphasized lateral planation by rivers, which developed a more or less uniform surface of overlapping rock-cut straths as lateral divides were cut through. Crickmay could cite no well-developed examples of panplains but thought that pediments and pediplains approached them closely.

These special cases may lead to some confusion in the field, but are, by definition, clearly limited to special settings. They will not be treated further in this volume.

REFERENCES

Baulig, H., 1939. Sur les gradins du piedmont. J. Géomorph., 2, p. 281–304.

Crickmay, C. H., 1933. The later stages in the cycle of erosion. Geol. Mag., 70, p. 337–347.

Daly, R. A., 1905. The accordance of summit levels among Alpine mountains. J. Geol., 13, p. 105–125.

————, 1889. Topographic development of the Triassic formations of the Connecticut Valley. Amer. J. Sci., 3rd ser., 37, p. 430.

————, 1896. Plains of marine and sub-aerial denudation. Bull. Geol. Soc. America, 7, p. 377–398; also in Geographic essays, p. 323–349, Ginn and Co., Boston.

————, 1922. Peneplains and the geographic cycle. Bull. Geol. Soc. America, 33, p. 587–598.

Dutton, C. E., 1882. Tertiary history of the Grand Canyon District. U.S. Geol. Surv. Monograph 2.

Gilbert, G. K., 1880a. Report on the geology of the Henry Mountains, 2nd ed. U.S. Geogr. Geol. Surv. Rocky Mt. Region.

————, 1880b. Contributions to the history of Lake Bonneville. U.S. Geol. Surv. 2nd Ann. Rept., p. 167–200.

Penck, W., 1924. *Die morphologische Analyse*. Engelhorn, Stuttgart, 283 p.

Ramsey, A. C., 1846. The denudation of South Wales. Mem. Geol. Surv. Gt. Brit., 1, p. 197–235, H.M. Stationery Office, London.

Schumm, S. A., and M. P. Mosley, 1973. *Slope Morphology*. Dowden, Hutchinson & Ross, Inc., Stroudsburg, Pa.

Simons, M., 1962. The morphological analysis of landforms: a new review of Penck's work. Inst. Brit. Geogr. Trans. Papers, 31, p. 1–14.

Part I
PENEPLAINS

Editor's Comments
on Paper 1

1   DAVIS
The Geographical Cycle

In one of his fuller statements on his cycle of erosion, Davis brings insight to the idea of progressive landscape development (youth–maturity–old age) and to the general processes of stream erosion, transportation, and deposition. Clearly written and easy to comprehend, this paper should be read carefully and perhaps returned to for critical review after reading later papers.

1

Reprinted from *Geogr. J.*, **14(A)**, 481–484, 485–503 (1899)

THE GEOGRAPHICAL CYCLE.

By WILLIAM M. DAVIS, Professor of Physical Geography in Harvar
University.

THE GENETIC CLASSIFICATION OF LAND-FORMS.—All the varied forms of
the lands are dependent upon—or, as the mathematician would say, are
functions of—three variable quantities, which may be called structure,
process, and time. In the beginning, when the forces of deformation
and uplift determine the structure and attitude of a region, the form of
its surface is in sympathy with its internal arrangement, and its height
depends on the amount of uplift that it has suffered. If its rocks were
unchangeable under the attack of external processes, its surface would
remain unaltered until the forces of deformation and uplift acted again ;

and in this case structure would be alone in control of form. But no rocks are unchangeable ; even the most resistant yield under the attack of the atmosphere, and their waste creeps and washes downhill as long as any hills remain ; hence all forms, however high and however resistant, must be laid low, and thus destructive process gains rank equal to that of structure in determining the shape of a land-mass. Process cannot, however, complete its work instantly, and the amount of change from initial form is therefore a function of time. Time thus completes the trio of geographical controls, and is, of the three, the one of most frequent application and of most practical value in geographical description.

Structure is the foundation of all geographical classifications in which the trio of controls is recognized. The Alleghany plateau is a unit, a "region," because all through its great extent it is composed of widespread horizontal rock-layers. The Swiss Jura and the Pennsylvanian Appalachians are units, for they consist of corrugated strata. The Laurentian highlands of Canada are essentially a unit, for they consist of greatly disturbed crystalline rocks. These geographical units have, however, no such simplicity as mathematical units ; each one has a certain variety. The strata of plateaus are not strictly horizontal, for they slant or roll gently, now this way, now that. The corrugations of the Jura or of the Appalachians are not all alike ; they might, indeed, be more truly described as all different, yet they preserve their essential features with much constancy. The disordered rocks of the Laurentian highlands have so excessively complicated a structure as at present to defy description, unless item by item ; yet, in spite of the free variations from a single structural pattern, it is legitimate and useful to look in a broad way at such a region, and to regard it as a structural unit. The forces by which structures and attitudes have been determined do not come within the scope of geographical inquiry, but the structures acquired by the action of these forces serve as the essential basis for the genetic classification of geographical forms. For the purpose of this article, it will suffice to recognize two great structural groups : first, the group of horizontal structures, including plains, plateaus, and their derivatives, for which no single name has been suggested ; second, the group of disordered structures, including mountains and their derivatives, likewise without a single name. The second group may be more elaborately subdivided than the first.

The destructive processes are of great variety—the chemical action of air and water, and the mechanical action of wind, heat, and cold, of rain and snow, rivers and glaciers, waves and currents. But as most of the land surface of the Earth is acted on chiefly by weather changes and running water, these will be treated as forming a normal group of destructive processes ; while the wind of arid deserts and the ice of frigid deserts will be considered as climatic modifications of the norm, and set apart for particular discussion; and a special chapter will be needed to explain the action of waves and currents on the shore-lines at the edge of the lands. The various processes by which destructive work is done are in their turn geographical features, and many of them are well recognized as such, as rivers, falls, and glaciers; but they are too commonly considered by geographers apart from the work that they do, this phase of their study being, for some unsatisfactory reason, given over to physical geology. There should be no such separation of agency and work in physical geography, although it is profitable to give separate consideration to the active agent and to the inert mass on which it works.

TIME AS AN ELEMENT IN GEOGRAPHICAL TERMINOLOGY.—The amount of change caused by destructive processes increases with the passage of time, but neither the amount nor the rate of change is a simple function of time. The amount of change is limited, in the first place, by the altitude of a region above the sea; for, however long the time, the normal destructive forces cannot wear a land surface below this ultimate baselevel of their action; and glacial and marine forces cannot wear down a land-mass indefinitely beneath sea-level. The rate of change under normal processes, which alone will be considered for the present, is at the very first relatively moderate; it then advances rather rapidly to a maximum, and next slowly decreases to an indefinitely postponed minimum.

Evidently a longer period must be required for the complete denudation of a resistant than of a weak land-mass, but no measure in terms of years or centuries can now be given to the period needed for the effective wearing down of highlands to featureless lowlands. All historic time is hardly more than a negligible fraction of so vast a duration. The best that can be done at present is to give a convenient name to this unmeasured part of eternity, and for this purpose nothing seems more appropriate than a "*geographical cycle.*" When it is possible to establish a ratio between geographical and geological units, there will probably be found an approach to equality between the duration of an average cycle and that of Cretaceous or Tertiary time, as has been indicated by the studies of several geomorphologists.

"THEORETICAL" GEOGRAPHY.—It is evident that a scheme of geographical classification that is founded on structure, process, and time, must be deductive in a high degree. This is intentionally and avowedly the case in the present instance. As a consequence, the scheme gains a very "theoretical" flavour that is not relished by some geographers, whose work implies that geography, unlike all other sciences, should be developed by the use of only certain ones of the mental faculties, chiefly observation, description, and generalization. But nothing seems to me clearer than that geography has already suffered too long from

19

the disuse of imagination, invention, deduction, and the various other mental faculties that contribute towards the attainment of a well-tested explanation. It is like walking on one foot, or looking with one eye, to exclude from geography the "theoretical" half of the brain-power, which other sciences call upon as well as the "practical" half. Indeed, it is only as a result of misunderstanding that an antipathy is implied between theory and practice, for in geography, as in all sound scientific work, the two advance most amiably and effectively together. Surely the fullest development of geography will not be reached until all the mental faculties that are in any way pertinent to its cultivation are well trained and exercised in geographical investigation.

[*Editor's Note:* Several paragraphs are omitted dealing with exhortations to "geographers" to adopt the explanatory method.]

THE IDEAL GEOGRAPHICAL CYCLE.—The sequence in the developmental changes of land-forms is, in its own way, as systematic as the sequence of changes found in the more evident development of organic forms. Indeed, it is chiefly for this reason that the study of the origin of land-forms—or geomorphogeny, as some call it—becomes a practical aid, helpful to the geographer at every turn. This will be made clearer by the specific consideration of an ideal case, and here a graphic form of expression will be found of assistance.

The base-line, $a\omega$, of Fig. 1 represents the passage of time, while verticals above the base-line measure altitude above sea-level. At the epoch 1, let a region of whatever structure and form be uplifted, B representing the average altitude of its higher parts, and A that of its lower parts; thus AB measuring its average initial relief. The surface rocks are attacked by the weather. Rain falls on the weathered surface, and washes some of the loosened waste down the initial slopes to the trough-lines where two converging slopes meet; there the streams are formed, flowing in directions consequent upon the descent of the trough-lines. The machinery of the destructive processes is thus put in motion, and the destructive development of the region is begun. The larger rivers, whose channels initially had an altitude, A, quickly deepen their valleys, and at the epoch 2 have reduced their main channels to a moderate altitude, represented by C. The higher parts of the inter-stream uplands, acted on only by the weather without the concentration of water in streams, waste away much more slowly, and at epoch 2 are reduced in height only to D. The relief of the surface has thus been increased from AB to CD. The main rivers then deepen their channels very slowly for the rest of their life, as shown by the curve CEGJ; and the wasting of the uplands, much dissected by branch streams, comes to be more rapid than the deepening of the main valleys, as shown by comparing the curves DFHK and CEGJ. The period 3–4 is the time of the most rapid consumption of the uplands, and

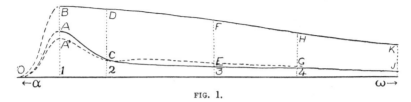

FIG. 1.

thus stands in strong contrast with the period 1–2, when there was the most rapid deepening of the main valleys. In the earlier period, the relief was rapidly increasing in value, as steep-sided valleys were cut beneath the initial troughs. Through the period 2–3 the maximum value of relief is reached, and the variety of form is greatly increased by the headward growth of side valleys. During the period 3–4 relief is decreasing faster than at any other time, and the slope of the valley sides is becoming much gentler than before; but these changes advance much more slowly than those of the first period. From epoch 4 onward the remaining relief is gradually reduced to smaller and smaller measures, and the slopes become fainter and fainter, so that some time after the latest stage of the diagram the region is only a rolling lowland, whatever may have been its original height. So slowly do the later changes advance, that the reduction of the reduced relief JK to half of its value might well require as much time as all that which has already elapsed; and from the gentle slopes that would then remain, the further removal of waste must indeed be exceedingly slow. The frequency of torrential floods and of landslides in young and in mature mountains, in contrast to the quiescence of the sluggish streams and the slow movement of the soil on lowlands of denudation, suffices to show that rate of denudation is a matter of strictly geographical as well as of geological interest.

It follows from this brief analysis that a geographical cycle may be subdivided into parts of unequal duration, each one of which will be characterized by the strength and variety of relief, and by the rate of change, as well as by the amount of change that has been accomplished since the initiation of the cycle. There will be a brief youth of rapidly increasing relief, a maturity of strongest relief and greatest variety of form, a transition period of most rapidly yet slowly decreasing relief, and an indefinitely long old age of faint relief, on which further changes are exceedingly slow. There are, of course, no breaks between these subdivisions or stages; each one merges into its successor, yet each one is in the main distinctly characterized by features found at no other time.

THE DEVELOPMENT OF CONSEQUENT STREAMS.—The preceding section gives only the barest outline of the systematic sequence of changes that run their course through a geographical cycle. The outline must be at once gone over, in order to fill in the more important details. In the first place, it should not be implied, as was done in Fig. 1, that the forces of uplift or deformation act so rapidly that no destructive changes occur during their operation. A more probable relation at the opening of a cycle of change places the beginning of uplift at O (Fig. 1), and its end at 1. The divergence of the curves OB and OA then implies that certain parts of the disturbed region were uplifted more than others, and that, from a surface of no relief at sea-level at epoch O, an upland having AB relief would be produced at epoch 1. But even during uplift, the streams that gather in the troughs as soon as they are defined do some work, and hence young valleys are already incised in the trough-bottoms when epoch 1 is reached, as shown by the curve OA'. The uplands also waste more or less during the period of disturbance, and hence no absolutely unchanged initial surface should be found, even for some time anterior to epoch 1. Instead of looking for initial divides separating initial slopes that descend to initial troughs followed by initial streams, such as were implied in Fig. 1 at the epoch of instantaneous uplift, we must always expect to find some greater or less advance in the sequence of developmental changes, even in the youngest known land-forms. "Initial" is therefore a term adapted to ideal rather than to actual cases, in treating which the term "sequential" and its derivatives will be found more appropriate. All the changes which directly follow the guidance of the ideal initial forms may be called consequent; thus a young form would possess consequent divides, separating consequent slopes which descend to consequent valleys; the initial troughs being changed to consequent valleys in so far as their form is modified by the action of the consequent drainage.

22

THE GRADE OF VALLEY FLOORS.—The larger rivers soon—in terms of the cycle—deepen their main valleys, so that their channels are but little above the baselevel of the region ; but the valley floor cannot be reduced to the absolute baselevel, because the river must slope down to its mouth at the sea-shore.   The altitude of any point on a well-matured valley floor must therefore depend on river-slope and distance from mouth. Distance from mouth may here be treated as a constant, although a fuller statement would consider its increase in consequence of delta-growth.   River-slope cannot be less, as engineers know very well, than a certain minimum that is determined by volume and by quantity and texture of detritus or load.   Volume may be temporarily taken as a constant, although it may easily be shown to suffer important changes during the progress of a normal cycle.   Load is small at the beginning, and rapidly increases in quantity and coarseness during youth, when the region is entrenched by steep-sided valleys ; it continues to increase in quantity, but probably not in coarseness, during early maturity, when ramifying valleys are growing by headward erosion, and are thus in-creasing the area of wasting slopes ; but after full maturity, load continually decreases in quantity and in coarseness of texture ; and during old age, the small load that is carried must be of very fine texture or else must go off in solution.   Let us now consider how the minimum slope of a main river will be determined.

In order to free the problem from unnecessary complications, let it be supposed that the young consequent rivers have at first slopes that are steep enough to make them all more than competent to carry the load that is washed into them from the wasting surface on either side, and hence competent to entrench themselves beneath the floor of the initial troughs,—this being the condition tacitly postulated in Fig. 1, although it evidently departs from those cases in which deformation produces basins where lakes must form and where deposition (negative denudation) must take place, and also from those cases in which a main-trough stream of moderate slope is, even in its youth, over-supplied with detritus by active side streams that descend steep and long wasting surfaces ; but all these more involved cases may be set aside for the present.

If a young consequent river be followed from end to end, it may be imagined as everywhere deepening its valley, unless at the very mouth. Valley-deepening will go on most rapidly at some point, probably nearer head than mouth.   Above this point the river will find its slope increased ; below, decreased.   Let the part up-stream from the point of most rapid deepening be called the headwaters ; and the part down-stream, the lower course or trunk.   In consequence of the changes thus systematically brought about, the lower course of the river will find its slope and velocity decreasing, and its load increasing ; that is, its ability to do work is becoming less, while the work that it has to do is becoming greater. The original excess of ability over work will thus in time be corrected, and when an equality of these two quantities is brought about, the river is *graded*, this being a simple form of expression, suggested by Gilbert, to replace the more cumbersome phrases that are required by the use of " profile of equilibrium " of French engineers. When the graded condition is reached, alteration of slope can take place only as volume and load change their relation ; and changes of this kind are very slow.

In a land-mass of homogeneous texture, the graded condition of a river would be (in such cases as are above considered) first attained at the mouth, and would then advance retrogressively up-stream. When the trunk streams are graded, early maturity is reached ; when the smaller headwaters and side streams are also graded, maturity is far advanced ; and when even the wet-weather rills are graded, old age is attained. In a land-mass of heterogeneous texture, the rivers will be divided into sections by the belts of weaker and stronger rocks that they traverse ; each section of weaker rocks will in due time be graded with reference to the section of harder rock next down-stream, and thus the river will come to consist of alternating quiet reaches and hurried falls or rapids. The less resistant of the harder rocks will be slowly worn down to grade with respect to the more resistant ones that are further down stream ; thus the rapids will decrease in number, and only those on the very strongest rocks will long survive. Even these must vanish in time, and the graded condition will then be extended from mouth to head. The slope that is adopted when grade is assumed varies inversely with the volume ; hence rivers retain steep headwaters long after their lower course is worn down almost level ; but in old age, even the head-waters must have a gentle declivity and moderate velocity, free from all torrential features. The so-called " normal river," with torrential head-waters and well-graded middle and lower course, is therefore simply a maturely developed river. A young river may normally have falls even in its lower course, and an old river must be free from rapid movement even near its head.

If an initial consequent stream is for any reason incompetent to carry away the load that is washed into it, it cannot degrade its channel, but must aggrade instead (to use an excellent term suggested by Salisbury). Such a river then lays down the coarser part of the offered load, thus forming a broadening flood-land, building up its valley floor, and steepening its slope until it gains sufficient velocity to do the required work. In this case the graded condition is reached by filling up the initial trough instead of by cutting it down. Where basins occur, consequent lakes rise in them to the level of the outlet at the lowest point of the rim. As the outlet is cut down, it forms a sinking local baselevel with respect to which the basin is aggraded ; and as the lake is thus destroyed, it forms a sinking baselevel with respect to which the tributary streams grade their valleys ; but, as in

24

the case of falls and rapids, the local baselevels of outlet and lake are temporary, and lose their control when the main drainage lines are graded with respect to absolute baselevel in early or late maturity.

THE DEVELOPMENT OF RIVER BRANCHES. — Several classes of side streams may be recognized. Some of them are defined by slight initial depressions in the side slopes of the main river-troughs: these form lateral or secondary consequents, branching from a main consequent; they generally run in the direction of the dip of the strata. Others are developed by headward erosion under the guidance of weak substructures that have been laid bare on the valley walls of the consequent streams: they follow the strike of the strata, and are entirely regardless of the form of the initial land surface; they may be called subsequent, this term having been used by Jukes in describing the development of such streams. Still others grow here and there, to all appearance by accident, seemingly independent of systematic guidance; they are common in horizontal or massive structures. While waiting to learn just what their control may be, their independence of apparent control may be indicated by calling them "insequent." Additional classes of streams are well known, but cannot be described here for lack of space.

RELATION OF RIVER ABILITY AND LOAD.—As the dissection of a landmass proceeds with the fuller development of its consequent, subsequent, and insequent streams, the area of steep valley sides greatly increases from youth into early and full maturity. The waste that is delivered by the side branches to the main stream comes chiefly from the valley sides, and hence its quantity increases with the increase of strong dissection, reaching a maximum when the formation of new branch streams ceases, or when the decrease in the slope of the wasting valley sides comes to balance their increase of area. It is interesting to note in this connection the consequences that follow from two contrasted relations of the date for the maximum discharge of waste and of that for the grading of the trunk streams. If the first is not later than the second, the graded rivers will slowly assume gentler slopes as their load lessens; but as the change in the discharge of waste is almost infinitesimal compared to the amount discharged at any one time, the rivers will essentially preserve their graded condition in spite of the minute excess of ability over work. On the other hand, if the maximum of load is not reached until after the first attainment of the graded condition by the trunk rivers, then the valley floors will be aggraded by the deposition of a part of the increasing load, and thus a steeper slope and a greater velocity will be gained whereby the remainder of the increase can be borne along. The bottom of the V-shaped valley, previously carved, is thus slowly filled with a gravelly flood-plain, which continues to rise until the epoch of the maximum load is reached, after which the slow degradation above stated is entered upon. Early maturity may therefore witness a slight shallowing of the main valleys, instead of the slight deepening (indicated by the dotted line CE in Fig. 1); but late maturity and all old age will be normally occupied by the slow continuation of valley erosion that was so vigorously begun during youth.

THE DEVELOPMENT OF DIVIDES.—There is no more beautiful process to be found in the systematic advance of a geographical cycle than the definition, subdivision, and rearrangement of the divides (water-partings) by which the major and minor drainage basins are separated. The forces of crustal upheaval and deformation act in a much broader way than the processes of land-sculpture; hence at the opening of a cycle one would expect to find a moderate number of large river-basins, somewhat indefinitely separated on the flat crests of broad swells or arches of land surface, or occasionally more sharply limited by the raised edge of faulted blocks. The action of the lateral consequent streams alone would, during youth and early maturity, sharpen all the vague initial divides into well-defined consequent divides, and the further action of insequent and subsequent streams would split up many consequent drainage slopes into subordinate drainage basins, separated by subdivides either insequent or subsequent. Just as the subsequent valleys are eroded by their gnawing streams along weak structural belts, so the subsequent divides or ridges stand up where maintained by strong structural belts. However imperfect the division of drainage areas and the discharge of rainfall may have been in early youth, both are well developed by the time full maturity is reached. Indeed, the more prompt discharge of rainfall that may be expected to result from the development of an elaborate system of subdivides and of slopes from divides to streams should cause an increased percentage of run-off; and it is possible that the increase of river-volume thus brought about from youth to maturity may more or less fully counteract the tendency of increase in river load to cause aggradation. But, on the other hand, as soon as the uplands begin to lose height, the rainfall must decrease; for it is well known that the obstruction to wind-movement caused by highlands is an effective cause of precipitation. While it is a gross exaggeration to maintain that the quaternary Alpine glaciers caused their own destruction by reducing the height of the mountains on which their snows were gathered, it is perfectly logical to deduce a decrease of precipitation as an accompaniment of loss of height from the youth to the old age of a land-mass. Thus many factors must be considered before the life-history of a river can be fully analyzed.

The growth of subsequent streams and drainage areas must be at the expense of the original consequent streams and consequent drainage areas. All changes of this kind are promoted by the occurrence of inclined instead of horizontal rock-layers, and hence are of common occurrence in mountainous regions, but rare in strictly horizontal plains. The changes are also favoured by the occurrence of strong contrasts in the resistance of adjacent strata. In consequence of the migration of divides thus caused, many streams come to follow valleys that are worn down along belts of weak strata, while the divides come to occupy the ridges that stand up along the belts of stronger strata; in other words, the simple consequent drainage of youth is modified by the development of subsequent drainage lines, so as to bring about an *increasing adjustment of streams to structures*, than which nothing is more characteristic of the mature stage of the geographical cycle. Not only so: adjustments of this kind form one of the strongest, even if one of the latest, proofs of the erosion of valleys by the streams that occupy them, and of the long continued action in the past of the slow processes of weathering and washing that are in operation to-day.

There is nothing more significant of the advance in geographical development than the changes thus brought about. The processes here involved are too complicated to be now presented in detail, but they may be briefly illustrated by taking the drainage of a denuded arch, suggested

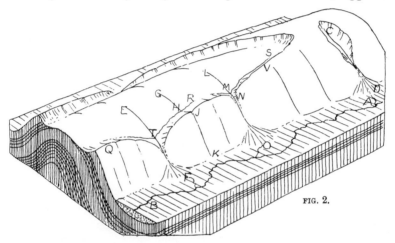

FIG. 2.

by the Jura mountains, as a type example. AB, Fig. 2, is a main longitudinal consequent stream following a trough whose floor has been somewhat aggraded by the waste actively supplied by the lateral consequents, CD, LO, EF, etc. At an earlier stage of denudation, before the hard outer layer was worn away from the crown of the mountain arch, all the lateral consequents headed at the line of the mountain crest. But, guided by a weak under-stratum, subsequent streams, TR, MS, have been developed as the branches of certain lateral consequents, EF, LO, and thus the hard outer layer has been undermined and partly removed, and many small lateral consequents have been beheaded. To-day, many of the laterals, like JK, have their source on the crest of the lateral ridge VJQ, and the headwaters, such as GH, that once belonged to them, are now diverted by the subsequent streams to swell the volume of the more

27

successful laterals, like EF.   Similar changes having taken place on the further slope of the mountain arch, we now find the original consequent divide of the arch-crest supplemented by the subsequent divides formed by the lateral ridges. A number of short streams, like JH, belonging to a class not mentioned above, run down the inner face of the lateral ridges to a subsequent stream, RT. These short streams have a direction opposite to that of the original consequents, and may therefore be called obsequents. As denudation progresses, the edge of the lateral ridge will be worn further from the arch-crest; in other words, the subsequent divide will migrate towards the main valley, and thus a greater length will be gained by the diverted consequent headwaters, GH, and a greater volume by the subsequents, SM and RT.   During these changes the inequality that must naturally prevail between adjacent successful consequents, EF and LO, will eventually allow the subsequent branch, RT, of the larger consequent, EF, to capture the headwaters, LM and SM, of the smaller consequent, LO. In late maturity the headwaters of so many lateral consequents may be diverted to swell the volume of EF, that the main longitudinal consequent above the point F may be reduced to relatively small volume.

THE DEVELOPMENT OF RIVER MEANDERS.—It has been thus far implied that rivers cut their channels vertically downward, but this is far from being the whole truth.   Every turn in the course of a young consequent stream causes the stronger currents to press toward the outer bank, and each irregular, or, perhaps, subangular bend is thus rounded out to a comparatively smooth curve.   The river therefore tends to depart from its irregular initial path (background block of Fig. 3) towards a serpentine course, in which it swings to right and left over a broader belt than at first.   As the river cuts downwards and outwards at the same time, the valley-slopes become unsymmetrical (middle block of Fig. 3), being steeper on the side toward which the current is urged by centrifugal force.   The steeper valley side thus gains the form of a half-amphitheatre, into which the gentler sloping side enters as a spur of the opposite uplands.   When the graded condition is attained by the stream, downward cutting practically ceases, but outward cutting continues; a normal flood-plain is then formed as the channel is withdrawn from the gently sloping side of the valley (foreground block of Fig. 3). Flood-plains of this kind are easily distinguished in their early stages from those already mentioned (formed by aggrading the flat courses of incompetent young rivers, or by aggrading the graded valleys of over-loaded rivers in early maturity) ; for these occur in detached lunate areas, first on one side, then on the other side of the stream, and always systematically placed at the foot of the gentler sloping spurs. But, as time passes, the river impinges on the up-stream side, and withdraws from the down-stream side of every spur, and thus the spurs are gradually consumed ; they are first sharpened, so as better to observe their name; they are next reduced to short cusps; then they are worn back to blunt salients; and finally, they are entirely consumed, and the river wanders freely on its open flood-plain, occasionally swinging against the valley side, now here, now there. By this time the curves of youth are changed into systematic meanders, of radius appropriate to river volume; and, for all the rest of an undisturbed life, the river persists in the habit of serpentine flow. The less the slope of the flood-plain becomes in advancing old age, the larger the arc of each meander,

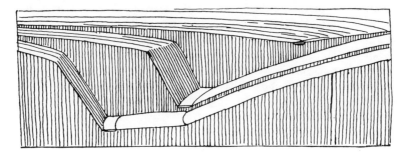

FIG. 3.

and hence the longer the course of the river from any point to its mouth. Increase of length from this cause must tend to diminish fall, and thus to render the river less competent than it was before; and the result of this tendency will be to retard the already slow process by which a gently sloping flood-plain is degraded so as to approach coincidence with a level surface; but it is not likely that old rivers often remain undisturbed long enough for the full realization of these theoretical conditions.

The migration of divides must now and then result in a sudden increase in the volume of one river and in a correspondingly sudden decrease of another. After such changes, accommodation to the changed volume must be made in the meanders of each river affected. The one that is increased will call for enlarged dimensions; it will usually adopt a gentler slope, thus terracing its flood-plain, and demand a greater freedom of swinging, thus widening its valley. The one that is decreased will have to be satisfied with smaller dimensions; it will wander aimlessly in relatively minute meanders on its flood-plain, and from increase of length, as well as from loss of volume, it will become incompetent to transport the load brought in by the side streams, and thus its flood-plain must be aggraded. There are beautiful examples known of both these peculiar conditions.

THE DEVELOPMENT OF GRADED VALLEY SIDES. —When the migration of divides ceases in late maturity, and the valley floors of the adjusted streams are well graded, even far toward the headwaters, there is still to be completed another and perhaps even more remarkable sequence of systematic changes than any yet described : this is the development of graded waste slopes on the valley sides. It is briefly stated that valleys are eroded by their rivers ; yet there is a vast amount of work performed in the erosion of valleys in which rivers have no part. It is true that rivers deepen the valleys in the youth, and widen the valley floors during the maturity and old age of a cycle, and that they carry to the sea the waste denuded from the land ; it is this work of transportation to the sea that is peculiarly the function of rivers ; but the material to be transported is supplied chiefly by the action of the weather on the steeper consequent slopes and on the valley sides. The transportation of the weathered material from its source to the stream in the valley bottom is the work of various slow-acting processes, such as the surface wash of rain, the action of ground water, changes of temperature, freezing and thawing, chemical disintegration and hydration, the growth of plant-roots, the activities of burrowing animals. All these cause the weathered rock waste to wash and creep slowly downhill, and in the motion thus ensuing there is much that is analogous to the flow of a river. Indeed, when considered in a very broad and general way, a river is seen to be a moving mixture of water and waste in variable proportions, but mostly water ; while a creeping sheet of hillside waste is a moving mixture of waste and water in variable proportions, but mostly waste. Although the river and the hillside waste-sheet do not resemble each other at first sight, they are only the extreme members of a continuous series ; and when this generalization is appreciated, one may fairly extend the " river " all over its basin, and up to its very divides. Ordinarily treated, the river is like the veins of a leaf ; broadly viewed, it is like the entire leaf. The verity of this comparison may be more fully accepted when the analogy, indeed, the homology, of waste-sheets and water-streams is set forth.

In the first place, a waste-sheet moves fastest at the surface and slowest at the bottom, like a water-stream. A graded waste-sheet may be defined in the very terms applicable to a graded water-stream ; it is one in which the ability of the transporting forces to do work is equal to the work that they have to do. This is the condition that obtains on those evenly slanting, waste-covered mountain-sides which have been reduced to a slope that engineers call " the angle of repose," because of the apparently stationary condition of the creeping waste, but that should be called, from the physiographic standpoint, " the angle of first-developed grade." The rocky cliffs and ledges that often surmount graded slopes are not yet graded ; waste is removed from them faster than it is supplied by local weathering and by creeping from still higher slopes, and hence the cliffs and ledges are left almost bare ;

they correspond to falls and rapids in water-streams, where the current is so rapid that its cross-section is much reduced. A hollow on an initial slope will be filled to the angle of grade by waste from above; the waste will accumulate until it reaches the lowest point on the rim of the hollow, and then outflow of waste will balance inflow; and here is the evident homologue of a lake.

In the second place, it will be understood, from what has already been said, that rivers normally grade their valleys retrogressively from the mouth headwards, and that small side streams may not be graded till long after the trunk river is graded. So with waste-sheets; they normally begin to establish a graded condition at their base, and then extend it up the slope of the valley side whose waste they "drain." When rock-masses of various resistance are exposed on the valley side, each one of the weaker is graded with reference to the stronger one next downhill; and the less resistant of the stronger ones are graded with reference to the more resistant (or with reference to the base of the hill): this is perfectly comparable to the development of graded stretches and to the extinction of falls and rapids in rivers. Ledges remain ungraded on ridge-crests and on the convex front of hill spurs long after the graded condition is reached in the channels of wet-weather streams in the ravines between the spurs; this corresponds nicely with the slower attainment of grade in small side streams than in large trunk rivers. But as late maturity passes into old age, even the ledges on ridge-crests and spur-fronts disappear, all being concealed in a universal sheet of slowly creeping waste. From any point on such a surface a graded slope leads the waste down to the streams. At any point the agencies of removal are just able to cope with the waste that is there weathered *plus* that which comes from further uphill. This wonderful condition is reached in certain well-denuded mountains, now subdued from their mature vigour to the rounded profiles of incipient old age. When the full meaning of their graded form is apprehended, it constitutes one of the strongest possible arguments for the sculpture of the lands by the slow processes of weathering, long continued. To look upon a landscape of this kind without any recognition of the labour expended in producing it, or of the extraordinary adjustments of streams to structures, and of waste to weather, is like visiting Rome in the ignorant belief that the Romans of to-day have had no ancestors.

Just as graded rivers slowly degrade their courses after the period of maximum load is past, so graded waste-sheets adopt gentler and gentler slopes when the upper ledges are consumed and coarse waste is no longer plentifully shed to the valley sides below. A changing adjustment of a most delicate kind is here discovered. When the graded slopes are first developed, they are steep, and the waste that covers them is coarse and of moderate thickness; here the strong agencies of removal have all they can do to dispose of the plentiful supply of coarse waste from the strong ledges above, and the no less plentiful supply of waste that is weathered from the weaker rocks beneath the thin cover of detritus. In a more advanced stage of the cycle, the graded slopes are moderate, and the waste that covers them is of finer texture and greater depth than before; here the weakened agencies of removal are favoured by the slower weathering of the rocks beneath the thickened waste cover, and by the greater refinement (reduction to finer texture) of the loose waste during its slow journey. In old age, when all the slopes are very gentle, the agencies of waste-removal must everywhere be weak, and their equality with the processes of waste-supply can be maintained only by the reduction of the latter to very low values. The waste-sheet then assumes a great thickness—even 50 or 100 feet—so that the progress of weathering is almost *nil;* at the same time, the surface waste is reduced to extremely fine texture, so that some of its particles may be moved even on faint slopes. Hence the occurrence of deep soils is an essential feature of old age, just as the occurrence of bare ledges is of youth. The relationships here obtaining are as significant as those which led Playfair to his famous statement concerning the origin of valleys by the rivers that drain them.

OLD AGE.—Maturity is past and old age is fully entered upon when the hilltops and the hillsides, as well as the valley floors, are graded. No new features are now developed, and those that have been earlier developed are weakened or even lost. The search for weak structures and the establishment of valleys along them has already been thoroughly carried out; now the larger streams meander freely in open valleys and begin to wander away from the adjustments of maturity. The active streams of the time of greatest relief now lose their headmost branches, for the rainfall is lessened by the destruction of the highlands, and the run-off of the rain water is retarded by the flat slopes and deep soils. The landscape is slowly tamed from its earlier strength, and presents only a succession of gently rolling swells alternating with shallow valleys, a surface everywhere open to occupation. As time passes, the relief becomes less and less; whatever the uplifts of youth, whatever the disorder and hardness of the rocks, an almost featureless plain (a pene-plain) showing little sympathy with structure, and controlled only by a close approach to baselevel, must characterize the penultimate stage of the uninterrupted cycle; and the ultimate stage would be a plain without relief.

Some observers have doubted whether even the penultimate stage of a cycle is ever reached, so frequently do movements in the Earth's crust cause changes in its position with respect to baselevel. But, on the other hand, there are certain regions of greatly disordered structure, whose small relief and deep soils cannot be explained without supposing them to have, in effect, passed through all the stages above described—and doubtless many more, if the whole truth were told—before reaching the

penultimate, whose features they verify. In spite of the great disturbances that such regions have suffered in past geological periods, they
have afterwards stood still so long, so patiently, as to be worn down to
pene-plains over large areas, only here and there showing residual reliefs
where the most resistant rocks still stand up above the general level.
Thus verification is found for the penultimate as well as for many earlier
stages of the ideal cycle. Indeed, although the scheme of the cycle is
here presented only in theoretical form, the progress of developmental
changes through the cycle has been tested over and over again for many
structures and for various stages; and on recognizing the numerous
accordances that are discovered when the consequences of theory are
confronted with the facts of observation, one must feel a growing belief
in the verity and value of the theory that leads to results so satisfactory.

It is necessary to repeat what has already been said as to the
practical application of the principles of the geographical cycle. Its
value to the geographer is not simply in giving explanation to landforms; its greater value is in enabling him to see what he looks at, and
to say what he sees. His standards of comparison, by which the unknown are likened to the known, are greatly increased over the short
list included in the terminology of his school-days. Significant features
are consciously sought for; exploration becomes more systematic and
less haphazard. "A hilly region" of the unprepared traveller becomes
(if such it really be) "a maturely dissected upland" in the language of
the better prepared traveller; and the reader of travels at home gains
greatly by the change. "A hilly region" brings no definite picture
before the mental eyes. "A maturely dissected upland" suggests a
systematic association of well-defined features; all the streams at grade,
except the small headwaters; the larger rivers already meandering on
flood-plained valley floors; the upper branches ramifying among spurs
and hills, whose flanks show a good beginning of graded slopes; the
most resistant rocks still cropping out in ungraded ledges, whose
arrangement suggests the structure of the region. The practical value
of this kind of theoretical study seems to me so great that, among
various lines of work that may be encouraged by the Councils of the
great Geographical Societies, I believe there is none that would bring
larger reward than the encouragement of some such method as is here
outlined for the systematic investigation of land-forms.

Some geographers urge that it is dangerous to use the theoretical
or explanatory terminology involved in the practical application of the
principles of the geographical cycle; mistakes may be made, and harm
would thus be done. There are various sufficient answers to this objection. A very practical answer is that suggested by Penck, to the effect
that a threefold terminology should be devised—one set of terms being
purely empirical, as "high," "low," "cliff," "gorge," "lake," "island;"
another set being based on structural relations, as "monoclinal ridge,"

" transverse valley," " lava-capped mesa ; " and the third being reserved
for explanatory relations, as " mature dissection," " adjusted drainage,"
" graded slopes." Another answer is that the explanatory terminology
is not really a novelty, but only an attempt to give a complete and systematic expansion to a rather timid beginning already made ; a sand-dune
is not simply a hillock of sand, but a hillock heaped by the wind ; a delta
is not simply a plain at a river mouth, but a plain formed by river action ; a
volcano is not simply a mountain of somewhat conical form, but a mountain
formed by eruption. It is chiefly a matter of experience and tempera-
ment where a geographer ceases to apply terms of this kind. But little
more than half a century ago, the erosion of valleys by rivers was either
doubted or not thought of by the practical geographer ; to-day, the
mature adjustment of rivers to structures is in the same position ; and
here is the third, and to my mind the most important, answer to those
conservatives who would maintain an empirical position for geography,
instead of pressing forward toward the rational and explanatory geography of the future. It cannot be doubted, in view of what has already
been learned to-day, that an essentially explanatory treatment must in
the next century be generally adopted in all branches of geographical
study ; it is full time that an energetic beginning should be made
towards so desirable an end.

INTERRUPTIONS OF THE IDEAL CYCLE.—One of the first objections that
might be raised against a terminology based on the sequence of changes
through the ideal uninterrupted cycle, is that such a terminology can
have little practical application on an Earth whose crust has the habit
of rising and sinking frequently during the passage of geological time.
To this it may be answered, that if the scheme of the geographical
cycle were so rigid as to be incapable of accommodating itself to the
actual condition of the Earth's crust, it would certainly have to be
abandoned as a theoretical abstraction ; but such is by no means the
case. Having traced the normal sequence of events through an ideal
cycle, our next duty is to consider the effects of any and all kinds of
movements of the land-mass with respect to its baselevel. Such movements must be imagined as small or great, simple or complex, rare or
frequent, gradual or rapid, early or late. Whatever their character,
they will be called "interruptions," because they determine a more or
less complete break in processes previously in operation, by beginning
a new series of processes with respect to the new baselevel. Whenever
interruptions occur, the pre-existent conditions that they interrupt can
be understood only after having analyzed them in accordance with the
principles of the cycle, and herein lies one of the most practical applications of what at first seems remotely theoretical. A land-mass,
uplifted to a greater altitude than it had before, is at once more intensely
attacked by the denuding processes in the new cycle thus initiated ; but
the forms on which the new attack is made can only be understood by

considering what had been accomplished in the preceding cycle previous to its interruption. It will be possible here to consider only one or two specific examples from among the multitude of interruptions that may be imagined.

Let it be supposed that a maturely dissected land-mass is evenly uplifted 500 feet above its former position. All the graded streams are hereby revived to new activities, and proceed to entrench their valley floors in order to develop graded courses with respect to the new baselevel. The larger streams first show the effect of the change; the smaller streams follow suit as rapidly as possible. Falls reappear for a time in the river-channels, and then are again worn away. Adjustments of streams to structures are carried further in the second effort of the new cycle than was possible in the single effort of the previous cycle. Graded hillsides are undercut; the waste washes and creeps down from them, leaving a long even slope of bare rock; the rocky slope is hacked into an uneven face by the weather, until at last a new graded slope is developed. Cliffs that had been extinguished on graded hillsides in the previous cycle are thus for a time brought to life again, like the falls in the rivers, only to disappear in the late maturity of the new cycle.

The combination of topographic features belonging to two cycles may be called "composite topography," and many examples could be cited in illustration of this interesting association. In every case, description is made concise and effective by employing a terminology derived from the scheme of the cycle. For example, Normandy is an uplifted peneplain, hardly yet in the mature stage of its new cycle; thus stated, explanation is concisely given to the meandering course of the rather narrow valley of the Seine, for this river has carried forward into the early stages of the new cycle the habit of swinging in strong meanders that it had learned in the later stages of the former cycle.

If the uplift of a dissected region be accompanied by a gentle tilting, then all the water-streams and waste-streams whose slope is increased will be revived to new activity; while all those whose slope is decreased will become less active. The divides will migrate into the basins of the less active streams, and the revived streams will gain length and drainage area. If the uplift be in the form of an arch, some of the weaker streams whose course is across the axis of the arch may be, as it were, "broken in half;" a reversed direction of flow may be thus given to one part of the broken stream; but the stronger rivers may still persevere across the rising arch in spite of its uplift, cutting down their channels fast enough to maintain their direction of flow unchanged: and such rivers are known as "antecedent."

The changes introduced by an interruption involving depression are easily deduced. Among their most interesting features is the invasion of the lower valley floors by the sea, thus "drowning" the valleys to a certain depth, and converting them into bays. Movements that tend to produce trough-like depressions across the course of a river usually give birth to a lake of water or waste in the depressed part of the river valley. In mountain ranges frequent and various interruptions occur during the long period of deformation; the Alps show so many recent interruptions that a student there would find little use for the ideal cycle; but in mountain regions of ancient deformation, the disturbing forces seem to have become almost extinct, and there the ideal cycle is almost realized. Central France gives good illustration of this principle. It is manifest that one might imagine an endless number of possible combinations among the several factors of structure, stage of development at time of interruption, character of interruption, and time since interruption; but space cannot be here given to their further consideration.

ACCIDENTAL DEPARTURES FROM THE IDEAL CYCLE.—Besides the interruptions that involve movements of a land-mass with respect to baselevel, there are two other classes of departure from the normal or ideal cycle that do not necessarily involve any such movements: these are changes of climate and volcanic eruptions, both of which occur so arbitrarily as to place and time that they may be called "accidents." Changes of climate may vary from the normal towards the frigid or the arid, each change causing significant departures from normal geographical development. If a reverse change of climate brings back more normal conditions, the effects of the abnormal "accident" may last for some small part of a cycle's duration before they are obliterated. It is here that features of glacial origin belong, so common in north-western Europe and north-eastern America. Judging by the present analysis of glacial and interglacial epochs during quaternary time, or of humid and arid epochs in the Great Salt Lake region, it must be concluded that accidental changes may occur over and over again within a single cycle.

In brief illustration of the combined interruptions and accidents, it may be said that southern New England is an old mountain region, which had been reduced to a pretty good peneplain when further denudation was interrupted by a slanting uplift, with gentle descent to the south-east; that in the cycle thus introduced the tilted peneplain was denuded to a sub-mature or late mature stage (according to the strength or weakness of its rocks); and that the maturely dissected region was then glaciated and slightly depressed so recently that little change has happened since. An instructive picture of the region may be conceived from this brief description.

Many volcanic eruptions produce forms so large that they deserve to be treated as new structural regions; but when viewed in a more general way, a great number of eruptions, if not the greater number, produce forms of small dimensions compared to those of the structures on which

they are superposed : the volcanoes of central France are good instances
of this relation.   Thus considered, volcanoes and lava-flows are so
arbitrarily placed in time and space that their classification under the
head of "accidents" is warranted.   Still further ground for this classification is found when the effects of a volcanic eruption on the pre-
existent processes of land-sculpture are examined.   A valley may be
blockaded by a growing cone and its lava-flows ; lakes may form in the
up-stream portion of such a valley, even if it be mature or old.   If the
blockade be low, the lake will overflow to one side of the barrier, and
thus the river will be locally displaced from its former course, however
well adjusted to a weak structure that course may have been.   If the
blockade be higher than some points on the headwater divides, the lake
will overflow " backwards," and the upper part of the river system will
become tributary to an adjacent system.   The river must cut a gorge
across the divide, however hard the rocks are there ; thus systematic
adjustments to structure are seriously interfered with, and accidental
relations are introduced.   The form of the volcanic cone and the sprawl-
ing flow of its lava-streams are quite out of accord with the forms that
characterize the surrounding region.   The cone arbitrarily forms a
mountain, even though the subjacent rocks may be weak ; the lava-flows
aggrade valleys that should be degraded.   During the dissection of the
cone, a process that is systematic enough if considered for itself alone,
a radial arrangement of spurs and ravines will be developed ; in long
future time the streams of such ravines may cut down through the
volcanic structures, and thus superpose themselves most curiously on
the underlying structures.   The lava-flows, being usually more resistant
than the rocks of the district that they invade, gain a local relief as the
adjoining surface is lowered by denudation ; thus an inversion of topography is brought about, and a "table-mountain" comes to stand where
formerly there had been the valley that guided the original course of
the lava-flow.   The table-mountain may be quite isolated from its
volcanic source, where the cone is by this time reduced to a knob or
"butte."   But although these various considerations seem to me to
warrant the classification of volcanic forms as "accidental," in contrast
to the systematic forms with which they are usually associated, great
importance should not be attached to this method of arrangement ; it
should be given up as soon as a more truthful or more convenient classification is introduced.

THE FORMS ASSUMED BY LAND WASTE.—An extension of the subject
treated in the section on Graded Valley Sides, would lead to a general
discussion of the forms assumed by the waste of the land on the way to
the sea ; one of the most interesting and profitable topics for investigation that has come under my notice.   Geographers are well accustomed
to giving due consideration to the forms assumed by the water-drainage
of the land on the way to the sea, and a good terminology is already in

use for naming them ; but much less consideration is given to the forms assumed by the waste that slowly moves from the land to the sea. They are seldom presented in their true relations ; many of them have no generally accepted names—for example, the long slopes of waste that reach forward from the mountains into the desert basins of Persia ; forms as common as alluvial fans are unmentioned in all but the most recent school-books ; and such features as till plains, moraines, and drumlins are usually given over to the geologist, as if the geographer had nothing to do with them ! There can be no question of the great importance of waste-forms to the geographer, but it is not possible here to enter into their consideration. Suffice it to say that waste-forms constitute a geographical group which, like water-forms, stand quite apart from such groups as mountains and plateaus. The latter are forms of structure, and should be classified according to the arrangement of their rocks, and to their age or stage of development. The former are forms of process, and should be classified according to the processes involved, and to the stage that they have reached. The application of this general principle gives much assistance in the description of actual landscapes.

[*Editor's Note:* The paper ends with a brief discussion of shoreline development not pertinent to planation surfaces.]

Editor's Comments
on Papers 2 and 3

2 **TARR**
   The Peneplain

3 **DAVIS**
   The Peneplain

EARLY CHALLENGES TO THE EXISTENCE OF PENEPLAINS

Tarr (Paper 2), like Davis, was disturbed by the proliferation of peneplains of doubtful authenticity, and decided to question the idea of recognizing peneplains solely from accordant hill summits in erosional topography. Tarr ended by virtually denying the existence of peneplains, citing as one reason the alleged absence of active examples. The reader may wish to refer ahead to Paper 31, in which the authors describe what they accept as a virtually undissected peneplain in Australia.

Students of plate tectonics and their eustatic effects (Larsen and Pitman, 1972) suggest that the last peak episode of continental splitting and sea-floor spreading occurred in the Upper Cretaceous, about 80 million years ago. At that time, the world sea level was at least 500 m above that of today. Since then, episodic and discontinuous emergence of all continents has been postulated; thus a universal mechanism has become available that would lead to progressive lowering of base levels and thus to systematic dissection and destruction of Mesozoic peneplains and pediplains. The remnants that have been preserved are those far from the sea or partly preserved by duricrusts (R. W. Fairbridge, personal communication). The general absence of contemporary peneplains is a point well taken by Tarr, but completely understandable today, whereas in Tarr's frame of reference 75 years ago an explanation was not available.

Other objections to the reality of peneplains are answered by

Davis in Paper 3. Davis brings to bear the weight of authority, the advantage of wide experience, and the use of adroit reasoning to refute alternative explanations of the landscape offered by Tarr in Paper 2. Let the reader weigh the arguments for himself.

Davis's reply to Tarr is organized around the three allegations regarding peneplains: (1) they are unreal (i.e., they do not exist); (2) they are improbable (i.e., they are not likely to be formed); and (3) they are unnecessary (i.e., landscapes can be explained adequately without them). Many subsequent workers have found Davis's refutation of these allegations unconvincing.

REFERENCE

Larsen, R. L., and Pitman, W. C., 1972. World wide correlation of Mesozoic magnetic anomalies and its implications. Bull. Geol. Soc. Amer., 83, p. 3645–3662.

2

Reprinted from *Amer. Geol.*, **21**, 351–370 (June 1898)

THE PENEPLAIN.

By R. S. TARR, Ithaca, N. Y.

Reasons for the Paper:—Five years ago doubts concerning the value of the evidence of peneplains, which had previously come to my mind, were distinctly strengthened as the result of study in the highlands of New Jersey. I was, therefore, led to call in question the explanation which even then was being quite generally accepted. So widespread was the adoption of the idea that I hesitated to publish these doubts and decided to give the matter more thought. After two or three years a paper was prepared stating my objections, and sent to Prof. W. M. Davis for his consideration. It did not convince him, nor did his comments upon the paper convince me that the objections were unsound.

Nevertheless, the failure to convince Prof. Davis induced me to give the question still more study, with the result that the longer I have thought upon the matter, and the more extended my field observations have become, the stronger grows the conviction that the peneplain explanation is in error. Therefore, notwithstanding the fact that nearly all American geologists have adopted the peneplain explanation, and that no one has publicly questioned it, I have decided at last to state my objections in print.*

I have been led to this decision in the belief that it should be done. Every month, and sometimes oftener, one finds a statement concerning a newly discovered peneplain. They are being found nearly everywhere. Indeed they are announced upon the most meagre evidence, and oftentimes with no statement of evidence whatever. Frequently a new peneplain is mentioned as one might state the discovery of a delta

---

*I am indebted to Prof. J. C. Branner, Prof. I. C. Russell, Prof. A. C. Gill and Mr. J. B. Woodworth for kindly reading and commenting upon this paper.

or a fossil vertebrate; and not only are single peneplains found in a given district, but oftentimes several of different ages.

It is perfectly certain that many of the so-called peneplains have been announced without any semblance of proof. But it is not against these that I write, for the author of the peneplain idea has himself urged more careful study before the announcement of a peneplain,—advice which has not been generally followed. The literature of geology is becoming overburdened with peneplains, and the geological history and geography of the past are often interpreted upon the basis of these. If any of the peneplains are well founded, their discovery and correct interpretation form an important factor in geological investigation. On the other hand, if they are wrongly interpreted, and the entire idea is incorrect, geological literature is becoming seriously confused, as it has been at times in the past, when erroneous ideas have prevailed in large measure as the result of authority. Should this be the case with the peneplain, the time has long since passed when the error should have been detected. Believing as firmly as I do that the peneplain explanation is incorrect, I feel that I should do wrong to longer delay the publication of my reasons.

At the same time, while I have a firm belief, as stated, doubts concerning the validity of this position cannot help arising, for the views that I hold seem opposed to those of the larger number of leading American geologists. I may be wrong, and the weight of authority would seem to indicate that I am. I hope that my paper will call out a discussion and that if I am wrong, the case will be proved beyond question. Even if this is the outcome of this paper, the discussion may perhaps have a salutary effect in putting a stop to the reckless announcement of unproved peneplains, and should lead all geologists to give a more careful study before they put forward the announcement.

Definition of a Peneplain.—A peneplain is "a nearly featureless plain" produced by subaerial denudation.* These are not true plains, but "nearly always possess perceptible inequalities, amounting frequently to two or three hundred

---

*Davis: Am. Journ. Sci., 1889, ser. III., vol. XXXVII., p. 430.

feet."* This levelness is in spite of irregularity of rock "structure."† No extensive peneplains are known to exist at the present time in any part of the earth, but many are inferred from the crest lines of old mountains, which are believed to represent the remnants of dissected ancient peneplains, produced during some previous geographic cycle. It is this conception of a peneplain which is discussed here, and for typical illustrations the peneplains of New England and New Jersey are selected, because they have been most fully studied and discussed, and rest upon the firmest basis.

General Acceptance of the Peneplain Idea.—Few new theories have been so rapidly and uniformly accepted in this country as that of the peneplain suggested by Prof. W. M. Davis about nine years ago. ‡ Indeed its acceptance has become so universal and indiscriminate that the author of the explanation has found it necessary to caution his followers against rashness of conclusion, and to call for a more careful study of specific cases.§ As in the case of most new ideas the followers have gone beyond the originator, and it is perfectly apparent that a great many of the so-called peneplains which have been described rest upon very much less secure basis than the types to which Prof. Davis has called especial attention. In this country many have evidently accepted Prof. Davis' explanation without question, and applied it to very doubtful cases.‖

Improbability of the Peneplain Explanation.—So far there has been no extensive peneplain of recent date, nor even an approximation to one, found on the earth's surface in regions of folded rocks. Yet if we may judge from the evidence adduced by the modern workers in physiographic geology, peneplains have been produced again and again at various times in the past. That is to say, during some past times there

---

*Davis: Bull. Geol. Soc. Amer., 1896, vol. VII., p. 393.

†Davis: Am. Journ. Sci., 1889, ser. III., vol. XXXVII., p. 430; Proc. Boston Soc. Nat. Hist., 1889, vol. XXIV., p. 373; Nat. Geog. Monog., 1896, vol. I., p. 271.

‡Am. Journ. Sci., 1889, ser. IV., vol. XXXVII., p. 430.

§ Bull. Geol. Soc. Amer., 1896, vol. VII., pp. 377-398.

‖ I feel free to speak upon this point, since I have been guilty of the same error, having described as a peneplain in Texas something which may perhaps be a plain of marine denudation. See Proc. Phila. Acad. Nat. Sci., 1893, p. 317.

have been periods of sufficient land rest to allow mountain masses to be worn down to very near the base level. This means relative quiet, or fluctuations about an average level, for a sufficiently long period of time to admit of the slow process of approximate base leveling. Therefore, in accepting the peneplain theory, we need, as a fundamental assumption, to believe that during a part of the remote past, the conditions have been different from those that have prevailed in any portion of the known earth during the present and immediate past.

Few American geologists will be found who will deny the possibility of base leveling,—that, given time, the surface of the land will be leveled to the condition of a peneplain. Such a principle may readily be given a place in an ideal cycle of land development but there should be some real evidence before applying the ideal to the interpretation of existing conditions.

The wearing down of elevated mountains to those of moderate relief may be granted, and the theoretical possibility of their further reduction to the base level may also be accepted. But when the stage of maturity has been reached, the further process of down-wearing must become progressively slower. This will be so, partly because decreased relief of land diminishes the power of the agents of denudation, and partly in a more indirect manner, by furnishing to the undulating surface a capping of residual soil which protects the rock from the action of many of the agents of weathering. It seems impossible to state just what would be the curve of rapidity of denudation with diminishing altitude, but it is evident that the rate diminishes so rapidly with decreasing slope, that, before the condition of the peneplain is really reached, the rate of down wearing must become exceedingly slow.

In the summer of 1897 I spent a month among the mountains of central Maine, the larger part of the time being in the Penobscot drainage area. When I started upon the ascent of Mt. Katahdin, there had been five days of very heavy rain, so that the mountain trails were transformed to brooks, and the East Branch of the Penobscot had risen a number of feet, almost to the level of the spring freshets. The trail up the mountain led across this river, which was fed by mountain torrents, having their source from 2,000 to 5,000 feet above the main river. The Penobscot was not even clouded with sediment. The mountain torrents and the smaller branches from the primeval forests were doing little more work of transportation than that of carrying their slight load of dissolved mineral. This period represents one of the three or four annual freshets when the greatest amount of work of destruction is done in the drainage area. But, even at such a time, the work done was marvelously slight in amount. During the remainder of the year, still less is done. Yet this is a mountainous region where denudation is certainly much more active than it would be in a more reduced area approaching the peneplain stage. At this rate how long will it take to reduce Mt. Katahdin, from its elevation of about a mile, and its neighbors, only slightly lower, to the condition of a peneplain?

During all the time necessary to reduce a hilly country to the condition of a "nearly featureless plain," time to be counted in immense ages, the land must remain nearly at one level; for if it is elevated, the task is increased, and the time needed for reduction correspondingly lengthened; if much depressed a part of the lowered region is submerged, and the work checked, or perhaps even lengthened by the deposit of a load of sediment upon it, which must be removed before further lowering can be accomplished.

The belief in the reduction of a country to the condition of a peneplain rests upon an assumption very difficult to realize, but which could be granted if the peneplain were proved to represent a real condition, and this to be the sole explanation. This assumption of immense periods of time, with relative land quiet during certain periods of the earth's history, conflicts so markedly with what we know of the present and past, both immediate and remote, that its acceptance means no less than the belief that at some periods of the past the conditions have been different from those of the present, and from those of that portion of the past whose history has been worked out by purely stratigraphic methods.

Add to this the fact that the extensive peneplains so far discovered are all of the past, and that no part of the earth reveals even an approximation to this supposed condition, and it seems fair to call for evidence of the most convincing nature before accepting the ancient peneplains. So far as I can judge the evidence is not of this nature.

Lack of Evidence of Ancient Peneplains.*—Several observers, both in this country and abroad, have called attention to the fact that if a person stands upon a high hill in certain regions he looks over a vista of apparently level-topped crests, even though between the hills there are many deep valleys. This appearance has been described with especial fullness by Prof. Davis† and others for the New Jersey and the New England highlands. It is argued that these even-crested ridges and hills occur in regions of complex rock structure, and hence that the explanation cannot be the same as that for plateau crests capped by hard rock in a horizontal position. From this condition an ancient plain is affirmed, and the American school explains it by subaerial denudation, while the British school has advocated marine denudation.

It is true that as one stands upon an eminence and looks over the surface of the surrounding region, the hill tops in these places appear to be quite level. But it is equally true that the same appearance will be produced even where the ridges reach a quite different level. The appearance to the eye may be most deceptive. In order to see exactly what the conditions are, I have made a careful examination of the topographic sheets covering the highlands of New Jersey and Connecticut, and have constructed a series of profiles across the former regions.

In the case of New Jersey, leaving out of consideration all the valleys and all of the lower hills, there is a range of fully 500 feet in the elevation of the higher crests, and there is about the same range along the crest line of the very even-topped Kittatinny mountains. There is a difference of fully 900 feet between the crests of these two neighboring highlands. A series of nine parallel profiles from east to west in this region shows a very distinct lack of uniformity in the elevation of the upland crests, even if all the lower hills are eliminated.

---

*It is to be understood that this refers only to the best established peneplains, not to those whose proof is most doubtful.

†See previous references.

A similar method was followed in Connecticut; and in each sheet where it was possible only those high hills whose elevations were especially marked by the U. S. Geological Survey were chosen, an especial effort being made to select only those hills which could be considered to be a part of the supposed ancient peneplain. On the westernmost sheet selected, the Cornwall, the range is between 1,787 and 1,215 feet. From this sheet eastward a strip five miles wide was followed; and upon the Winsted sheet, next east from the Cornwall, the range is between 1,600 and 1,160 feet; on the Granby between 1,240 and 720 feet; on the Hartford sheet the prevailing condition is lowland; on the Tollard sheet the range is between 985 and 660 feet, and on the Woodstock between 761 and 540 feet. The distance between the edges of the sheets, in an easterly direction, is 91 miles, in which space the total range of the elevation of the "peneplain" crests is 1,247 feet, while in each sheet the distance to the sea shore in the southern part of Connecticut is practically the same. Upon most of these sheets a range of several hundred feet between the most uniform of the crests may be found at places not more than a mile or two apart.

In Maine, New Hampshire, Vermont, western Massachusetts and the Adirondack region, with similar structure to that of the region above mentioned, and so near them that they must have been subjected to the same general degradation, the lack of uniformity of upland crests is very much more marked. One standing upon the crest of one of the mountains of central Maine would hardly find the evenness sufficient to give the appearance of levelness even to the eye, unless he were upon the top of a mountain rising well above the lower peaks, so that the differences in level disappear from view.

This unevenness of crests in the ancient peneplain remnant has not been overlooked by the advocates of this explanation; but to account for it two assumptions have been made. In New England there is an increasing elevation of the upland crests from south to north and from east to west, a difference which, in the section of Connecticut above considered, amounts to over 1,200 feet in 91 miles, measured east and west. Near the sea coast the hill tops, the supposed remnants of the ancient peneplain, stand but slightly above the sea level, while in Maine, New Hampshire, Vermont and western Massachusetts, their elevation is 2,000 to 4,000 feet above the sea. This average rise is taken as evidence that the peneplain has been uplifted, and that in the uplifting it has been so tilted that it slopes at about the rate here indicated. So far as I can find out this is an assumption rendered necessary to explain the difference in elevation of the supposed peneplain; but I fail to find that there is any evidence to prove it, unless the peneplain be previously accepted as a condition covering this entire region. While near the coast there is a certain semblance of levelness, I am utterly unable to find even the appearance of uniformity in the more elevated sections of New England.

In addition to this general deviation from the level condition of the supposed ancient peneplain, there are certain peaks which rise distinctly above the average level of the surrounding crests which are supposed to be remnants of this peneplain. Of this mount Monadnock, in southern New Hampshire, is selected as typical, and the members of this class of hills have been given the name monadnock by Prof. Davis. There are innumerable smaller monadnocks scattered about in New England; and some, like Mt. Katahdin in Maine, are even higher than the type. This class of irregularity is explained by a second assumption,—that they now rise above the general level, somewhat as they did before the peneplain was uplifted, because they were residuals that had not been lowered to the peneplain level, probably either because of the greater durability of their rocks or some other accidental reason, such as greater original elevation. In this case also I am unable to find that there is any other proof that this interpretation is correct than that which comes from the necessity of such an explanation, made necessary by first accepting the existence of the peneplain.

Granting these two assumptions for the purpose of examining the evidence of a peneplain, I have taken several of the sheets of Connecticut and New Jersey, disregarding the gradual north and south rise of the crests, as well as eliminating the hills that rise well above what might be considered the average crest line, and hence which might be called monadnocks. Then, drawing a section on scale, I have computed the area occupied by the crests, reaching an elevation within 300 feet of one another and compared this area, which is all that is now left of the "peneplain," with that above or below the level. In each case the area occupied by the crests of this elevation is less than 25 per cent of the entire area examined, and is generally about 10 per cent. Assuming this to be the remnant of the ancient peneplain, as has been done by the advocates of the theory, and reconstructing this supposed ancient plain, by filling in the valleys and raising the lower hills, we have a peneplain constructed of which 75 to 90 per cent has been gratuitously supplied because of the moderate uniformity of crests whose total area is from 10 to 25 per cent of the whole. Moreover, this uniformity of crests has been obtained only after making use of two assumptions, and by means of them somewhat arbitrarily, disregarding those irregularities which are explained as monadnocks and the result of tilting.

As the result of these considerations, I cannot but believe that the basis upon which the peneplain theory is supported is not altogether solid. In point of fact there seems to be very little real evidence upon which to construct the ancient peneplain, and I am led to raise the query whether, even granting in its entirety the evidence claimed, we would be warranted in drawing so broad a conclusion from so small a basis of fact.

The second fundamentally important point in the peneplain explanation is the claim that we get this uniformity notwithstanding the complexity of the rock "structure." That is to say, I suppose, there is a lack of sympathy between the level-topped hills and the rock texture and position,* excepting possibly where the residual monadnock rises above the ancient plain. That the stratigraphy of the region here considered is complex, and the rocks variable in texture and attitude, is evident, but I question whether there is after all such a lack of sympathy between topography and rock structure as would

*From Prof. Davis' paper it is evident that attitude of the rocks is considered as the main element under "structure," and I am not quite certain whether he meant to include texture under the term, as is so commonly done. Whether he did or not, the consideration of this point is warranted, since some have certainly considered structure as synonymous with both texture and attitude.

be inferred from the statement, "Not less notable than the former continuity of the dissected upland is the want of sympathy between its surface and the structure of the rock masses of which the region is composed."*

Concerning the rock texture in Connecticut and Massachusetts I know very little from direct observation; but for two seasons I worked in the type region of the highlands of New Jersey, climbing from valley to hilltop, and collecting rocks from all portions in a part of the western Highlands. While there are many low hills of hard rock, and possibly some of the higher ones composed of the less durable gneiss, there is, in that region, a very evident general sympathy between the present topography and the rock texture. Where limestone or non-resistant schist and gneiss exist, there are low hills and valleys, while the coarser and more durable gneiss quite generally makes the crests of the high hills. In other words, the apparent remnants of the New Jersey highland peneplain in Sussex county are really somewhat irregular hill tops composed of durable, coarsely crystalline gneisses that have a general uniformity of texture and power of resistance to subaerial denudation. So far as my observation goes, the conditions in Maine and Massachusetts are the same.

Hence the small portion of the so-called peneplain still existing, and the only part upon which the argument for its former greater extension can be based, namely from 10 to 25 per cent of the total area, is that in which the rocks are hard and rather uniform in durability. Therefore, although the rocks are complex in kind and position, they now stand in very general harmony with topography. To say that because of their hardness they now stand up at this level, while the remainder, being softer, has been lowered from the former condition of the peneplain, should be prefaced by proof that this minute fraction of the supposed whole, in reality represents the remnant of such a plain. Neither the area occupied by the remnants, nor the nature of their rocks, seems to bear evidence of a conclusive kind in favor of the peneplain theory.

It may be argued that one of the strongest bits of evidence in favor of the peneplain is not here considered. I refer to

*Davis: Nat. Geog. Monog., 1896, vol. I., p. 271.

the fact that the region is a lowered mountain mass, evidently once of very rugged topography, but now much reduced and traversed by drainage lines of a somewhat mature form, the result of elevation. These facts I accept; but I believe that they admit of a much simpler explanation, which is proposed in a later part of the paper where this point can be considered more appropriately than here.

Evidence Against the Peneplain Theory.—It is difficult to find positive evidence against this explanation. In fact it would seem hardly necessary to do so, since the burden of proof should rest with the advocates of the explanation. At present it is generally accepted, not as a theory to be considered possible, but as a fact amply proved and to be accepted without question. Convincing evidence in its favor, which would place it upon a more solid foundation than that of a mere theory, I am unable to find, though I have searched carefully for it. I should like to have a statement of the evidence which gives the explanation any other rank than that of a hypothesis.

The point already dwelt upon concerning the improbability of the explanation seems to bear evidence against the peneplain. The length of time required to reduce a surface to this condition, and the necessity of assuming a moderate uniformity of level, or an oscillation about an average level during all this time, argues against the explanation. The time since the glacial period, by many believed to be represented in from 5,000 to 10,000 years, though lately multiplied several times this by some, has not been sufficient to strip off the till left by the ice upon the hillsides, nor to notably modify the very perfect form of drumlins, eskers and deltas formed when the ice was here. Yet mountains have been reduced to the condition of a plain so uniform that, from the remnants left us, one has but to climb to the hilltop to see the proof of a plain, needing only a power of imagination sufficient to fill up the valleys between the hills and thus build up the former undulating surface. When we see the slowness of denudation in a hilly country even a single peneplain seems most difficult to conceive; but when three, four, or even five successive peneplains are argued, as if it were as natural to grind down mountains to form plains as it would be to clip off the edges of these pages, one may feel distinctly skeptical; and when this is argued in spite of the fact that the land is apparently so unstable, one may well demand that evidence of the best and most satisfactory kind be adduced. The instability of the land, both present and past, combined with the slowness of denudation even in distinctly upland regions, and its rapidly increasing slowness as these are lowered, appear to be evidence against the peneplain of such strength that only the most convincing proof that such plains have really existed can offset it.

Then the very irregularity of the surface, let us say of New England and the neighboring regions, argues against the peneplain so strongly that here also convincing proof of the peneplain should be necessary to offset this. The type feature of New England is not the peneplain remnant, but the low mountain. It is only in the lower portions, not far removed from the sea, that there is any semblance of a dissected peneplain. Much more than one-half of New England is distinctly mountainous and irregular. There are single isolated peaks, isolated groups of peaks, and entire mountain masses. Where will one go in the White mountains to find evidence of a former plain? or where in northwestern Massachusetts and Vermont, or in the Adirondacks? Last summer I stood upon the crests of several of the higher peaks of Maine and looked in vain for any series of peaks that even to the eye appeared uniform in level. The region is essentially that of mature mountains somewhat roughened by recent elevation. Less markedly is the same true of the coast of Maine. Mt. Desert, Blue hill and many other peaks in that neighborhood contrast very strongly with their neighbors, some of which are half as high, others a quarter, and still others mere low hills or even reefs in the sea. A model of New England large enough to really show the differences in elevation would reveal a very irregular surface, not merely where incised by valleys cut during the Tertiary uplift, but among those uplands which should represent the ancient peneplain. Unless the evidence of the New England peneplain is of the very strongest kind, this irregularity would seem to stand forth in positive testimony against the belief in the former reduction of this region to anything approaching a plain. To attempt to account for this by exceptional conditions seems an admission of a weakness in the explanation.

So-called monadnocks appeal to me as proof against the peneplain theory. Grant for the moment the destruction of a mountainous surface to the condition of a plain under subaerial denudation, and this reduction must certainly call for a very great lapse of time. During such reduction it may be admitted that the soft rock will be much more reduced than the harder ones, and that the latter may stand well above the general level as residuals; but are the monadnock rocks essentially harder than the other hilltop rocks of the neighborhood? I know of no evidence that has been made public that Mts. Monadnock and Washington are made of harder rock than many of the much lower hills within a radius of twenty miles from them. I know of no proof that they are more resistant than the Blue hills near Boston, nor that these are harder than the lower granite hills of Essex county, Massachusetts, a few miles away. Is the rock of Mt. Washington more durable than that of Essex county, Massachusetts, or the rock of Mt. Katahdin or Blue hill, in Maine, harder than that of scores of lower hills not far away? I believe that I am correct in saying that there are no very distinct differences between the rocks of the monadnocks and the lower hills, in point of durability, while there is a difference in elevation of more than a mile, and, even in short distances, of 2,000 or 3,000 feet. In some of these cases it is certain that the rocks of low and high hills are not markedly different.

Without the existence of very notable differences in power of resistance, is it probable that a hill would stand several thousand feet above a plain which stretched all around its base, and which has been reduced to this condition by the slow process of subaerial denudation? After the region surrounding the monadnock had been reduced to the condition of a low, undulating hilly country, all the time required to plane it down to the condition of a peneplain has not been able to reduce the elevation of the higher, and hence more rapidly destroyed part, to approximately the same level!

It must be confessed that this opposing evidence is based purely upon my own ability to conceive of the processes involved. In so far as this power of conception is strong or weak, this part of the argument is good or bad. I would put it forward with more hesitation if there did not appear to be

53

good evidence that the rate of denudation is exceedingly slow in a forest covered country, and that the land is and has been far from stable, when long periods of time have been involved.

Alternate   Hypotheses.—Two hypotheses have been suggested to explain the facts considered above: (1) that of marine denudation; (2) that of subaerial denudation. Although American geologists in general consider marine denudation possible only in rather restricted areas, this hypothesis is certainly not an improbability. So, also, subaerial denudation, if continued long enough, with land level maintained somewhat uniformly throughout, would undoubtedly reduce any area to a level condition, though the places most likely to be so reduced are those near the sea, or those in which the rocks are soft, or the elevation slight. The possibility of these two causes for reduction is not questioned, although the probability of general reduction by such causes is called in question. This doubt is still further strengthened by the belief that the evidence of ancient peneplains, upon which the entire argument of former base-leveling is founded, is far from convincing.

Summarizing this evidence it seems that the ancient plain is constructed upon the basis of the existence of moderately uniform hill crests, whose total area is not more than 10 to 25 per cent of the entire area. Even among these, the hill tops reach to considerably different elevations. Of the remaining 75 to 90 per cent the greater part is sunk below this upland level. The hill tops are mainly of hard rock, while the valleys are mainly located in the areas of less resistant beds, and the streams are, in general, in quite close accord with the rock structure. There are, moreover, numerous localized elevations, called monadnocks, reaching well above the upland level; and, in the western and northern portions of New England, at no very great distance from the coast, the region is elevated and very irregular and mountainous, as for instance in Maine, the White mountains, Green mountains, Berkshire hills and Adirondacks. These greater elevations do not correspond with marked differences in rock structure. The advocates of the peneplain admit past irregularity, in the form of monadnocks rising above the ancient peneplain, and they also admit present irregularity in a marked degree, but ex-

plain it by one of three conditions, either post-peneplain denudation, or ancient irregularities upon this peneplain surface,
or differential elevation of the peneplain since its formation.

In the present condition of New England and New Jersey,
I am unable to see any evidence that the region was ever reduced further than the condition of full maturity of topography,—that is, a region of hills and valleys of considerable variety, and, away from the sea shore, of rounded but considerably
elevated mountains. That this mature mountain region has
been subjected to later elevation, which has rejuvenated the
rivers, seems certain. According to this, the present New
England topography is mainly one of reduced mountains,
lowered to the stage of full maturity, then elevated and made
more rugged. By this explanation it is held that the region
was never reduced to the peneplain stage, but has always been,
as it still is, a mountainous section, though once less mountainous than now, because of the recent uplift.

So far as I can see, the facts in the field are in fuller harmony with this explanation than with that of the peneplain.
The present marked irregularity of surface is explained without other assumption than that certain places were formerly,
as now, either high or low, as they would naturally be in a region of mature mountains. It does away with the necessity
of assuming long periods of time during which the land remained at approximately one level. To reduce a mountainous
region to the stage of maturity is an easy task compared with
the reduction of a mature mountain region to a peneplain. It
would be impossible to state what the ratio of time is, but it
is certain that to lower a mature mountain to a peneplain must
take many times as long as to reduce a mountainous area to
that of maturity. Moreover, much more variation in elevation
is possible under the explanation here proposed than under
that of the peneplain. While the mountains were being lowered to the stage of maturity, there might be very much fluctuation of level without marked interference with the continuation of the process of production of mature forms. Besides
this, while there are no existing peneplains, there are at present many regions of reduced mountains approaching the stage
of maturity—witness the very regions under consideration.

It may be argued that the number of hills reaching to a moderately uniform elevation which are found in southeastern New England, and in New Jersey, cannot be accounted for by this hypothesis. When we take into account their present irregularity of surface, this asserted uniformity does not appear so marked. Upon my mind the impression of irregularity is produced much more strongly than that of regularity, particularly when the monadnocks and higher irregularities of the northern and western part of New England are included. It is true that near the coast the uniformity is more marked than in the interior; but here, of course, the mountains would have been more lowered than in the interior, and, in the coastal region, there may well have been an approach toward the condition of a local peneplain. Yet, when we consider such isolated elevations as that of the Blue hills, near Boston, and the ruggedness of the Maine coast and of Nova Scotia, as well as of the region farther north in Labrador, even here, where the peneplain condition should have been most fully reached, the regularity of level can be urged only when numerous local exceptions are eliminated.

However, it is necessary, if this proposed hypothesis is to be accepted, to account for even the measure of uniformity that exists, even though it is really less marked than some believe. In the reduction of a mountain mass toward base level, long before the peneplain stage is reached it seems certain that there would be a uniformity of level among the mountain crests fully as marked as that now found in New England, and that this uniformity would naturally be greater near the sea, where development would have been most advanced.

Given a mountain region of marked irregularity, such as New England must have been during the Paleozoic, the rocks from place to place varied greatly in hardness and in attitude, while the peaks in different sections naturally reached to very different altitudes. If we should select from these, two neighboring peaks or ridges of approximately the same texture attitude, and altitude, it would follow that, since they were exposed to the same climatic conditions, their downwearing toward base level would be continued at about the same rate, as a general proposition, though, of course there might, in any selected place, be accidents of variations which would interfere. This rate of denudation among these neighbors would at first be rapid; for, in the inception of the work, the elevation was great, the slope steep, and the rocks were exposed to strong winds and powerful frost action, while they were not protected by trees. The rate of downwearing, as we see upon similar peaks in the higher mountains of the present, would have been much more rapid than at any later stage, decreasing in rapidity as they were lowered; though still being worn down rapidly until the zone of the timber line was reached. Then conditions of an entirely new kind would have been introduced, and, from that line downward, the rate of denudation of the peaks would greatly decrease, partly because of the lessened slope, but chiefly because of the protection of the forest, which holds the disintegrated pieces in place, and helps make a protection of residual soil. As the forest covering became greater, and the slope less, the soil covering would become deeper, and the rocks more and more protected, until denudation had become exceedingly slow. These two peaks, starting at the same level, having the same kind of rock throughout, and exposed to the same conditions, would reach this stage of development (namely their crests at approximately the timber line), at about the same time; and, as their crests sank lower below this level, the peaks would still stand at about the same elevation. In an extensive mountainous region there may have been a number of such cases.

But there would not be many such peaks of the same hight or so similar that they would be reduced at nearly the same rate. Some would be of easily denuded rock, and, in time, these would be very much lowered, while the harder ones stood well above the base level. There would at first be very marked ruggedness, partly the result of difference in original elevation, and partly the result of the effects of subaerial denudation upon the much elevated and differentiated surfaces. One peak, perhaps of slightly less durable rock than a neighbor, would be lowered at a very much more rapid rate than its neighbor. But there would come a time when this difference in rapidity would be very much diminished, even if the rock of the two peaks were quite different. This time would come when the zone of trees was reached; and the difference in rate of downwearing would even more rapidly diminish as soon as a soil covering became possible. In the meantime, a higher or more durable neighbor might still be sinking more rapidly, and, in time, might almost catch up with a more favorably situated and lower peak. The curves of the rate of denudation in the two cases would approach and finally almost coincide; and, unless the rock differences were marked, the two peaks would proceed to be lowered at about the same rate. If the rock differences were very marked, there would be no exact approach; but, according to this view of the method of denudation, even though there was originally a marked difference in altitude, all peaks whose rocks were approximately the same in power of resistance would in time approach each other in altitude, the one originally higher catching up with the other whose rate of lowering was becoming rapidly diminished because of decreased elevation. It must be granted that in such a mountainous country as that of New England down below the surface there are extensive beds of rock of approximately the same hardness. That this is so is proved by the abundance of durable gneiss and granite in most low mountainous areas, as for instance in New England and New Jersey.

By this there would be a beveling of the hill tops, the highest area of beveling being that part of the tree zone in which, because of lessened slope, the rock was protected by trees and by a residual soil blanket. Down to this zone denudation would be relatively rapid, below it much slower, and increasingly slower as the beveling continued still further. In a mature mountain region so developed there would be some peaks not yet lowered to this area, and there would be great valleys depressed below it. But would it be incorrect to assume that in a given area where most lowered, from 10 to 25 per cent of the reduced mountain tops would probably have reached a fair uniformity of level? This beveling of the hill tops would be very much further advanced near the coast than in the interior, thus coinciding with the conditions found in New England.

According to this view, by the time maturity of topographic form has been reached, there will be a beveling of hill tops where the harder gneissic and granitic rock exists, the stream valleys standing near the base level and hills of softer strata standing at levels still lower than those in which the rock is harder. Areas originally distinctly higher or harder than usual, or more unfavorably situated, may be less lowered and more irregular than the surrounding region, though still engaged in an approach to this lower level. A well matured surface would then present three intergrading stages in different places and under different conditions. (1) Local base levels in the valleys; (2) general well matured topography with many hills reaching to approximately the same general level, but with some distinct and many indistinct "monadnocks"; (3) exceptional and localized *early* maturity, found particularly in the interior. The further the topographic development had gone toward old age, the greater would be the extent of the first two areas. Can any evidence be adduced to show that New England has ever. advanced further in development than this stage?

Granting such a reduction, with many hills of hard rock standing at a moderately regular level if an elevation succeeds, while the valleys will be deepened and the hills lowered, the rate of lowering of the hills will be so nearly uniform, since the climate and rock are so nearly alike, that the measure of uniformity of upland level will in part be maintained.

Conclusion—The questions raised in this paper are not against the great importance of subaerial denudation, which few American geologists are inclined to underestimate. The stamp of the genius of Powell, Gilbert, Davis and others is too plainly marked upon the minds of American geologists for any underestimation of the importance of this. The question I raise is whether far too much importance has not been assigned to this great work. The facts and assumptions upon which the peneplain theory is based are also called in question, and an attempt is made to show that all the phenomena believed to indicate the existence of peneplains in New England and New Jersey can best be explained without assuming the reduction of a high mountainous country to the condition of old age, a condition now nowhere found on the earth.

The alternate hypothesis of beveling down to mature form is advanced. This hypothesis requires no long periods of relative quiet, and no assumptions to explain the irregularities of the surface, which, by the peneplain theory, call for special causes whose operation is apparently not otherwise proved, and which, in part, appear to be hardly probable. The theory of the peneplain calls for a "nearly featureless plain"; the alternate hypothesis of beveling calls merely for a greatly reduced, but still markedly irregular surface. To some the difference between these two hypotheses may seem slight, but really it is great; for, after the rounded features of maturity are reached, the advance to such old age topographic features as the peneplain demands, calls for immense periods of time with land standing at nearly the same level, conditions which seem at variance with the facts which geologists have been collecting in the last half century.

3

Reprinted from *Amer. Geol.*, **22(B)**, 207–222, 223–224, 224–227, 228–232, 234–239
(Apr. 1899)

THE PENEPLAIN.

By W. M. DAVIS, Cambridge, Mass.

Had it not been for the distractions of foreign travel during a year of absence from college duties, I should have sooner written a reply to Professor Tarr's article on "The Peneplain" that appeared in the Geologist for June, 1898. The delay has not, however, been a disadvantage on my part, for it has enabled me to talk over the problem with a number of English and French geologists and geographers who are interested in such matters, and thus to free my reply somewhat from individual prejudices. The discussion that Professor Tarr's article should awaken will be a welcome one, for as he has well said, the peneplain is too important a matter to gain an accepted position without close scrutiny. The courteous and earnest tone of Professor Tarr's essay will, I hope, determine the style of those that follow it.

At the outset, allow me to correct the implication that the "peneplain idea" was original with me. The name is of my invention, and, as has sometimes happened, the introduction of a definite name for a thing previously talked about only in general terms has promoted its consideration:—witness the name, antecedent, for rivers that hold their courses against mountains uplifted beneath them. The idea of antecedent rivers had occurred to several observers who gave it no name, and unnamed it gained no general currency; but it became popular when Powell named it. Moreover, the ideas of antecedence and peneplanation were ripe in many minds about the time the names were suggested, and it is chiefly for that reason, as it seems to me, that antecedent rivers have been so frequently mentioned in the last thirty years, and peneplains in the last ten.

It was in Powell's "Exploration of the Colorado River" (1875) that the "peneplain idea," along with a number of other important facts and principles, first came to my notice. The idea is not stated categorically, but when describing the even surface of deformed rocks beneath the horizontal Carboniferous strata in the Colorado canyon, Powell said that "aerial forces carried away 10,000 feet of rocks by a process slow yet unrelenting, until the sea again rolled over the land," and the evenly denuded surface is referred to as "the record of a long time when the region was dry land" (p.212). In a later work, the same author writes:—"Mountains cannot long remain as mountains; they are ephemeral topographic forms. Geologically speaking, all existing mountains are recent; the ancient mountains are gone" (Geology of the Uinta Mountains, 1876, 196). Again, "in a very low degree of declivity approaching horizontality, the power of transporting material is also very small. The degradation of the last few inches of a broad area of land above the level of the sea would require a longer time than all the thousands of feet which might have been above it, so far as this degradation depends on mechanical process—that is, driving or flotation; but here the disintegration by solution and the transportation of material by the agency of fluidity come in to assist the slow processes of mechanical degradation, and finally perform the chief part of the task" (Ibid., 196). Dutton referred to Powell's having given precision to the idea of baselevel, an idea probably known previously in a general way to many geologists. "All regions"—Dutton says—"are tending to baselevels of erosion, and if the time be long enough each region will, in its turn, approach nearer and nearer, and at last sensibly reach it" (U. S. G. S. Monogr. II, 76).* In Great Britain, where the litera-

---

*I had expected to find some similar sentences in Gilbert's "Geology of the Henry Mountains," but discover instead the following statement:—"It is evident that if steep slopes are worn more rapidly than gentle, the tendency is to abolish all differences of slope and produce uniformity. The law of uniformity of slope thus opposes diversity of topography, and if not complemented by other laws, would reduce all drainage basins to plains. But in reality it is never free to work out its full results; for it demands a uniformity of conditions which nowhere exists" (p. 115).

ture very generally indicates a belief in plains of marine abrasion, a number of geologists have, without public announcement in any formal manner, gradually enlarged the share of work attributed to subaerial forces, until, as some of them have lately assured me, the "peneplain idea" has come to be for a number of years as familiar to them as to most American geologists; and some of them certainly entertained it before the term, peneplain, was suggested. Several examples of the recognition of the "peneplain idea" by continental geologists might be given if time and space permitted, but I will here refer only to the essay by Penck, quoted below (B 4).

Professor Tarr argues, in effect, that certain regions instanced as dissected peneplains have never really been lowlands of faint relief; that the process of peneplanation is in itself an extremely unlikely one; and that the so-called peneplains, all of which are now more or less dissected, are capable of other explanation; in brief, that peneplains are (A) unreal, (B) improbable, and (C) unnecessary. Several subdivisions of each of these headings will be made in replying to them. Page numbers refer to Professor Tarr's original article.

A 1. *Certain regions show no trace of peneplanation.* It is stated that "one standing upon the crest of one of the mountains of central Maine would hardly find the evenness [of the sky line] sufficient to give the appearance of levelness even to the eye" (p. 357). But no one, so far as I know, has thought that the mountain tops of Maine mark the remnants of a peneplain. The mountains there are probably of the nature of monadnocks; it is only the general upland surface above which the mountains rise that can be regarded as a peneplain, uplifted and dissected, if the features that I have seen about Portland and at some other points along the coast may be extended inland. The White mountains have been, in my mind, tentatively classed as a group of monadnocks; they do not, as far as I have seen them in brief excursions, stand upon any distinct basement comparable to that of the uplands of New England further south; but Mr. Philip Emerson, master of the Cobbett school, Lynn, tells me that he has in summer excursions traced what he thinks may be regarded as the extension of the more southern uplands around

the White mountains on the east, north, and west. Northern
New England is not to-day well enough mapped or studied to
give either decided support or disproof to the theory of peneplains. Its ruggedness is generally so great that it is quite
possible that the peneplain explanation does not apply to the
greater part of the area. Little wonder that an observer
whose attention is given to this mountainous district, under
the impression that its mountain tops represent the remnants
of a peneplain, should come to discredit such an explanation.

A 2. *The uplands of southern New England and of northern New Jersey are not of uniform altitude.* It is urged that
a careful examination of the topographic maps of these regions disproves the accordance claimed for their upland altitudes. In answer, I should say that the lack of uniformity
among the uplands—a fact perfectly familiar to those who accept the peneplain idea—is partly the result of titling, as will
be further considered below (see A 4); and that for the rest
the unevenness of the uplands of to-day is a natural result of
imperfect peneplanation followed by submature dissection.
The examination of the peneplain remnants by means of topographic maps is not a new method of investigation, as it was
employed for New Jersey in 1888-89, and for southern New
England a few years later; but, like observation out doors, it
seems to lead different investigators to different results. Considerable as the inequalities of altitude are, frequent study of
the maps and repeated views of the uplands from various hill
tops impress me much more with the relative accordance of
their altitudes than with their diversity. I cannot admit that
the appearance of accordance from hill top to hill top is an
optical deception. There is an important matter of fact behind the appearance.

The comparative evenness of the uplands in Connecticut
was recognized and well described by Percival over half a
century ago. The state being divided into eastern and western areas of primary rocks by the trough of Triassic strata,
he said:—"the eastern and western primary may both be regarded as extensive plateaus, usually terminating abruptly toward the larger secondary basin, but sinking more gradually
toward the south, on the sound. These plateaus present, when
viewed from an elevated point on their surface, the appear-

ance of a general level, with a rolling or undulating outline, over which the view often extends to a very great distance, interrupted only by isolated summits or ridges, usually of small extent. These plateaus are also intersected by valleys and basins, which serve to mark the arrangement of their surface even more definitely than the elevations. This arrangement will be found to correspond very exactly with that of the geological formation, indicating that it was caused essentially by the original form of the surface of these formations, and not by any subsequent denudation" (Geol. Conn., 1842, 477). "The western primary . . . forms, within the limits of this state, a wide plateau . . . of so uniform an elevation that from many points, the view extends across its entire width, and to a great distance north and south" (Ibid., 478.) "The eastern primary, viewed from its more elevated points, presents the same general appearance as the western; that of an extensive undulating surface of nearly uniform elevation, diversified by detached summits" (Ibid., 482). The peculiar conclusion of the first of the above quotations is interesting in contrast to modern views.

In eastern Massachusetts, dissection has gone so far that it would be difficult to discover an uplifted peneplain on local evidence alone; but in the central and western parts of the state, the uplands are generally so well defined and so accordant that I am at loss to understand why Professor Tarr should say:—"While near the coast there is a certain semblance of levelness, I am utterly unable to find even the appearance of uniformity in the more elevated sections of New England" (p. 358). Looking eastward from the Berkshire hills across the Connecticut valley lowlands in northern Massachusetts, the skyline of the central plateau is to my eyes astonishingly uniform, though its altitude is over 1,000 feet. Hence it must be agreed that with the same facts before us, both out doors and on maps, our descriptions and interpretations of them do not correspond; one of us being impressed with the diversity of the upland altitudes, and the other with their accordance.

A 3. *The remains of certain peneplains are fragmentary.* It is urged that ten per cent of the original area of the supposed peneplain, now preserved in the uplands of Connecticut, is too small a fraction to serve as a basis of reconstruction.

This does not strike me as a serious or a novel difficulty. Geologists are often compelled to work on fragmentary evidence; they are satisfied if the fragments can be logically built up into the complete structure. In most parts of the world, rock outcrops occupy less than ten per cent—often less than one per cent—of the land surface; yet no field geologist hesitates to "color in" a formation over an area where scattered outcrops give reasonable proof of its occurrence. The surface area thus colored in is often but a small part of the entire body of the original formation, which may be largely covered by later deposits or destroyed by erosion; but the covered and eroded portions are reasonably inferred, and a formation thus established is a stock subject in historical geology. It is therefore not so much a high percentage of direct observation, as a logical method of reconstructing the unseen whole from the observed parts that is necessary. Here the dissected peneplain seems to me to stand on a par with many other things. Its fragmental condition is most natural; its discovered parts are connected and the lost peneplain is restored by a line of argument that is perfectly reasonable in itself, and that is objected to only because it runs counter to certain views that are held by Professor Tarr to be established principles in the science of geology. These views will be considered below (B 1, 2).

A 4. *Certain so-called peneplains are now inclined.* Professor Tarr says:—Uplift or tilting "is an assumption rendered necessary to explain the difference in elevation of the supposed peneplain; but I fail to find that there is any evidence to prove it, unless the peneplain be previously accepted as a condition covering this entire region" (p. 358). Here we fully agree. I have repeatedly insisted that it was only by recognizing the existence of a peneplain that uplift or deformation could be determined in certain cases; and that only in this way could certain stages of geological history be discovered, in the absence of what might be called orthodox geological evidence in the form of marine deposits. For example, it is by the remnants of an uplifted, inclined, and warped peneplain in the even crest lines of the Pennsylvania Appalachians that the post-Cretaceous uplift of the mountain belt has been determined: it was formerly supposed that the existing ridges were the unconsumed remnants of the ancient Appalachians, and by implication, that no uplift of the region had occurred since the mountains were crushed, folded, and upheaved. So in southern New England: there was no means of determining the date of uplift, as a result of which the existing valleys were eroded, until the peneplain of the uplands was recognized and dated. Twenty or thirty years ago, it was not uncommon to meet the suggestion that the valleys might be of glacial origin, so little understanding had then been reached of the geographical development of the region. Those who believe in the verity of peneplains will infer uplift, where they see a high-standing and dissected peneplain, as confidently as the geologists of the end of the eighteenth century inferred uplift when they found marine fossils in stratified rocks far above sea level.

But it does not seem warranted to conclude that the peneplain theory is invalidated because certain peneplains are now uplifted on a slant, although this is implied in Professor Tarr's argument (p. 359). It is no objection to the peneplain idea to say that the crest of Kittatinny mountain is higher than the upland surface of the New Jersey highlands (p. 356), or that the crest of the palisades is lower. It would be as extraordinary to find no slanting peneplains as to find no inclined strata. Warped and faulted peneplains are no more unlikely products of crustal deformation than warped and faulted sedimentary formations; witness the dislocations of the plateaus trenched by the Colorado canyon, the plateau surface having been worn down to "a very flat expanse" before the uplift and displacements that have determined the altitudes and forms of to-day.

A 5. *Objections based on the fragmentary condition of certain peneplains further considered.* If the best preserved peneplains were not less fragmentary than the ones that Professor Tarr has discussed quantitatively, the theory of peneplains might perhaps be overthrown: but when the imperfect peneplains of New England and New Jersey are considered in connection with many more nearly perfect peneplains elsewhere, the series becomes so well graded, from better to worse, that the theory seems to me unassailable. A few examples of the better preserved peneplains may therefore be now considered.

The Piedmont belt of Virginia has been described by a number of observers in recent years. McGee writes:—"The plain is not monotonously smooth; here it undulates in graceful swells, there it dips into rocky river gorges, winding across its width. . . . Such is the Piedmont plain within view of Monticello and such is the province throughout its extent from New York to Alabama" (Nat. Geogr. Mag., viii, 1896, 261). The Piedmont rivers "rush through narrow, rockbound gorges. . . . All the Piedmont rivers, large and small, are incessantly corrading their beds" (Ibid., 262). The plain "must be regarded as the basal portion of a vast mass of inclined rocks of which an unmeasured upper portion has been planed away" (Ibid., 263). In describing the same region, Darton writes:—"The Piedmont plateau is a peneplain of Tertiary age . . . the plain has been deeply trenched by drainage ways, but wide areas are preserved on the divides" (Chicago Journ. of Geol., ii, 1894, 570). He believes that this peneplain, AB, fig. 1, continues across the inner strata, BC, of the coastalplain, and that it should therefore be distinguished from an earlier peneplain carved on the same ancient rocks, part of which is BE, preserved beneath the strata of the coastal plain, and part of which, DB, is generally hereabout destroyed by erosion. It is upon the older peneplain that the Potomac formation, with its fossil terrestrial flora, directly rests. Keith gives an elaborate account of a part of the Piedmont plain in his "Geology of the Catoctin Belt" (14th Ann. Rep. U. S. G. S.), and discusses its relations to various members of the coastal plain series.

Any one who will follow up the foregoing references, or who will, better still, look over the region on the ground, will find a decidedly larger portion of the peneplain surface preserved than is the case in New England or New Jersey; and this is most natural, for the Virginia Piedmont plain is of distinctly later origin than the peneplain of the uplands further north: the latter corresponds to the earlier peneplain, DBE, in Virginia. But it is not only the comparative continuity of the Piedmont plain that makes it a valuable example: the deep soils of the upland plain and the rocky walls of the narrow steep-sided valleys are as important witnesses to the once lower position of the plain and to the uplift by which its present altitude has been gained as are the forms of the upland and the valleys. To explain this point more fully, a brief digression may be allowed.

The peneplain is only one element in the theory of the geographical cycle. The systematic sequence in the development of land forms through the cycle is a much larger and more important principle than the penultimate development of a peneplain, considered alone; for the former includes the latter. One of the elements of the cycle is the development of the graded condition of streams of water during maturity, whereby an essential agreement is brought about between the ability of a stream to do work, and the work that it has to do. Another element, less generally recognized, is the development of the graded condition in the streams and sheets of rock waste or soil on sloping surfaces, where no running streams of water occur. By following out the ideal scheme thus suggested, it must result that just as the graded condition of water streams is normally propagated from the mouth towards the head, and in time reaches the source of all the branches, so the graded condition of soil-covered slopes is in time extended all over a land surface, from the valley floors to the divides. The supply of waste by the disintegration of the sub-soil rock is then everywhere essentially equal to its removal by all available agents of transportation. In a late stage of a cycle, when the surface slopes are small, agents of transportation are weak; hence the supply of waste must then be slow and the waste to be removed must be of fine texture. In order that the supply shall be slow, the waste comes to have a great depth,* and the upper parts greatly protect the rock beneath from the attack of the weather. At the same time,

---

*It is not at first sight clear why the depth of soil should increase on a graded waste slope, if the supply and removal of rock waste are (essentially) equal. As a matter of fact, the supply exceeds the removal by a quantity of the second order. Then as the slope decreases and the agencies of removal weaken, the depth of soil increases by just such a measure as will suffice to reduce the agencies of weathering (supply) to equality with the waning agencies of transportation (removal). This is only one of the many natural examples of an (essential) equilibrium, maintained between varying forces and resistances. In elementary presentation, when the condition of a graded waste slope at a given stage of development is considered, and attention is not directed to the variation of the forces and the resistances with the advance of the cycle, the equilibrium may be announced without qualification: in more advanced presentation, the hedge-word, essential, is an assistance to clear understanding.

transportation is facilitated by the refinement of the surface soil during its long exposure to the weather. Hence, under ordinary climatic conditions, normal peneplains must have deep local soils of fine texture at the surface, and grading into firm rock at a depth of 30, 50, or more feet. Moreover, it is only on a lowland surface of small slope that such a depth and arrangement of local soil can be normally produced.

In contrast to the deep soil of a peneplain, the steep sides of young valleys, whose graded waste sheets are not yet developed, must frequently reveal bare, rocky ledges. Only as the valleys widen and their side slopes become somewhat more gentle, will the ledges disappear; and even then the rock will be covered only by a relatively thin and coarse sheet of rapidly creeping waste. It therefore follows that the uplands of the Piedmont belt, with their deep soil, are of an essentially different cycle of development from the narrow valleys, with their bare ledges. The two elements of form remain mutually inconsistent, until reconciled by the postulate of an uplift of the region between their developments. But if this postulate is accepted, the plain is shown to have been a lowland of faint relief before the existing narrow valleys were cut in it. It is this double line of argument, based on deep soil and bare ledge, as well as undulating plain and narrow valley, that has convinced various observers of the verity of the peneplain in the Piedmont belt.

The Great Plains of eastern Montana include an area of nearly horizontal Cretaceous strata on either side of the Missouri river, regarding which the evidence of peneplanation seems to me beyond dispute. Here and there, volcanic buttes, dikes, and mesas surmount the plain by several hundred feet; on the south, the Highwood mountains, a network of dikes among nearly horizontal shales and sandstones, rise in still stronger relief. Hence there can be no question that strata, measuring hundreds if not thousands of feet in thickness, have been broadly removed from the region by denudation. Yet the surface between the various eminences that rise above it is a true geographical plain. It is not absolutely level, but broadly undulating, with a sky line almost as even as that of the ocean itself. In this plain, the Missouri river and its chief branches have cut narrow, steep-sided valleys, several hun-

dred feet below the general upland level. These valleys are
so young that the Missouri itself has not yet developed an
even slope; witness its several leaps at Great Falls. Innumerable wet-weather side-streams are cutting sharp ravines in the
larger valley sides. It does not seem possible to avoid concluding that the upland plain is to-day in process of destruction by an agency that could not have been in operation while
the finishing touches were given to its production. It was
upon this peneplain in 1883 that the necessity of believing in
penultimate denudation was first strongly impressed upon me.
Dr. Waldemar Lindgren, now of the U. S. Geological Survey, who was with me in the field, may recall how the conviction grew upon our minds; if I am not mistaken, he accepted it before I did. A brief account of the region is published in volume XV of the Tenth. U. S. Census Reports.

The extended plains of Central Russia, as lately described
by Philippson (Zeitschr. Ges. f. Erdk. Berlin, xxxiii, 1898,
37-68, 77-110), have a gently undulating surface at a height of
200 or 300 meters, broadly continuous, but here and there
dissected by relatively narrow, steep-sided, young valleys. The
upland surface is not a structural plain, for it bevels across
formations of very different ages: it is therefore a plain of
erosion. In the south, there is a partial covering of loess, a
thin veneer often absent and leaving the rock surface visible
over large areas. In the north, the drift cover is heavier and
more continuous; but the plateau surface is still the continuation of the same plain of erosion as in the south. There is
no record of marine action on the great plain, hence its erosion is ascribed to the lateral swinging of the lower courses of
large rivers, but the origin of the rivers is unknown: it can
only be said that when the erosion was going on, the Russian
"Scholle" must have stood 200 meters lower than to-day. The
narrow valleys have been cut since the uplift of the plain and
are older than the glacial period (see especially p. 38-42, 54,
55, 62, of the above-cited essay). This is the largest peneplain of which I have found any account.

A 6. *The asserted discordance of peneplain surface and
rock structure is open to question.* It is said to be question-
able "whether there is after all such a lack of sympathy between topography and rock structure" as has been repre-

sented (p. 359), and in evidence of this doubt it is said that in the Highlands of New Jersey there is "a very evident general sympathy between the present topography and the rock texture" (p. 360)   There is some danger that our discussion may here run into cross purposes, for this objection to peneplanation does not meet the arguments advanced in its favor. Disregarding the weaker structures, which are now worn down beneath the inferred surface of the peneplain, it seems to me undeniable that the peneplain surface was strongly discordant with the hard structures that still preserve its remnants.   It is a matter of necessity that the present topography of an uplifted and dissected peneplain should exhibit sympathy between form and structure, for where should better accordance of form and structure be expected than in such a region of adjusted drainage; but this is a matter quite apart from the present discussion.

Various gneisses, sandstones, and trap sheets, standing in a more or less inclined position, are truncated with good appearance of system by the gently slanting surface of the peneplain of northern New Jersey.   The following description of the region is taken from one of Cook's reports.   "The Highland mountain range consists of many ridges which are in part separated by deep valleys and in part coalesce, forming plateaus or table-lands of small extent.   .   .   .   A characteristic feature is the absence of what might be called Alpine structure [form?] or scenery.   There are no prominent peaks or cones.   The ridges are even-topped for long distances and the average elevation is uniform over wide areas.   Looking at the crests alone and imagining the valleys and depressions filled, the surface would approximate to a plane gently inclined toward the southeast and toward the southwest" (Geol. Surv. N. J., Ann. Rep. 1883, 27.   See also p. 28, 29, 60, 61). It is this indifference of the peneplain to the various structures that it systematically truncates that has always been the chief argument of those who thought they saw traces of a former lowland where there is to-day a dissected highland, whether they believed in marine abrasion or in subaerial denudation.

Special mention may be made here of certain features that will be referred to more briefly in a later section (C 4).   Descending the Hudson from Haverstraw to Jersey City, one may see a gradual decrease of altitude in the palisades, a ridge formed on a monoclinal sheet of dense intrusive trap, from a hight of about 600 feet in the north, to sea level in the south. There is no corresponding variation in the thickness of the trap sheet. The uplands of schists and gneisses on the east of the Hudson have a similar descent from the Highlands to Long Island sound. In Connecticut, the view from East Rock, New Haven, discloses the extraordinarily even crest line of Totoket mountain, the edge of a strongly warped sheet of extrusive trap. The crest line slowly descends southward and is continued by the somewhat lower crest line of Pond mountain, of similar structure. Furthermore, the descent of these crest lines agrees very well with the descent of the crystalline uplands next on the east. The systematic relation of these and many other crest lines and uplands suggests a peneplain, and the peneplain thus inferred is strikingly indifferent to the structures that it truncates. It might be urged that the observed discordance of form and structure is of some other origin than peneplanation; but the discordance does not seem open to question.

A 7. *The rocks of monadnocks are not proved to be more resistant than those of the adjoining peneplain.* It is urged that there is no other proof of the durability of the rocks of monadnocks other "than that which comes from the necessity of such an explanation, made necessary by first accepting the existence of the peneplain" (p. 358). As far as my own work is concerned, there is some ground for this objection. I have as a rule given no particular attention to the composition of the monadnock rocks; indeed, it has generally seemed to me reasonable to infer their greater resistance on account of their form. But so far as attention has been given directly to this phase of the problem, the inference based on the peneplain theory is borne out by petrographic study. The buttes and mesas that surmount the plains of the upper Missouri are maintained by dense igneous rocks. The monadnocks of the Virginia piedmont belt "are ribbed with siliceous schists or quartzites or other rocks that resist well the work of the weather . . . while the rocks underlying the fertile fields of the plain are softer schists, easily weathered and worn away" (McGee, l. c., 262, 263). Near Atlanta, Ga., the Piedmont area is a well-finished peneplain, rather strongly dissected, with deep soils overlying the uplands of gneiss and schist. Stone mountain, a superb monadnock of abrupt form, consists of fine homogeneous granite, quite unlike the rocks of the peneplain (Purington, Amer. Geol., xiv, 1894, 105-108). Van Hise, in describing the uplands of the ancient disordered and indurated rocks in north-central Wisconsin, says that they constitute "as nearly perfect a baselevel plain as it has been my good fortune to see. * * * Above the valley of the Wisconsin river, an almost perfect plain is seen * * * large areas of which are but little dissected by any of the tributary streams of the Wisconsin." The upland plain is surmounted by Big Rib hill, a monadnock of exceedingly resistant quartzyte (Science, N. S., iv, 1896, 57-59). The upland of the Slate mountains in western Germany is a wonderfully fine peneplain of broad and gentle undulations, now undergoing active dissection by the branches of the Rhine, Mosel, and other strong rivers which have eroded their steep-sided valleys deep beneath its even surface. The upland is surmounted by several ridges or elongated monadnocks; and some of these at least are composed of a very resistant quartzyte. In New England, the type Monadnock is, if my memory serves me, largely composed of an andalusite schist, which certainly has every appearance of being a resistant rock. Yet it must be freely admitted that, as far as I know, no artificial test has been made of its resistance as compared with that of many apparently resistant rocks around its base. It may be added that an appropriate test would be difficult to devise, inasmuch as exposure for ages to the weather would certainly be the best means of discovering the way in which long ages of weathering will affect a rock. In view of this difficulty, I hope that those who regard the peneplain explanation as compulsory will not be left alone to devise appropriate tests to determine the resistance of monadnock **rocks**, but that Professor Tarr and any others who are interested in the development of land forms, but who feel no such compulsion from the peneplain theory, will nevertheless turn their ingenuity in this direction. It is the truth of the matter that we are all striving for, not the maintenance of this theory or that; and it seems unfriendly if not unscientific to say that "the burden of proof should rest with the advo-

cates of the explanation." I am no more willing to be considered **an** advocate than I suppose Professor Tarr is to be
thought an enemy of peneplanation. We are not retained to
argue for or against the theory; each of us follows the guidance of the best evidence he can find. I trust that Professor
Tarr is just as much interested to discover whether monadnock rocks are resistant as I am to discover whether his theory
will account for uplands with even skylines. It is of course
difficult to avoid the appearance and even the style of the advocate or the enemy when writing earnestly in expression of
one's convictions, but for my part I cannot say too emphatically
that the peneplain idea shall find no "defense" from me. Let
us all set forth the *pros* and *cons* to the best of our ability, and
then the peneplain idea must look out for itself, and stand or
fall according to its value.

The objections thus far discussed relate to actual examples
of supposed peneplains. Attention may be next turned to a
group of objections based on general considerations, leading
to the belief that the production or occurrence of peneplains
is improbable or even impossible.

B. 1. *No peneplains are now found standing close to baselevel.* It is stated that "no extensive peneplains [not uplifted
or dissected] are known to exist at the present time in any
part of the earth," although "peneplains have been produced
again and again in the past, * * * Therefore, in accepting the peneplain theory, we need, as a fundamental assumption, to believe that during a part of the remote past, the
conditions have been different from these that have prevailed
in any part of the known earth during the present and immediate past" (p. 353, 354). Here Professor Tarr and I are in
essential accord, although I should prefer to replace "fundamental assumption" by "necessary corollary." As far as my
own understanding of the problem is concerned, it was not at
all as a fundamental assumption, but as a very surprising corollary that I came upon the difference between the present and
certain parts of the past, with respect to peneplanation. This
aspect of the question has often been discussed with my advanced classes, but it has not yet received the attention that it
deserves, and I am obliged to Professor Tarr for bringing it
clearly forward.

Although agreeing in the belief that the theory of peneplains involves a certain difference between the past and the
present, we do not agree as to the bearing of this belief on the
theory. Professor Tarr implies that the past, "whose history
has been worked out by purely stratigraphic methods," is
proved to be so like the present that the theory of peneplains
must be wrong because it involves a past that is in some ways
unlike the present. My opinion is that stratigraphic methods
do not always disclose a past closely like the present (see B4);
and that, even at their best, stratigraphic methods are not so
complete in their revelations but that all other lines of evidence
concerning the nature of the past should have a careful hear-
ing.

[*Editor's Note:* The illustrations and general remarks found
on pp. 222 and 223 are omitted as being nonessential to
the argument and as having been discussed earlier.]

B 2. *The earth's crust will not stand still long enough for
the slow process of denudation to produce a peneplain.* It is
justly urged that according to theory the later stages of peneplanation are much longer than the early stages of dissection
(p. 354), as Powell clearly pointed out some years ago; and
it is inferred that the earth's crust will not stand still long
enough for even penultimate denudation to be accomplished
(p. 362). But the stability or instability of the earth's crust
can be learned only by comparing the consequences reasonably deduced from one condition or the other with observed facts.
It seems to me a prejudgment of the case to enter it with the
conclusion that the lands do not stand still long enough for
peneplanation. Certainly they do not stand still long enough
in certain regions; witness the manifest effects of uneasiness
in the varied and unconformable stratified deposits, or in the
repeated renewals of dissection in the Alps. But the opposite
conclusion is enforced by both lines of evidence in the Piedmont region of Virginia.

[*Editor's Note:* An additional example has been omitted.]

But it has never been my intention to imply an absolute still-stand of the crust during a cycle of denuda-

tion. Any sort of movement that does not cause a distinct dissection of the surface below the peneplain level is admissible. Well-preserved peneplains, now dissected only by young, narrow valleys, give assurance that no significant valley-cutting below the peneplain level was permitted before the uplift by which the erosion of the existing valleys was initiated. Even in so uneven a region as southern New England, the gradual decrease of relief on approaching the coast makes it extremely probable that the deep valleys of the interior were not cut till after the peneplain was essentially finished. Any other supposition involves special conditions of oscillation and tilting that I believe are less probable than those involved in peneplanation, as may be seen by drawing a series of diagrams to represent the successive attitudes assumed by the land under different hypotheses.

It is sometimes suggested that before peneplanation, but after valleys like those of southern New England had been excavated, economy of work and time would be served by postulating a depression and a truncation of so much of the mountains as then remained above baselevel. The truncated surface thus produced would, under this supposition, correspond to the New England uplands. This truly effects an economy of work, measured by area of baselevelling, but it effects no important economy of time; for it will require essentially as long a time to truncate or baselevel a large cone as a small cone, structure and slope being equal. Moreover, unless very special suppositions were made as to the attitude of the land before, during, and after such a truncation of its mountains, the existing forms of southern New England could not be explained. During a submergence long enough to truncate the mountains remaining above baselevel, many shallow valleys would be filled with marine deposits; and after elevation, the streams might frequently abandon their former valleys for new, superposed courses. The narrow new valleys excavated on such courses, and the former valleys in which remnants of marine deposits might long linger, are not represented in southern New England.

Yet any supposition or process that will aid in the destruction of a land mass must be welcomed by those who believe that land masses have been destroyed, close down to baselevel.

The lateral swinging of large rivers, occasional incursions of the sea, changes of climate, anything that will contribute to the end is a pertinent part of the theory of peneplanation. Still my own opinion is that, of all processes, subærial denudation is the most important. This is not simply an opinion of preference; it is an opinion based on the arrangement of rivers in uplifted and dissected peneplains (Bull. Geol. Soc. Amer., vii, 1896, 377-398). Such rivers frequently exhibit adjustments that they could not have gained from a disordered arrangement during the present cycle of denudation alone; adjustments that could have been gained for the most part only during the cycle of peneplanation, and that would have been lost if the rivers had wandered far, or if the sea had abraded much of the land during the later stages of that cycle. It is of course perfectly possible that a peneplain should be smoothed off by the sea after it had been worn down under the air: such appears to have been the case with the Cambro-Silurian plain of northwest England (see B3). But it is not reasonable to suppose that every uplifted and dissected peneplain was thus smoothed before it was uplifted, although this supposition finds much favor with certain English geologists.

As to the arguments based on the slow progress of denudation during a brief period of observation (p. 355) or during post-glacial time (p. 361), I can only reply that a geographical cycle must be so enormously longer than either of these intervals that their evidence is not of value. Truly denudation is retarded when a capping of waste protects the rocks from the attack of the weather, but rather than side with De Luc, who concluded that waste-covered mountains are practically protected from further change, I should prefer to side with Hutton, who maintained that even the slow denudation of waste-covered slopes could produce great changes of form. As to the time that has elapsed during the denudation or dissection of peneplains, there is apparently no way of measuring it but by the work done. Hence the question returns to the verity of the peneplains; whether much or little time is needed to produce them is a secondary matter. Above all, a preconception as to the insufficience of geological time should not in this day be urged (p. 361) as a reason for not believing in the possibility of peneplanation. One sometimes hears a student say:

"I should think that drumlin ought to be more eroded if it has stood there unprotected since the ice sheet disappeared." Evidently such an opinion is based on a preconception of too long a post-glacial interval; for how can the interval be measured except by what has happened to the drumlin during its passage! How can past time be estimated except by studying what has happened during the progress of its ages!

B 3. *No part of the earth reveals even an approximation to a peneplain.* It is contended that as all the reputed peneplains now known are of the past, and that as all are now more or less fragmentary, "no part of the earth reveals even an approximation to this supposed condition" (p. 355). This seems to me an over strong statement in view of the form of such districts as the Piedmont belt, above referred to; but leaving aside even the best examples of well-finished and slightly dissected peneplains, let us consider some examples of peneplains that were submerged and unconformably buried after their surface had been reduced to faint relief, and that are now more or less visible where valleys are cut into the compound mass in consequence of uplift. It is true that these peneplains are not today standing in the position in which they were formed, that they make a very small part of the earth's actual surface, and that they are imperfectly open to observation; but it seems to me that they give strong evidence of the verity of peneplains, and that they certainly suffice to set against the strong assertion quoted at the opening of this paragraph.

[*Editor's Note:* Davis cites several examples where this is the case, namely the unconformable contact between the pre-Carboniferous and Carboniferous of the Grand Canyon of the Colorado, and a similar case in England. (Other cases are discussed in later papers in this volume.) Davis himself likes the cases cited in B4 as definitely requiring subaerial erosion.]

A buried plain of remarkably even form underlies the heavy Carboniferous limestones of northwestern England. It has been repeatedly described and figured by English geologists. An official report states:—"It is evident that these [Carboniferous] beds were deposited on an uneven floor of the Silurian rocks, for the line dividing the two formations runs sharply up or down 20 or 30 feet in places, while the bedding of the limestone keeps nearly horizontal. In other places . . . . Silurian grit sticks up in a boss, against the west side of which limestone has been laid down in horizontal strata" (Mem. Geol. Surv. Gr. Britain, Geol. of the Country around Ingleborough, 1890, 23). The inequalities of the floor here referred to are very small in comparison to the hights that the Silurian strata must have reached after their great deformation, for the sections represent the contact surface by an essentially even line, parallel to the limestone beds, and so it may be seen on various valley sides; for example in upper Ribblesdale. The actual contacts displayed in certain hillside quarries on Moughton fell are extraordinarily clear; one of them is well reproduced in the frontispiece to Bird's "Geology of Yorkshire." The heavy Silurian flagstones are so evenly truncated that a single layer of limestone stretches smoothly over them for a hundred feet or more across the quarry; the same limestone bluff may be traced for two or three miles around the side of the fell, close above the uppermost outcrops of the flagstones; and the same general division of the Carboniferous formation lies on the denuded surface over tens or scores of miles. As there is no residual soil on the firm rocks of the denuded plain, and as the overlying strata are heavy marine limestones (excepting local deposits of pebbly beds, one or two feet thick), the floor must have been swept and worn by the sea before Carboniferous deposition began. There seems to be no way of determining how much work was thus done by the sea, and how much had been previously done by subærial agencies; but whatever the proportions, a well finished plain of denudation, hundreds of square miles in area, had taken the place of a vigorous mountain range, before the deposition of the limestones began.

Goodchild has repeatedly referred to this ancient plain of denudation, and to two others of later date in northwest England. He says that when the deformed Cambrian and Silurian rocks "were brought within the destroying action of the waves . . . . the end of it was that the whole surface of the country was shorn off to one general uniform level; depressions and elevations there were, beyond a doubt, just as there are both depths and islands left on a modern plain of denudation; but in the main the surface was tolerably uniform" (Trans. Cumberland and Westmoreland Assoc., xiii, 1888, 92, 93). Many mountain slopes in the Lake district consist of re-exposed areas of this ancient plain, from which the weaker covering strata have been worn off again (Ibid., xiv, 1889, 76). The plain extends, locally, with marvellous evenness of contour, across the edges of quite five miles of strata" in the Lake district alone (Geol. Assoc., London, 1889, 45).

B 4. *Stratigraphic evidence is against the occurrence of peneplains.* It is said that quiescence sufficient for peneplanation requires conditions different from "those of that portion of the past whose history has been worked out by purely stratigraphic methods" (p. 355). Although I am not sure of just what is meant by "purely stratigraphic methods," it seems fair to regard the two examples given in the preceding section as dependent on at least a mixed stratigraphic argument. These examples were, however, associated with marine strata, and hence the subærial origin of the buried plains is not assured, although there can be little doubt as to the actual occurrence of the plains themselves. The following examples are more pertinent to the present discussion, for they point chiefly to the action of subærial denudation in the production of peneplains.

The Central plateau of France is a part of the ancient Hercynian mountain system of post-Carboniferous deformation, that once stretched across west-central Europe. Judging by the strength of its foldings, its altitude may for a time have rivalled that of the Alps of to-day. The mountains were greatly denuded during secondary time, as is shown by the comparatively even overlaps of Jurassic and Cretaceous strata on the flanks of the central plateau and elsewhere; but of these buried portions no more need be said at present. Continued denudation at last reduced the region of the central area itself to a surface of moderate relief, and it was upon a surface thus prepared that several brackish Tertiary lakes, communicating with the sea on the north and south, laid down their sediments. Since then, the region as a whole has been much uplifted, its southern and eastern parts have been irregularly dislocated, volcanic action has diversified parts of the surface, and denudation has effected important changes in the complex uplifted mass: but the northwestern part is free from dislocation and from volcanic action, and there the uplifted surface of denudation is well displayed in an even plateau (abstracted chiefly from Depéret, Ann. de Géogr., i, 1892, 369-378). If stripped of its volcanic cones and flows and 'unfaulted', the plateau would have the form of a vast inclined plane, highest in the southeast, and descending very slowly to the northwest (Boule, in Joanne's Dictionnaire géogr. et admin. de la France, iv, 1895, 2538). The northwestern part of the plateau, unaffected by volcanic action, and not covered by the lacustrine formations that elsewhere rest upon it, exhibits a surface of crystalline rocks interrupted only by closely folded troughs of coal measures, whose outcrops are sharply cut across at plateau hight, as if the whole structure had been rubbed down by a great levelling machine. The perfect regularity of the uplands between Montluçon and Creuse is an excellent representation of the form that the whole extent of the plateau region must have had about the beginning of Tertiary time, before it was uplifted and dislocated. Long continued erosion had then reduced the region to a plain close to sea level, thus destroying a great mountain chain and leaving in its place a lowland composed chiefly of long belts of granitic rocks (Vélain, "Auvergne et Limousin, géographie physique," in " L' Itinéraire Miriam," Paris, 1897, 10).

These extracts make it clear that French observers regard the stratigraphic evidence of the Tertiary lake deposits as confirming the conclusion reached from the study of form alone: both lines of evidence show that the uplifted, dislocated, and dissected plateau region of to-day was a lowland of denudation in Tertiary time.

The highlands of the Ardennes along the border of France and Belgium is another part of the ancient Hercynian range, greatly denuded. It descends southward, where it is overlapped by Mesozoic formations, among which the Cretaceous strata are of special interest in the present connection. A belt of coal measures extends from the Ardennes southwestward under the Cretaceous; shafts have been sunk through the Cretaceous to the coal measures at many points, and thus the form of the buried denuded surface has been determined with much accuracy. Gosselet's elaborate "Mémoire sur l'Ardenne"* (Paris, 1888) gives much information concerning both the buried and the unburied portions of the denuded mountains. The frontispiece shows the valley of the Meuse incised in the plateau, "everywhere leveled to the same altitude." Many sections in the text show the Mesozoic strata lying on the deformed and denuded Paleozoic rocks, but the basal deposits beneath the Cretaceous are usually not marine. Even under the Jurassic, there is a ferruginous clay with limonite concretions, thought to be of terrestrial origin (l. c., p. 802). Under the Cretaceous strata, the most general deposit is a layer of black pyritous clay with vegetal remains, taken to represent the soil of a pre-Cretaceous land surface. Fluviatile and lacustrine deposits are also recognized. The Carboniferous limestone is often pitted, and the pits contain non-marine materials and fossils. Where the intermediate deposits are wanting, the ancient rocks are perforated by boring mollusks and strewn with shells of oysters and serpulæ (p. 808, 810). On the uplands at a considerable altitude, and far beyond the main overlap of the Mesozoic cover, there are scattered remnants of Cretaceous and Tertiary deposits, and these are all regarded as of earlier date than the elevation and dissection of the plateau (p. 831). Fuller details as to the composition and distribution of these deposits are given in other papers by Gosselet (Ann. Soc. géol. du Nord, vii, 1879, 100) and Barrois (Ibid., vi, 1879, 340). Bertrand says that the buried pre-Cretaceous surface is a denuded plain, and that its existing irregularities are due, at least in great

---

*Ardenne is the name of a department of France; Ardennes, of the Highlands themselves.

part, to subsequent movements that the chalk also has suffered (Ann. des Mines, Jan., 1893, 36).

The different parts of the ancient mountains of the Ardennes overlapped by Triassic, Jurassic, Liassic, Cretaceous, and Tertiary formations, were doubtless exposed to denudation for different periods of time, and successively submerged in encroaching seas: it is quite possible that the dissected uplands of today were peneplained at a distinctly later date than was the floor beneath the Jurassic strata, and that the relation of the two is similar to the relation stated by Darton for the two parts (AB and BE, Fig. 1) of the ancient rocks of the Piedmont belt in Virginia. French writers do not seem to have occupied themselves especially with this question, either in the Ardennes or in the Central plateau. But on the other hand there seems to be no question that the stratigraphy of both the marine and the terrestrial deposits proves the existence in northern France of a denuded surface of small relief, whose larger part is now buried, and whose smaller part is elevated and more or less dissected.

Bohemia offers another remarkably good example for citation, as summarized by Penck and here freely rendered. A great mountain range once rose there, probably reaching an altitude of 5,000 meters. It was worn down to a comparatively even lowland before the incursion of the Cretaceous sea, by whose deposits it is now thinly covered, for freshwater formations are everywhere found under the Cretaceous strata. This relation is repeated in many other parts of Europe, especially where truncated old mountains are found. Terrestrial formations are their first cover, and upon these rest the later marine deposits. It follows from this that the truncated mountains of Europe were not denuded by the surf of ancient seas, eating into their hights and gradually wearing them away; for before the sea rolled over the old mountains they were already laid low and covered with terrestrial formations (Ueber Denudation der Erdoberfläche, Vienna, 1887, 23, 24).

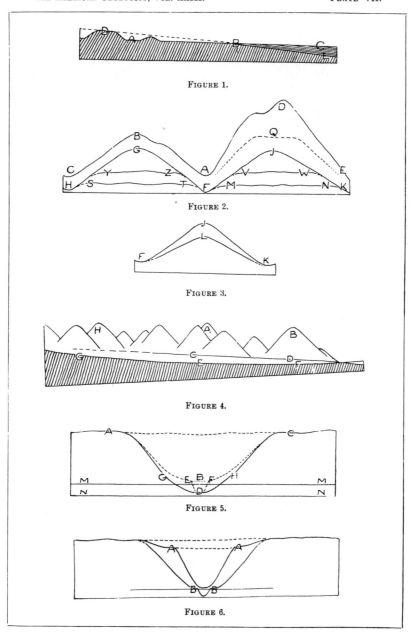

FIGURE 1.

FIGURE 2.

FIGURE 3.

FIGURE 4.

FIGURE 5.

FIGURE 6.

[*Editor's Note:* Material not essential to the argument has been omitted.]

C 2. *The subequality of mountain hights.* The development of a rough equality in the hight of mountain peaks by the faster destruction of the higher summits, on account of the greater violence of weather changes and of the absence of tree and soil covering at great hights, has been announced by Penck (Morphologie der Erdoberfläche) and by Dawson (Geology of the Kamloops area, Geol. Surv. Canada, 1894) and is now independently suggested by Professor Tarr. ABC and ADE, fig. 2, are in this way changed to FGH and FJK. The processes appealed to are not identical in the explanations of the three authors, but not having the writings of the two former at hand, I cannot give details. The approach to equal hight of many peaks of different structure in a given mountain group may be thus explained, as in the Alps, or in the Rocky mountains of Colorado: if any one felt *per contra* convinced that this subequality might be the result of the deep dissection of a greatly elevated peneplain, the discussion would doubtless be interesting.

The subequality of mountain hights being once gained, all

the mountains are then under essentially uniform climatic conditions, and for the rest of their lives difference of structure will determine their rate of decay. All changes would truly be very slow, but small differences in rate of wasting would suffice to develop distinct differences of altitude while the mountains were worn down from the tree line to the farm line. The uneven hill and mountain districts about lake Winnipiseogee, N. H., might be evolved in this way; but it seems to me very doubtful whether any such equality of hight as prevails in central Massachusetts can be explained as an inheritance from an equality determined by climatic control when the region had a much greater elevation, for to-day these uplands are about 3000 feet below the tree line, and their structure is by no means uniform.

C 3. *The beveling of hill tops.* .It is said that "by the time maturity of topographic form has been reached, there will be a beveling of hill tops where the harder gneissic and granitic rocks exist" (p. 368). I am not sure whether this and other references to beveling are correctly interpreted in what here follows, but the intended meaning seems to be that the hill tops would be somewhat flattened so as to imitate the broad uplands that are often found in (so-called) uplifted and dissected peneplains. That is, the conical form, FJK, would be changed to the beveled or truncated form, FVWK. Beveled forms of this kind are certainly common in central and western Massachusetts and in northern New Jersey; they are much better developed in such districts as the Piedmont belt of Virginia, the western part of the central plateau of France, and the uplands of the Slate mountains in western Germany. The broadly beveled upland, FMNK, would represent the type form in some of these examples. Not only are the ancient mountains beveled, but their beveled uplands fall in a systematic and accordant manner closely into a single plane, usually an inclined plane. There can be no question of the fact of such beveling, so far as existing form is concerned, in many well preserved peneplains; but no sufficient explanation of the process of beveling is given in Professor Tarr's article. It is briefly asserted. An explanation would be hard to find, unless it involved some especially active process, a function of altitude and climate, such as Richter has suggested for the beveling of the upland fjelds of Norway by means of the broadening of *kahre* floors under the action of local glaciers (Sitzungsber. Akad. Wien, 1896?); but it is manifest that all such processes are applicable only at altitudes above 5,000 or 6,000 feet in New England, as Q, fig. 2, and hence cannot be directly concerned in producing beveled hill tops at altitudes of 1,000 feet or less. In uplifted peneplains, where the uplands bevel across from valley to valley, it is manifest, as has already been pointed out, that existing processes are engaged in destroying, not in producing or perfecting the uplands. Indeed, the production of beveled hill tops, as here interpreted, seems to me so inherently impossible that I am for that reason persuaded that all this paragraph must be aside from the intention of Professor Tarr's theory. And yet, unless the systematic beveling of hill tops is some how accounted for, the theory must fail to explain the most essential features of those regions commonly regarded as uplifted and partly dissected plains of denudation.

Perhaps the change of form intended under the term, beveling, applies only to the mountain slopes, so that FJK, fig. 3, becomes FLK. If so, it does not materially modify the changes considered in the last paragraph of section C 2, and it leaves unexplained the important feature that I should call "beveling of hill tops," as considered just above. I hope that this important matter will be more fully presented by Professor Tarr.

C 4. *Denudation and beveling will be more advanced near the coast than in the interior.* It is urged that near the coast, "mountains would have been more lowered than in the interior, and, in the coastal region, there may well have been an approach to the peneplain condition" (p. 366); and again that "this beveling of the hill tops would be very much further advanced near the coast than in the interior, thus coinciding with the conditions found in New England" (p. 368). The brief statement allotted to these important propositions must leave the reader somewhat in doubt as to their explanation. As far as I can analyse them, they are not applicable even to the case of New England, much less to various other cases cited above.

The ancient mountain trends of southern New England are obliquely traversed by the shore line of Long Island sound.

The ancient mountain structures show no sign of weakening as they approach the shoreline, and we may fairly suppose that there has been about as much denudation to be done there as further north. From southern New Hampshire, across central Massachusetts and Connecticut, there is no indication of a weakening of the rocks; a good variety of more or less resistant gneisses and schists occur all across this district. Difference of climate cannot be appealed to as a cause of faster denudation and beveling near the coast, for the climate is more severe in the interior. The streams are larger and the valleys are necessarily lower and broader near the coast than in the interior, but the interstream uplands are under essentially similar conditions in the two regions. A rough equality of mountain hights being established, as H, A, B, fig. 4, there is then no reason for the greater action of the weather on a square foot of surface at B than at A. The master streams having gained a graded slope, GEF, there is no reason for the denudation of A to hesitate at the hight C, a thousand feet above the streams, while B is reduced to D, only a hundred feet or less above the streams. Hence I have to conclude that whatever strength there may be in the propositions quoted at the opening of this section, they are not presented in a way to make it apparent. The tilting of a previously denuded surface seems a relatively simple and safe way of accounting for the relation of the upland and stream profiles, CD and EF, as has been suggested above (A6); but tilting has no announced place in Professor Tarr's theory. The omission of so commonplace a movement is the more curious, inasmuch as frequent movement of some sort must be characteristic of a theory that "requires no long periods of relative quiet" (p. 369).

C 5. *New England and New Jersey as maturely disected mountains, rejuvenated.* It is maintained that the regions especially under discussion have been "lowered to the stage of full maturity, then elevated and made more rugged," and that although the surface has always been mountainous, it was "once less mountainous than now, because of the recent uplift" (p. 365). There are various other parts of the world to which a similar description might be applied: the indication of full maturity being found in the well-opened valleys of the larger streams, ABC, fig. 5, developed with respect to a former baselevel, MM; and the indication of "recent uplift" being no less apparent in the narrow gorges,D, incised in the floor of the maturely opened valleys, ABC,with reference to the new baselevel, NN. The essential evidence of such recent uplift and rejuvenation is found in benches E, F, on the compound valley slopes, AED and CFD. If rejuvenation were so long ago that this bench were destroyed by the development of continuous slopes, AGD and CHD, then the uplift could not be discovered by topographic evidence. It is on evidence of this kind that rejuvenation has been announced for the Susquehanna district of the Pennsylvania Appalachians, and for the Hudson valley; but no description has been published of any such evidence for the rejuvenation of New England. The deep valleys of many rivers in Massachusetts and Connecticut have no persistent benches on their slopes, and there is no visible reason for saying that any important pause was made in the elevation by which the former baselevel of the uplands, AC, was replaced by the present baselevel, NN, of the valleys, AGDHC. The rivers truly have many falls and rapids, due to their displacement from better graded courses by the irregular distribution of glacial drift, but this is quite another matter.

The valley-side benches are important items in the present discussion, if they exist at all; for when fully established they will define many interesting points. If high on the valley sides, as at A, fig. 6, (resembling the gorge of the Rhine) they would show that the previous cycle had gone far beyond maturity and well into old age before uplift occurred; if low down in the valley bottom, as at B (resembling the Frazer river valley in British Columbia), they are of trifling importance, as far as the present discussion is concerned. If the down-stream slope of the benches were about parallel to the present profile of the rivers, a uniform uplift would be suggested; if distinctly not parallel to the present stream profiles, an uneven uplift would be implied. All these points should receive specific attention if the new hypothesis, which "requires no long periods of relative quiet," is to be accepted. As now stated, the recent rejuvenation of New England seems to me very open to question. There is no evidence of recent rejuvenation in the occurrence of a young coastal plain along

the coastal border, such as ordinarily accompanies recent uplift, unless the postglacial coastal plain of Maine be so considered; and that would hardly be permissible, for the existing
valleys were eroded before the plain was formed.

[*Editor's Note:* Page 239 is omitted. This seems to repeat
earlier arguments.]

Editor's Comments
on Papers 4 Through 7

MODIFICATION AND APPLICATION OF THE PENEPLAIN CONCEPT

In 1913, Joseph Barrell, working in the Atlantic Piedmont, accepted the idea of the former existence of a broad peneplain, but suggested that the surface had been modified by a succession of marine advances and retreats into a series of terraces cut into the sloping peneplain. The abstracts (Papers 4A and 4B) express Barrell's general hypothesis. Comments by W. M. Davis, D. W. Johnson, and N. H. Darton, with a reply by Barrell, provide a lively discussion. A more lengthy paper by Barrell, which extends the marine terrace concept to New England, appeared posthumously in 1920. Modifications and criticisms can be found in articles by Johnson (1931), Rich (1938), and Adams (1945).

In the last decade, extensive borings and sedimentological studies on the Atlantic Coastal Plain from Virginia to Florida have shown many supposed "marine abrasion terraces" to be largely accumulations of lagoonal and of intertidal facies (Colquhoun, 1969). Although these studies do concern relatively young (Pleistocene) shorelines located for the most part in areas of rapid sediment accumulation, they point up the need for caution in the interpretation of terrace origins. Comparison is also invited with the *piedmont benchlands* idea introduced by Walther Penck in 1924.

In 1922, W. M. Davis published a short paper in which he indicated his comparative objectivity by making a strong plea for the modification, if necessary, of his cycle of erosion theory, rather than the use of piecemeal criticism to throw out the whole idea.

One of the first to do this was Walther Penck (1924). His compendious work, written in German, is difficult to understand, even for those well-versed in the language. Even Davis, who lectured and wrote in German, has been criticized for misrepresenting Penck's work. It was not until 1953 that this major work of Penck's was translated into English, by H. Czech and K. C. Boswell. Although no part of the original or the translation is reproduced here, a critique by M. Simon is excerpted in the following paragraphs.* The critique attempts to clarify some of the obscurities of the original work and to correct some of the misinterpretations that have arisen from it.

Basically, Penck offered two related ideas: (1) that a study of valley form would lead to an evaluation of vertical tectonic movements, and (2) that many planation surfaces are the products of stream erosion in regions undergoing slow, but accelerating, uplift. In this latter assumption he was accepting an alternative possibility already proposed by Davis, whose cycle theory typically depended on what happened *after* the uplift had virtually ceased.

Simons (1962), commenting on the Penckian view of the development of valley sides, says (pp. 3–4):

> Far from retreating parallel to itself, in Penck's view a valley side which begins as a smoothly convex surface above a stream which is eroding ever more rapidly, is transformed into a concave slope as erosion wanes, and thereafter becomes progressively flatter. If no rejuvenation takes place, the final result is a peneplain with flattish concave slope profiles, intersecting on the divides in obtuse convex angles which, Penck claimed, could be rounded to a very limited extent by weathering (M.A., 141–3).

*These selections are reprinted from *Inst. Brit. Geogr. Trans.*, **31**, 3–4, 9, 13 (1962). Copyright © 1962 by the Institute of British Geographers.

Simons continues with a discussion of three types of land-form associations presented by Penck either in *Die morphologische Analyse* (1924) or in "Piedmontflächen des südlichen Schwarzwaldes" ("Piedmont flats of the Black Forest; 1925). The first type is called by Simon the "great folding" and is described in the Black Forest paper as being a dome-like structure with a "single continuous peneplain summit-surface, tilting downward toward the margins, without any bench-like forms."

How the great folding type differs from the second type, or "crustal dome," is not too clear, except that it is associated with contraction not only in the massive sense but also in surface exposure. The crustal dome involves mass stretching and expansion of the surface area exposed to erosion. As Simons puts it:

> Describing the landforms on the domes, Penck observed that the summit-surface, once thought to be a single peneplain upraised and dissected, was actually a series of very broad, roughly concentric erosional benches arranged in descending, step-like order ("*Piedmonttreppen*"). Such landforms, he believed, are characteristic of domes which have risen, and expanded outwards, continuously.
>
> Each bench was said to have originated as a piedmont-flat ("*Piedmont-flache*") around the margin of the dome, for while the centre of the mass might be rising quite swiftly, the margin would ascend very slowly at first, and further out still would lie a region quite unaffected by the upheaval. Erosion at the edges of the dome would thus produce what was called a "*Primärrumpf*," a plain-like surface with very gentle convex slopes caused by the very slow waxing of erosion. Such a surface, Penck claimed, was the logical beginning of any erosion cycle.

Note here that the broad peripheral region of the dome, with its regional and local convexity, is taken to indicate the beginning of a cycle of erosion rather than the end (as Davis pictured it). The third type of landform association, the "inselberg," some-what anticipates pediplains in considering, at least incidentally, developments in more arid regions. Quoting from Simons:

> Processes of mass wastage alone, without river erosion, Penck claimed, could produce only concave slopes. Absence of all erosion could only result if a land mass remained stationary for a very long period. The result would be a landscape dominated by gently concave slopes with surviving, concave-sided residuals on the divides. This Penck termed the "Inselberg" form association, and in his view "it extends over very vast areas of the earth's crust" (M.A., 196). Inselbergs, he claimed, "are distinctive, not of any one climate, but of the continental

masses" (M.A., 196). If his general views on slope forms are even approximately correct, this conclusion is inescapable. However, in his elaborate discussion of slopes he seems to have ignored completely the possibility of abrupt changes of process, such as the transition from waste creep to sheetwash at the base of inselbergs, which has been described in semi-arid regions since his death. Penck stated somewhat rashly that "every observation made shows the error of the view that there is a sharp nick where the sides of the inselberg break off against the surrounding peneplane" and "the slope of all inselbergs is concave. A continuously concave curve always forms the foot slope . . ." (M.A., 191). The common error, that Penck regarded concave slope nicks as normal features of waning slopes in all climates, is due to a natural misunderstanding of his diagrams in *Morphological Analysis*. These are angular, but in a belated section entitled "Continuity of curvature of slopes" Penck explained that "the concave slopes of waning development are in reality, continuously curved." However, as has already been mentioned, sharp convex breaks of gradient, he claimed "play a highly significant part among the world's landforms" (M.A., 160).

The Penckian slope nick, therefore, is a convexity at the top of a valley side-slope, not, as so often stated, a concave angle at its foot.

Simons makes a final evaluation of Penck's work in his conclusions:

It has been possible in the foregoing paper to present only an outline of some aspects of Penck's system. There can be no doubt that he made some serious errors and contradicted himself on a number of occasions, thus casting grave doubts upon the reliability of his conclusions. None the less, there remains much in his work which is of value, and which will repay further study and development. In particular it seems very possible that the argument he used to demonstrate the dependence of stream erosion on crustal movements, though highly artificial, is capable of some limited applications. It may be true that the character of the uplift of a land mass is stamped on the landforms during the early stages of the erosion cycle at least, and may, thereafter, be recognizable.

Similarly, Penck's insistence that slope forms are a reliable record of erosional history does not seem unreasonable and certainly cannot be dismissed out of hand. In this respect, the validity of Penck's law of denudation, that, other things being equal, the rate of retreat of part of a slope is proportional to its steepness, remains to be tested. It may even be said that quantitative investigation of this principle is of first importance to geomorphology. For valley-side slopes to retreat parallel to themselves would be impossible if Penck's law is even approximately correct. At the same time, the rounding and flattening of valley shoulders envisaged by Davis and G. K. Gilbert is also contrary to Penck's law, the strict application of which would require valley shoulders to become less rounded and more angular early in the cycle, and later to become somewhat obtuse but nevertheless recognizable as distinct breaks of slope.

Since the inception of the planation idea, reference has been frequently made to those erosion surfaces buried under considerable thicknesses of younger sediments and constituting stratigraphic unconformities. After burial, these planation surfaces with their overlying sediments may be tilted, so that through subsequent erosion the line of unconformity may appear at the present topographic surface. Beveling across pre- and postburied surface rocks, this line may be the contact between the buried planation surface and another being formed. The general idea is important since it has a direct bearing on the dating of planation surfaces. It may also be possible to determine the kind of planation surface or surfaces involved.

An early paper by Wilson (1903), building on the work of others in the Canadian shield, describes the Laurentian upland as a typical peneplain developed mainly on crystallines but including a "sedimentary fringe" (Paleozoic), the sediments being underlain by an older surface that intersects the broad Laurentian upland. The older presedimentary peneplain is described as having the same hummocky surface as the glaciated Laurentian surface, which has interesting implications (including questions of exhumation, the relative ineffectivess of continental ice erosion, and the semi-permanence of hard-rock relief). This rather long paper is omitted in favor of two shorter ones that deal with better known areas. Paper 5 deals with the problem of passage from the "upland" (i.e., an unburied region) to the still buried but partly exposed surfaces. The essential question is whether the upland surface bends and passes beneath the sediments or whether a younger surface truncates an older one.

Paper 6 takes us to the Grand Canyon, Arizona, where two planation surfaces, both buried, can be seen to intersect along valley walls. Here the author deals more with the nature of the planation surface, discussing form, extent, and buried weathered surfaces.

Paper 7, which develops the theme of intersecting peneplains, is from the book by D. W. Johnson, *Stream Sculpture along the Atlantic Slope*. Only the introductory chapter is presented here; it reviews previous work on the Appalachian peneplains and sets the stage for Johnson's own regional study.

Douglas Johnson, a student of Davis, followed him in the acceptance of peneplains but emphasized the relationship between these surfaces and their possible more extensive covers, and the relation of these to regional superposition (Paper 7). Many questions raised by Johnson still remain unanswered, and many workers feel that his hypotheses lack firm support.

REFERENCES

Adams, G. F., 1945. Upland terraces in southern New England. J. Geol., 53, p. 289–312.

Barrell, J., 1920. The piedmont terraces of the northern Appalachians. Amer. J. Sci., 199, p. 227–258, 327–361, 407–428.

Colquhoun, D. J., 1969. Coastal plain terraces in the Carolinas and Georgia, U.S.A., in Quaternary geology and climate, Proc. Inst. Ass. Quat. Res., 7th Congress, 1965, 16, p. 150–162.

Czech, H., and Boswell, K. C., 1953. Morphological analysis of land forms. English translation of W. Penck's *Die morphologische Analyse*. St. Martin's Press, New York, 429 p.

Davis, W. M., 1922. Peneplains and the geographic cycle. Bull. Geol. Soc. America, 33, p. 587–598.

Johnson, D. W., 1931. Planes of lateral corrasion. Science, 73, 174–177.

Penck, W., 1924. *Die morphologische Analyse*. Engelhorn, Stuttgart, 283 p.

————, 1925. Piedmontflächen des südlichen Schwarzwaldes. Zeit. Ges. Erdkunde, Berlin, p. 81–108.

Rich, J. L., 1938. Recognition and significance of multiple erosion surfaces, Bull. Geol. Soc. America, 49, p. 1695–1722.

Simons, M. 1962. The morphological analysis of landforms: a new review of Penck's work. Inst. Brit. Geogr. Trans. Papers, 31, p. 1–14.

Wilson, A. W. G., 1903. The Laurentian peneplain. J. Geol., 11, p. 615–669.

4A

Reprinted from *Bull. Geol. Soc. America,* **24,** 688–690 (1913)

PIEDMONT TERRACES OF THE NORTHERN APPALACHIANS AND THEIR MODE OF ORIGIN

Joseph Barrell

(*Abstract*)

Physiographers are accustomed to build on the results of structural and historical studies in order to explain the origin of the present surface forms. Here the opposite method is followed—one which it is thought might be more widely adopted by historical geologists—building on a knowledge of the present relations of surface forms to rock structures in order to explain the past. In this first paper is given a description of the significant forms and the arguments as to mode of origin. In the second paper this data is combined with stratigraphic and structural evidence to elucidate the post-Jurassic history of the region.

Students of the strata of the Coastal Plain have discovered numerous surfaces of unconformity in those deposits, indicating repeated retreats of the sea; but they have drawn their shorelines close to the present outcrops, thus expressing the conclusion that the oscillations of the strand have been essentially confined to the present coastal plain, including the portions now submerged. The geologist who has carried the evidence of the sea farthest inland is W. M. Davis, who showed from the character of the superimposed drainage that the sea had at one time advanced over the trap ridges of New Jersey onto the border of the crystalline highlands. In the present paper it will be argued, however, from additional lines of evidence that earlier invasions of the sea advanced much farther inland, the farthest reaching in fact to central Pennsylvania and western Massachusetts.

As bearing on the nature of the evidence for determining the limits of former marine invasions the address of Gilbert is recalled, delivered to the Geological Society December 30, 1892. In this it is noted that the present limits of sedimentary formations are primarily determined by the length of time which their outlying portions have been above baselevel, secondarily on the thickness, resistance, and original limits. It is seen, therefore, that the mere absence of Coastal Plain deposits from the higher slopes of the Piedmont Plateau or seaward slopes of the Appalachians raises no presumption either for or against the former presence of the sea over that region. If they had once been present they would now be absent. The problem must, therefore, be approached through other lines of evidence.

Difficulties are then noted in the explanation of the higher surfaces planed across resistant formations over certain parts of Pennsylvania, New Jersey, and Connecticut as remnants of a single peneplain of subaerial denudation, commonly spoken of as the Cretaceous peneplain.

As a preliminary step for the reexamination of the problem a statement of criteria is then given for determining with some accuracy where the evidence is preserved, the elevation of former baselevels of subaerial and submarine erosion. In addition to other criteria, it is shown that wind-gaps of a certain character cut through hard ridges give the best record of subaerial baselevels. The submarine plains are best discriminated through a determination of their shores. For such plains as are shown by various lines of evidence to be of marine origin the ancient levels are best determined from the marked accordance and flat tops of the remnants of the more resistant formations.

To detect by means of these criteria the existence of marine plains which have been uplifted and very largely destroyed by subaerial erosion a method of projected profiles is employed. This is applied to western Connecticut and Massachusetts, and brings to light the fact that the plateau sloping from the Green Mountains and Taconic Range to Long Island Sound is not a simple uplifted and dissected peneplain, but consisted originally of a flight of rock terraces, themselves somewhat sloping. The lower of these are relatively narrow and poorly developed, but preserve a considerable approximation to the original surface. The higher terraces, originally wide and flat, were successively raised and warped so long ago that they have been very largely destroyed. Certain intermediate terraces are the most unmistakable in their character, having been rather well developed and still preserving well characterized remnants. This series of terraces is shown to have been developed on the eastern side of the Appalachians over the whole region studied, from northern Massachusetts to the Potomac River. They have been named by the author in descending order from localities in western Massachusetts and Connecticut, where they are preserved in recognizable state. Their names and the approximate elevation of their inner margins in their type localities are as follows:

	Feet
Becket terrace	2,400
Canaan terrace	2,000
Cornwall terrace	1,720
Goshen terrace	1,380
Litchfield terrace	1,140
Prospect terrace	920
Towantic terrace	730

Lower terraces of less perfect development are also noted at the following elevations:

	Feet
Appomattox (Lafayette) terrace	500 to 540
New Canaan terrace	340 to 380
Sunderland terrace	200 to 240
Wicomico terrace	80 to 120

In regard to mode of origin these terraces are shown to still preserve in a measure the forms resulting from wave planation, the features of cliff and terrace developed across various structures and rock formations at right angles to the drainage system indicating that they cannot be reasonably ascribed to river erosion or subaerial denudation. The latter agencies, however, worked preparatory to the marine planation, and have in later time through their destructive effects largely masked the evidence of the former invasions and retreats of the sea.

Certain independent tests of these conclusions are derived from the drainage of eastern Pennsylvania and the finding of scanty gravels remaining at an elevation of 700 to 730 feet in Maryland.

The demonstration of the theses of this paper depend very largely on the maps and profiles shown in the stereopticon slides.

4B

Reprinted from *Bull. Geol. Soc. America,* **24,** 690–696 (1913)

POST-JURASSIC HISTORY OF THE NORTHERN APPALACHIANS

Joseph Barrell

(Abstract)

The previous paper develops the evidence of an alternating series of uplifts and strand lines. The present paper seeks to determine their place in geologic time by correlating them with the strata of the Coastal Plain. An outline of the results is as follows:

The dates of the piedmont terraces may correspond with those of certain formations of the Coastal Plain. The dates of the uplifts, on the other hand, should be marked by the intervening planes of unconformity; or, if the uplift did not reach the Coastal Plain, should correspond to intervening formations. Warping is shown to have progressed variably, but without reversal, during the periods involved. Thus physiography, stratigraphy, and structural geology give three converging methods for testing sequence and age—the province of historical geology. The several divisions of the science are interlocked and render mutual support.

The sequence of regional events during post-Jurassic time is shown to be determined with considerable probability as follows:

Comanche.—Intermittent interior uplift reaching large amount. marginal downwarping, and continental sedimentation.

Cretaceous (Upper Cretaceous).—The sea invades the Appalachians for the first time since the Paleozoic, but now from the ocean instead of the continental interior. At the maximum of transgression its waves cut the Becket terrace against the southeast side of the Green Mountains and the south side of the Catskills. This was followed by emergence of several hundred feet, a crustal stability for a considerable period, and the cutting of the Canaan terrace. Both of these terraces appear to be distinctly older than the following series; so that, as the next is probably Oligocene, these may be tentatively regarded as probably middle and late Cretaceous respectively.

Eocene.—Emergence of several hundred feet and retreat of the shore to the region of the present Coastal Plain. Long period of subaerial erosion.

Oligocene.—Probable date of submergence and development of the Cornwall terrace.

Miocene.—Emergence of several hundred feet. Erosion of inner portion of Oligocene from the Coastal Plain, followed by submergence of present coastal plain and edge of the present Piedmont Plateau either at end of Miocene or early Pliocene. No terrace recognized.

Pliocene.—Intermittent uplift and retreat of the sea; cutting of the Goshen, Litchfield, and Towantic terraces. The Appomattox (Lafayette) terrace, still preserving in part its surface deposits, is 200 feet below the Towantic and is probably late Pliocene in age. Below the Appomattox, or Lafayette, are the fainter Pleistocene terraces.

An uplift of a few hundred feet was all that was necessary at any time to cause a retreat of the shore from these high-level terraces to the region of the present Coastal Plain, since the present great difference in elevation is mostly due to later progressive warping. Similarly, intervening phases of slight submergence and crustal rest are sufficient to account for the planing inland by the sea. The sequence of events thus indicates progressive but oscillatory uplift during Tertiary time, becoming more rapid in the Pliocene and Pleistocene. The conclusions, as in the previous paper, rest largely on an analysis of the profiles.

DISCUSSION

Prof. D. W. JOHNSON pointed out that if we have an uplifted and tilted peneplain on which there are monadnock residuals, it is possible to select positions in which terrace profiles can be drawn similar to some of those on which Professor Barrell bases his conclusions in favor of marine planation. According to the theory of marine planation, what are now generally regarded as monadnocks are to be interpreted as stacks on the wave-cut plain, and should still show traces of marine cliffs. So far as the speaker has observed, this is not the case. Water-worn pebbles on the upland surface may be of stream origin as well as of marine origin, and hence special care is necessary to eliminate the possibility of stream origin before such gravels are accepted as a confirmation of the theory of marine planation. The speaker would also like to know whether Professor Barrell found the terrace levels to be horizontal over wide areas.

Prof. W. M. DAVIS: The introduction of a new view by a critical investigator, who thus traverses an earlier view that had gained general acceptance, is likely to lead us nearer to the correct interpretation of natural phenomena, particularly if the investigator himself had previously accepted the earlier view which he is now obliged to give up; and, as in the present case, the new view is based on facts not previously recognized and, indeed, in the earlier lack of topographic maps hardly recognizable. Hence on general grounds Professor Barrell's new interpretation of the uplands of southern New England seems more likely to be correct than my own earlier one.

As an additional test for the occurrence of a series of ancient sea cliffs in these uplands, let me suggest a search for "cliffs of decreasing height," particularly along the latest-made (lowest-standing) shorelines; for if a sea cliff is cut back over a distance several times greater than the diameter of the hills into which the background is dissected, and if the shoreline is many times longer than a hill diameter, a certain number of the hills must be transected by the retreating shoreline on the inland side of their summits, so that the cliff height must decrease with further recession. The absence of such cliffs—indeed, the prevailing absence of any cliffs—along the inner hilly border of the even uplands of Devonshire-Cornwall and of Brittany has led me to regard these uplands as uplifted peneplains of normal erosion, in spite of their nearness to the sea, and in spite of the occurrence on them of gravel patches with marine fossils. These are explainable by a brief submergence of the peneplain after it was degraded and before it was uplifted and dissected. On the other hand, the occurrence of cliffs, and especially of cliffs of decreasing height, along the inner border of an even upland would go far toward proving its marine origin. However, if the hills of the background were elongated ridges, cliffs of decreasing height would be rare or absent. The long and narrow ridges, which the inclined trap sheets of the Triassic area would soon form in an uplifted platform of abrasion, would be reduced to a lower level by sea action south of a shoreline, while retaining their former level little reduced north of the shoreline. As there are many shorelines and many trap ridges, it would seem probable that some of the latter should be transected by some of the former, and at the point of transection a sudden increase of height should occur independent of the fault-lines which traverse the trap ridges.

As to the inland reach of a primitive Chesapeake Bay far northward in Pennsylvania, the transverse course of the Susquehanna, where it escapes from the syncline of the Wyoming Valley, ought not, in my opinion, to be regarded as an example of superposition, for a river must somehow escape from a synclinal valley by a transverse course, whether it is superposed or not. Likewise, the longitudinal course of a stream, pointed out by Professor Barrell along the northern side of the west end of the Wyoming syncline, is as easily explained by adjustment to structure as by indirect effect of marine action.

An important question on which Professor Barrell did not explicitly state his opinion is: What was the form of the land area in which the sea cut the broad platforms and low cliffs that he describes as occurring at successively lower and lower levels? The land seems to have had a hilly or mountainous surface in the background, as has long been recognized in the description which treats the uplands of southern New England as an uplifted and dissected peneplain; and in the foreground there was a buried land surface of small relief, which the Potomac or some other formation still covers unconformably, this buried surface being an element that is common to both the earlier and the later view. But what sort of surface existed in the broad belt of country between the background, not invaded by the sea, and the foreground, still covered by stratified deposits? It seems conceivable that this intermediate belt, as well as the buried foreground, may already have had a moderate or small relief more or less interrupted by monadnocks, and a general slope to the southeast, when it was exposed to the attack of the sea; it may, indeed, have served in this reduced condition as the floor on which a heavy subsiding marine strata were laid down during the advance of the sea on the subsiding continent; and the advancing sea may have even then trimmed and shaved the subsiding land at successively higher and higher levels; and in this case the retreating sea has had, during each pause in its retreat, to clear away the inner extension of the overlapping strata before it could bench the underlying floor. But all this may have happened without producing a great amount of change in the form of the preexistent surface of the intermediate belt; and if thus interpreted, this belt might be called a sea-benched peneplain. If, on the other hand, the work of the sea was more profound, the resulting surface would be best described as a series of broad sea benches or platforms, separated by low cliffs.

Let me turn to another aspect of this problem. In the outline of method written on the blackboard in explanation of my paper, that preceded Professor Barrell's, there were certain headings of which no mention was made in my remarks. One of these, in the column of safeguards, was the word hospitality, which indicated the attitude that an investigator should strive to maintain toward all new theories, particularly toward such as may replace his own preferred view; for in this way he can best guard against the serious danger of favoritism. I confess freely that it is difficult immediately to take a perfectly hospitable attitude toward a new theory that will very likely be substituted for an earlier one of my own which I have held for more than twenty years with growing faith in its competence, but I propose to do my best to assume the proper attitude toward the new view, and hope that the recognition of the difficulty of this duty will make it easier. With hospitality should be associated responsibility. It is not appropriate for one investigator to leave to others the duty of testing any theory that enters into the solution of his problem; but here I do not expect my practice to equal my principle. It is not likely that I shall take up new field study of this problem; and the admirably fair and competent treatment that it is receiving from Professor Barrell, along with my abundant occupation at present in other directions, is my excuse. Revision of opinion is another safeguard against error; that I shall hope to make whenever need be; and under the lead of this heading Professor Barrell's paper, when printed, will receive the most attentive consideration.

Further, the capacity to do critical work implies training, and among the elements of training I would place much importance on the unpleasant experience of being compelled by the force of new evidence to give up an opinion that one has held for a year or more with satisfaction; for this shows the subject that his mind is really open to change when good reasons for change are found. Experience of this kind is seldom provided in systematic instruction. It does not often come from one's own intention. It is best supplied by the work of one's colleagues, to whom one ought therefore to feel grateful as occasion offers. But gratitude of so dutiful a kind may well be accompanied by a certain amount of personal discomfort. It would not be sincere on my part to say that gratitude in this case is an unmixed feeling. The case resembles that of a patient who swallows his medicine, in spite of its unpleasant taste, because he has confidence in his physician. There is no one among our members from whom I, as a patient, would more confidently swallow an unpalatable dose in the expectation of its leading to a cure than from Dr. Barrell.

Mr. N. H. DARTON: I have given much attention to the physiographic development of the Piedmont slope in Maryland and Virginia and do not feel convinced that the region shows evidence of having been planed or terraced by the ocean. It seems more likely that the features described by Dr. Barrell are parts of slopes representing various stages of subaerial planation of Tertiary and later times. In a few places to the eastward there was some incursion of the sea during Chesapeake time, and there was extensive deposition of Lafayette and Pleistocene formation and later deposits, most of which have since been removed. I should mention in this connection that the Lafayette and Pleistocene deposits near the western margin of the Coastal Plain are not of marine origin, but are fluviatile and estuarine products.

In general it is unsafe to regard rounded hills of similar height as representing a terrace, for such hills are in process of degradation, and since Tertiary and early Pleistocene times have greatly decreased in altitude, and the approximate coincidence in height of scattered hills of hard and of soft rocks does not necessarily define a terrace.

Professor BARRELL replied as follows: Dr. Leverett has asked what relation the Hudson gorge may hold to the Sunderland terrace. In reply, I would say that this investigation shows no new facts on which to base a relationship. It is thought from glacial erratics dropped by floating ice on the Coastal Plain of Maryland that the Sunderland terrace was covered by the sea during an early epoch of glaciation. The terrace as shown on the Connecticut shore has been largely destroyed by erosion, yet fragments remain sufficiently intact to show level tops and integrate into an even sky line. Therefore a continental glacier was not here at the time of wave planation. The subaerial erosion features now submerged below present sealevel can be divided into the mature valleys which now constitute such drowned areas as Long Island Sound and Chesapeake Bay and the deep and narrow inner gorges. From the degree of their physiographic preservation these features do not impress one as older than the Sunderland terrace, so that the suggestion from this comparison is for a Pleistocene age.

Prof. D. W. Johnson has raised several points. First, in regard to the significance of gravels found on terraces. As Professor Johnson points out, the mere presence of such gravel is in no sense a criterion of marine as opposed to fluviatile origin. But when taken into connection with the size of pebble and the relations of the terrace to the highlands behind, then such gravels may become of significance. Bonney has shown that Alpine rivers of moderate length carry pebbles from four to six inches in maximum diameter up to distances of twenty miles and more from the Alps, over the Piedmont. On these Maryland terraces no such relations of mountains and grade existed as would give the rivers such carrying power. The gravels were found in scattered pebbles up to six inches in diameter on the highest remnants of a flat plain facing the sea and not within valley walls. Neither is the general height of the country behind such as could account for the presence of gravel of this size.

Professor Johnson speaks further of the possibility of fitting certain assumed terraces to any part of an irregular profile. Again, as an abstract proposition Professor Johnson makes a good point, but one which has been abundantly guarded against in this investigation and which is ruled out by the evidence. The terraces up to the 1,700-foot level are sufficiently well preserved over certain regions to give a markedly level sky-line, and these levels can be detected at intervals on resistant formations along hundreds of miles of the Atlantic slopes. The two higher terraces are less well preserved, and the conclusions in regard to them rest therefore not so much on their internal evidence as from the broader relations of these terraces, on the one hand, to the well-preserved ones at a lower elevation on the seaward side and, on the other hand, to the different character of the topography on the side of the mountains, especially the sharpness of that line diagonal to the structure which separates the highest terrace from the still higher and mountainous uplands.

Professor Johnson has spoken also of the distinction in form which should show between monadnocks and rock-stacks. These distinctions are largely features which would be lost long before the residuals had themselves disappeared. On the higher terraces the long exposure to subsequent subaerial activities has smoothed down all slopes, whatever their origin, to such grades as are in adjustment with the later history.

Mr. M. R. Campbell has spoken of the different degree of warping which would be implied in consequence of the acceptance of these terraces as surfaces of marine planation. It is true that if these be accepted, since they are greater in number than the baselevels previously used, it would mean a review of the correlation of the baselevels of the continental interior and would probably require the assumption of less warping in a new interpretation. On the interior slopes of the Appalachians the erosion has been presumably entirely subaerial, but the higher levels were largely developed on softer formations than where they are preserved on the Atlantic slope, with the result that subaerial peneplanation may there have been better developed, with the consequences in drainage history which have been previously held.

Mr. Darton has suggested that we do not know how much erosion has served to lower any particular hill and consequently that there is doubt thrown on the reconstruction of ancient terraces by the method of projected profiles. In answer, it should be stated that such a hypothesis must, of course, be in the geologist's mind in looking at every region, but that it may be readily tested in any region by the evidence of the field. When a plain is elevated and subjected to erosion, for a time there will be traceable a marked accordance between the level of the higher hilltops; but as all lose a part of their initial elevation, there will arise a maximum discordance of the higher summit levels. This will mark advanced maturity in the erosion cycle. In old age a new accordance will slowly become developed in relation to the lower and younger baselevel. The test, then, of the amount of loss by erosion suffered by the highest remnants of an old marine plain is the degree of accordance of the highest residuals—an accordance to be traced across the strike of resistant rocks and thus shown to be independent of structure. This criterion has been applied from New England to the Potomac River and indicates that it is still possible to rather closely determine the levels of even the terraces most destroyed.

Mr. Darton has mentioned that the Sunderland was largely an estuarine formation. It may also be so regarded in Connecticut, since Long Island must then have existed as a partial barrier to the open sea. Furthermore, all the Pleistocene terraces show the marked character of lapping around headlands, indicating that the land did not stand still sufficiently long after a stage of subsidence for shore erosion to produce a mature shoreline.

Professor Davis has made the point that the slopes of the hills partly consumed by sea erosion should be steeper facing the sea than on the landward side. It would seem that this criterion should be susceptible of application to the Pleistocene terraces, though the initial slopes may be somewhat masked by the effects of glaciation. The test must, however, be made in the field rather than from maps, since these New England maps are in general not faithful enough in regard to the details of slope to enable them to be used for such refined discriminations.

5

Reprinted from *Science,* **69**(1795), 544–545 (1929)

THE FALL ZONE PENEPLANE

Henry S. Sharp

WITH the passing of years the subject of the peneplanes of the eastern United States has not gained in
simplicity, and some may feel that the addition of
another name to those already proposed for such
features will needlessly complicate the existing confusion. Nevertheless, the erosional surface to which
it is proposed to apply the term Fall Zone peneplane
is of considerable importance to the geomorphologic
history of the region between Connecticut and
Georgia, although it has lacked proper recognition,
perhaps on account of the absence of a convenient
and suitable means of designation. In the following
paragraphs the author sets forth his reasons for
adopting the term "Fall Zone peneplane" in the hope
that others may find the name useful.

The recognition that the erosional surface underlying the Coastal Plain deposits is entirely distinct
in origin from the upland peneplanes of New
England and the Piedmont has been a matter of slow
development, which need not be discussed here. A
number of years ago Shaw[1] reviewed the literature
and emphasized the relationships for the region
southwest of the Hudson; more recently Renner[2] has
carried the discussion into southern New England.
He recognizes the presence of two peneplanes where
Davis and earlier writers had found only one: an
older steeply sloping surface exposed by the stripping

[1] E. W. Shaw, "Ages of Peneplanes of the Appalachian Province," *Bull. G. S. A.,* 29: 575–586, 1913.
[2] G. T. Renner, "The Physiographic Interpretation of
the Fall Line," *Geog. Review,* 17: 278–286, 1927.

away of the Coastal Plain to which he has tentatively applied the term Jurassic (?) peneplane, and the younger more gently sloping Upland peneplane to which he refers as the Tertiary (?). In his enlightening paper he assigns the origin of the falls and rapids of the Fall Line to the change in slope at the intersection of the two peneplanes.

Geologists in general have been slow to recognize this change in status of the Upland peneplanes, and they still are probably most commonly called the Cretaceous or Cretaceous (?) peneplanes, although as shown above it has become evident that the Coastal Plain basement and its uncovered extension is a distinct peneplane and not as formerly supposed the continuation of the Upland surfaces. Although of considerable extent and geologic importance this resurrected peneplane has had no more significant name than the Jurassic (?). In an unpublished report prepared in connection with a course in physiography at Columbia University, the writer, doubting the utility of age names of little accuracy and impressed by Renner's explanation of the origin of the Fall Line, first called this peneplane the Fall Line peneplane. Inasmuch as a peneplane is a surface the word line seemed misapplied, and at the suggestion of Professor Douglas Johnson the term Fall Zone peneplane was finally decided to be somewhat more appropriate. In fact, the peneplane stretches from Connecticut to Georgia in a long narrow belt or zone seldom fifteen miles wide between the more gently sloping Upland peneplanes on the west and the Coastal Plain on the east. It is considerably steeper than the Upland surfaces and perhaps averages fifty feet per mile in slope. Not only does the term Fall Zone avoid the implication of exact knowledge concerning the age of the peneplane, but it possesses the singular advantage of conveying a considerable amount of information about its location and significance in two words. For these and the following reasons the name is offered in the hope that its use may help to lessen the confusion surrounding the subject of the peneplanes of the eastern United States.

Although the Fall Zone surface has been called the Jurassic (?) peneplane, a moment's thought will show that this surface has the very structural relationship to the Cretaceous cover which the so-called Cretaceous peneplane, the Upland, was believed to have and which was the basis for dating the latter as Cretaceous. On correct evidence the resurrected peneplane, which it is proposed to call the Fall Zone, therefore has a better claim to the term Cretaceous than the Upland peneplane of New England or any surface correlated with the latter, although these have long been called Cretaceous because of a misconcep-

tion. Nevertheless any belated attempt to rectify the error by calling the Fall Zone peneplane the "Cretaceous peneplane" would surely add to the difficulty, since with the general though unjustifiable use of the term Cretaceous or even Cretaceous (?) for the Upland surface, its application to the Fall Zone peneplane would simply add a second Cretaceous peneplane to geologic literature.

In addition it may be doubted whether the Fall Zone peneplane could properly be called Cretaceous, even though there were no other surfaces of that name. It is surely younger than the Newark Series of the Upper Triassic, for it is known to bevel rocks of that age in Connecticut and New Jersey. If, as some believe, the top of the Newark is of Lower Jurassic age, its development started later than early Jurassic. The tectonic disturbances which resulted in the block faulting of the eastern Triassic areas are usually dated at the end of that period or during the early Jurassic, and the Fall Zone surface was developed after these movements. On the other hand, a progressive overlap toward the northeast covered it with successively younger beds of Comanchian and Cretaceous age, and at any given area it must be older than the oldest of these beds immediately above it, so that it varies in time of completion from place to place.

As suggested by the overlap, one part of the peneplane was being buried while another part further north was apparently being carried to a greater stage of completion by erosion. In places where covered by the Lower Comanchian its formation must have taken place entirely within the Jurassic, and it might there be called the Jurassic peneplane. Elsewhere the time of its formation included not only most of the Jurassic but all the Comanchian up to the base of the Cretaceous or higher. These parts might perhaps be called the Jura-Comanchian or the Jura-Cretaceous peneplane, but it is evident that these terms no more than the Jurassic can be applied to the surface as a whole. Different parts of the same erosional surface are materially different in age, and the term Fall Zone may be taken to indicate a surface unit although not the same time of formation for that unit. An analogous case is found in stratigraphy where a formation name may indicate a stratigraphic but not a chronologic unit.

For the several reasons stated it seems desirable to avoid implying a single precise date of origin for a peneplane of wide extent. In the case of the surface under discussion such an implication is conveniently and suitably avoided by calling it the Fall Zone peneplane.

6

Reprinted from *Bull. Geol. Soc. America*, **51**, 1236–1237, 1239, 1240, 1244–1245, 1264–1265 (1940)

EP-ARCHEAN AND EP-ALGONKIAN EROSION SURFACES, GRAND CANYON, ARIZONA

BY ROBERT P. SHARP

ABSTRACT

The buried Ep-Archean and Ep-Algonkian erosion surfaces of the Grand Canyon are outstanding examples of well-preserved peneplains. The maximum relief of the Ep-Archean surface is less than 50 feet, and monadnocks rise a maximum of 800 feet above the level of the Ep-Algonkian surface, about 95 per cent of which is relatively flat.

Residual regolith on these surfaces shows that they are of subaerial origin and that chemical weathering was dominant in the closing stages of the erosion cycle. The ultimate product of extended weathering consists of quartz, muscovite, clay, and hydrated iron oxides. In some places a residual enrichment of iron appears certain. The most maturely weathered materials contain an average of 88 per cent insoluble residue. Moderately fresh potash feldspar on the surfaces and in the overlying beds is attributed to removal of the more maturely weathered material by the encroaching seas.

Various lines of evidence are cited to show that climatic conditions toward the close of the Ep-Archean and Ep-Algonkian intervals were relatively humid. The most noticeable effects of marine processes are the reworking of the weathered mantle and erosion of monadnocks. A spectacular slide breccia resulting from such erosion is described. A calculation based on the best available evidence indicates approximately 100 million years as the length of the Ep-Algonkian interval, and the Ep-Archean interval may have been longer.

INTRODUCTION

INTRODUCTORY STATEMENT

Geomorphologists have been seriously handicapped in their study of peneplains because such surfaces are usually represented only by scattered remnants of limited extent and so modified by subsequent erosion that the details of form and mode of origin cannot be easily determined. At least a partial solution of this problem appears to lie in the study of buried erosion surfaces. On the steep walls of the Grand Canyon two such buried surfaces are wonderfully exposed. The extent and topographic features of these surfaces are described. The physical character and mineralogical and chemical composition of the overlying and underlying materials have been studied. An attempt is made to present a logical picture of the nature and mode of origin of these surfaces, of the climatic conditions under which they were formed, and of the modifications produced by marine processes when they were submerged.

[*Editor's Note:* Material has been omitted at this point.]

110

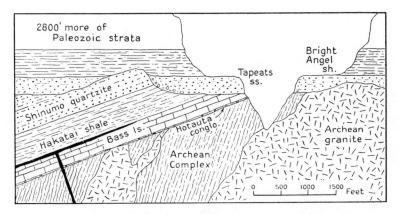

FIGURE 1.—*Diagrammatic cross section, inner part of Grand Canyon*
Hotauta, Bass, Hakatai, and Shinumo are Algonkian; Tapeats and Bright Angel are
Cambrian. No direction or elevations given because section is diagrammatic.

EP-ARCHEAN SURFACE

[*Editor's Note:* Material has been omitted at this point.]

TOPOGRAPHIC FEATURES

The extreme smoothness of this surface is its principal feature. Noble
(1910, p. 504; 1914, p. 42) notes a maximum relief of 20 feet in the
Shinumo quadrangle, and Hinds (1935, p. 9) reports a relief of less than
50 feet in the area investigated by his parties, chiefly the Bright Angel
and Vishnu quadrangles. Below Hance Rapids (Pl. 2, fig. 1), on Crystal
Creek (Fig. 3), in Hotauta Canyon, and at Granite Narrows (Pl. 2, fig. 2)
the Ep-Archean surface is extremely smooth with a relief of only a few
feet. Opposite the mouth of Shinumo Creek low rounded knobs of
Archean rock project 20 feet above the general level, and this was the
maximum relief observed.

The extremely low relief of this surface may mean that it approaches
the ultimate form of the erosion cycle. This ultimate form—the plain—
has probably never been recognized except in theory (Davis, 1899b,
p. 497); therefore, the Ep-Archean surface is usually referred to as a
peneplain. Its mode of origin is considered in a later section.

WEATHERING ON EP-ARCHEAN SURFACE

Depth and mineralogical changes.—The Archean rocks are noticeably
weathered to an average depth of 10 feet below the Ep-Archean surface,
although locally at the foot of Hance Rapids they are relatively fresh
just below the surface. This relation is thought to be due to removal of
weathered detritus by the Algonkian sea.

[*Editor's Note:* Material has been omitted at this point.]

FIGURE 2. EP-ARCHEAN SURFACE AT GRANITE NARROWS
Ep-Archean surface (dashed line) and Ep-Algonkian surface (dotted line). View westward downstream.

FIGURE 1. EP-ALGONKIAN SURFACE NEAR ZOROASTER CREEK
Tapeats sandstone wedges out both to the right and left. Monadnock of massive Archean granite on right.

EP-ALGONKIAN SURFACE

[*Editor's Note:* Material has been omitted at this point.]

TOPOGRAPHIC FEATURES

The relief of the Ep-Algonkian surface is controlled by the nature of the underlying rocks, for the highest monadnocks are composed of Shinumo quartzite (Pl. 7, fig. 1) and massive granite. The monadnocks underlain by the Shinumo quartzite are linear strike ridges (Noble, 1914, p. 40; McKee, 1933, p. 203-204; Hinds, 1935, p. 27; Wheeler and Kerr, 1936, p. 4-5) which form low hogbacks or cuestas on the Ep-Algonkian surface. The monadnocks of Archean rocks, chiefly massive granite, are lower, more rounded, and of no great lateral extent, as exemplified by a monadnock just west of Zoroaster Creek in the Bright Angel quadrangle (Pl. 4, fig. 1).

Hinds (1935, p. 27) reports a maximum relief of 250 feet in the Bright Angel and Vishnu quadrangles, and Noble (1914, p. 51) records a relief of 600 feet in Monadnock Amphitheater in the Shinumo quadrangle. A large monadnock on the north side of the Grand Canyon between Bright Angel and Ninetyone Mile creeks (Pl. 6, fig. 2) is almost 800 feet high (Wheeler and Kerr, 1936, p. 5), and this is the highest known point on the Ep-Algonkian surface in the Grand Canyon, where more than 95 per cent of the Ep-Algonkian surface has a relief of less than 150 feet and a large part is essentially flat (Pl. 3, fig. 2). Some students may wish to call such a surface an "old land" (Maxson and Anderson, 1935, p. 90-91), and others may prefer to call it a peneplain. In this paper the Ep-Algonkian surface will be referred to as a peneplain with the realization that not everyone will agree.

WEATHERING ON EP-ALGONKIAN SURFACE

Depth of weathering.—Extensive weathering of Archean rocks, chiefly gneissic granite 10 to 12 feet below the Ep-Algonkian surface, has been observed in many places, notably at the Garden Creek Indian ruins (Pl. 5, fig. 2), in Ninetyone Mile Canyon, in Elves Chasm (Fig. 4), and elsewhere. At the 96 and 117½ mile points Archean granite shows minor effects of weathering 20 to 25 feet below the surface.

[*Editor's Note:* Material has been omitted at this point.]

SUMMARY

The buried Ep-Archean and Ep-Algonkian erosion surfaces of the Grand Canyon are excellent examples of well-preserved peneplains. The maximum relief of the Ep-Archean surface is less than 50 feet, and this extreme smoothness is probably due to long erosion and to uniformity of the underlying rocks. At least 95 per cent of the Ep-Algonkian surface has a relief of less than 150 feet, and a large part is essentially flat. The remainder consists of monadnocks, underlain by massive Archean granite or Shinumo quartzite, which rise to a maximum height of 800 feet.

Both surfaces show evidence of subaerial weathering, and this weathered mantle has been studied in considerable detail. The greatest depth of weathering is 10 feet beneath the Ep-Archean surface; it is 50 feet beneath the Ep-Algonkian surface, but only the uppermost 5 to 10 feet is greatly altered. The following progressive changes with increasing intensity of weathering have been noted. Biotite changes first to a green chlorite and then to a colorless chloritelike material in which are dispersed small grains of iron oxide. Eventually, biotite is entirely eliminated, leaving hydrated iron oxide. Feldspars show some kaolinization even at considerable depth, and plagioclase is almost entirely decomposed in an early stage of the weathering. Potash feldspar has been more resistant, although in some places it too has been entirely decomposed. Quartz and muscovite have been relatively unaffected and show a residual enrichment. Quartz and muscovite with hydrated iron oxides and clay constitute most of the residual regolith produced by extended weathering; the amount of iron oxides ranges from 0.9 to 12.8 per cent. In a few places a residual concentration of iron oxide in the weathered zone appears certain, and it is suspected in other localities. Insoluble residues constitute an average of 88 per cent of the most maturely weathered material. The evidence of chemical weathering on these surfaces is far greater than the evidence of mechanical weathering.

Several lines of evidence offer proof that the weathering on the Grand Canyon surfaces is ancient. Perhaps the most convincing is the incorporation of the weathered detritus in the basal part of the overlying sedimentary beds.

The red color of the weathered detritus, the evidence of extensive chemical weathering, the high percentage of insoluble residues, the lack of soil caliche, and other types of evidence are believed to indicate a relatively humid climate during the closing phases of the Ep-Archean and Ep-Algonkian erosion intervals. The probable absence of plant life is admittedly a factor which cannot be easily evaluated.

The Grand Canyon surfaces, although buried beneath marine deposits, are believed to be the product of subaerial erosion and not marine planation, chiefly because of the evidence of subaerial weathering on both surfaces. No evidence of wind action was noted, and it seems likely that weathering, mass movements, and running water were the major factors producing the low relief of both surfaces.

Calculations are made showing that in some places at least 15,000 feet of rock has been eroded from the Ep-Algonkian surface. Using a composite figure for the rate of lowering of land by erosion, it is calculated that roughly 100 million years were required to produce the Ep-Algonkian surface.

It is concluded that marine processes have modified the topography of the surfaces only to a minor degree. The major effects of marine erosion have been (1) a reworking of part of the weathered material and its incorporation in the overlying beds, (2) a slight levelling of the surfaces, and (3) a more noticeable and spectacular steepening of the slopes of monadnocks on the Ep-Algonkian surface.. The face of at least one such monadnock has been so oversteepened by marine erosion that it has broken away in the form of a slide.

WORKS TO WHICH REFERENCE IS MADE

Campbell, Ian, and Maxson, J. H. 1938a *Geological studies of the Archean rocks at Grand Canyon Arizona* (Abstract), Carnegie Inst. Washington, Year Book 37, p. 359–364.

Davis, W. M. 1899b *The geographical cycle,* Geog. Jour., vol. 14, p. 481–504; also in *Geographical essays* (1909) Ginn and Co., p. 249–278.

Hinds, N. E. A. 1935 *Ep-Archean and Ep-Algonkian intervals in western North America,* Carnegie Inst. Washington, Publ. 463, p. 1–52.

Lawson, A. C. 1902 *The Eparchean interval,* Univ. Calif. Publ., Bull. Dept. Geol. Sci., vol. 3, p. 51–62.

Maxson, J. H., and Anderson, G. H. 1935 *Terminology of surface forms of the erosion cycle,* Jour. Geol., vol. 43, p. 88–96.

McKee, E. D. 1933 *Mountains at close of Algonkian Era,* Grand Canyon Nature Notes, vol. 8, no. 7, p. 202–204.

Noble, L. F. 1910 *Contributions to the geology of the Grand Canyon, Arizona—The geology of the Shinumo area,* Am. Jour. Sci., 4th ser., vol. 29, p. 369–386, 497–528.

———— 1914 *The Shinumo quadrangle, Grand Canyon district, Arizona,* U.S. Geol. Survey, Bull. 549, 100 p.

Wheeler, R. B., and Kerr, A. R. 1936 *Preliminary report on the Tonto Group of the Grand Canyon, Arizona,* Grand Canyon Nat. Hist. Asso., Bull. 5, p. 1–16.

Reprinted from *Stream Sculpture on the Atlantic Slope*, Douglas Johnson, Columbia University Press, New York, 1931, pp. 3–13 (and Hafner Publishing Co., Inc., 1967)

STREAM SCULPTURE ON THE ATLANTIC SLOPE

Douglas W. Johnson

CHAPTER ONE

PREVIOUS THEORIES OF APPALACHIAN HISTORY

No present study of Appalachian geomorphology can fail to rest heavily on the great body of facts and inferences set forth in the literature by many workers in this classic field of investigation. One who would, from a new point of view, examine the facts afresh, must feel his great indebtedness to those who have pointed the way for his particular studies. In suggesting a theory of evolution of Appalachian topography which differs in one or two essential points from those current in the literature, the writer acknowledges his debt to predecessors in this field whose works are too varied for all of them to receive specific mention in the brief review of earlier theories presented in the pages which follow.

So far as the writer is aware none of the many workers on Northern Appalachian geomorphology has considered the possibility that the transverse drainage of this region was superposed from a coastal plain extending far inland across the Appalachian ridges and resting on a peneplane older and higher than the highest surface (Schooley or Kittatinny) preserved in the present topography. It is true that more than one investigator has wondered where the pre-coastal plain surface would pass if projected inland. It is true that Davis[1] inferred a former moderate inland extension of the coastal plain, and he followed a suggestion of Tarr's[2] so far as to admit the possibility of a superposed origin for the lower portions of the Connecticut and Housatonic Rivers, and explained certain New Jersey drainage and the lower Susquehanna on the same basis. But he limited his restitution of the coastal plain cover to a belt from 25 to 40 miles broad in Connecticut and New Jersey, and possibly 80 miles broad in central Pennsylvania; and he imagined this extension of the coastal plain cover to rest upon the Schooley peneplane. In his earlier work Barrell [3] went a little farther in Pennsylvania by carrying the cover over the Kittatinny Mountain ridge instead of to the vicinity of the ridge as did Davis; while in New Jersey he likewise inferred a cover over the Kittatinny ridge,* although not over the crystalline Highlands farther southeast nor over that part of the Kittatinny ridge in New York. The extent of the coastal plain cover in New England is less clearly stated by Barrell, although he seems to have fixed a "Cretaceous shoreline" in northwestern Massachusetts and Vermont, over a hundred miles from the present shore. [4] We have no sufficiently clear statement as to Barrell's latest views regarding the relation of the coastal plain cover to the surface known as the Schooley peneplane. Apparently his earlier view was similar to that held by Davis — the coastal plain deposits rested on the fluvial "Kittatinny" or "Cretaceous" peneplane (the Schooley peneplane of Davis); but later he certainly inferred marine planation at various levels, interrupted more than once by wholesale fluvial erosion, possibly continued to the stage of planation. [5] Barrell follows Tarr and Davis in ascribing the southeast courses of the Connecticut and Housatonic Rivers to superposition, but apparently it was with marine invasions of post-Schooley date that were associated the marine deposits upon which these streams acquired the courses in question. [6]

If any one has derived Appalachian drainage by superposition from a coastal plain resting on a peneplane of pre-Schooley (pre-Kittatinny) date, the fact has escaped the writer's notice. In any case this theory has made no headway in the literature; neither have the facts which make it seem plausible, nor the

* Often called "Blue Ridge" by Barrell through confusion with "Blue Mountain," an alternative name for the Kittatinny ridge.

radical changes which the theory, if accepted, must introduce into our interpretations of later Appalachian history, received the attention they deserve. It seems pertinent, therefore, to outline the conception of Appalachian evolution which the writer and his students have been employing as a working hypothesis for some years;* to specify certain facts which seem to give a measure of support to this view; and to indicate briefly some of the consequences one must be prepared to accept in case one adopts it. If in the present volume the writer necessarily devotes much of his attention to the new interpretation being set forth, the reader must not from this fact infer that earlier theories of Appalachian evolution have definitely been discarded. He should rather, if the evidence and arguments seem to warrant, give the new theory hospitable welcome into the group of multiple hypotheses entitled to consideration in any effort to discover the true history of the Appalachians.

SOME ELEMENTS OF APPALACHIAN HISTORY

Recognition of two intersecting peneplanes on the crystallines. — It has long been recognized that the Northern Appalachians were (1) formerly reduced to a peneplane of remarkably low relief, (2) then partially overlapped by an encroaching coastal plain, and (3) later uplifted, the coastal plain cover largely stripped away, and the underlying folded structures deeply trenched by the superposed drainage. For Davis the widespread peneplane of this history was the Schooley surface, and he set moderate limits to the extent of coastal plain encroachment. Barrell apparently first accepted the idea of a fluvial peneplane, then inclined toward the conception of wholesale marine peneplanation; but he finally developed the view that an early fluvial peneplane of wide extent was later profoundly modified, and much of it completely destroyed, by

* For a brief preliminary statement of the theory, see program of the 1928 meeting of the Geological Society of America, published in *Bull.*, G. S. A., Vol. 40, 132–133, 1929.

marine agencies; and he assumed repeated marine transgressions which in New England were much greater than those recognized by Davis. We need not here concern ourselves with the age or the origin of the peneplane in question. But it is essential to our discussion to recall that Barrell [7] for New England, and Darton [8] and others for the Piedmont, some years ago recognized that the beveled crystalline surface underlying the Cretaceous coastal plain deposits along the Atlantic border, is steeper and may be older than the adjacent upland peneplane * — a view later set forth more fully by Shaw. [9]

This last point is so vital to the thesis of the present volume that it deserves some further discussion. Barrell's conception of marine erosion of the New England and Piedmont crystalline uplands rests largely on the recognition of what he considered to be wave-cut terraces on both surfaces. [10] The terraces were believed to trend roughly parallel with the present shore line, the descent from one terrace level to another being accomplished by a pronounced slope. Should the inclined crystalline surface under the Atlantic coastal plain represent one of these inter-terrace slopes, it would under Barrell's interpretation be *younger* than the adjacent upland, and certain of the deductions set forth in the following pages would be invalidated. Barrell does not discuss this possibility, although he does make a clear distinction between the inter-terrace slopes and the inclined floor upon which rest the coastal plain deposits. Doubtless some of the considerations set forth in later paragraphs seemed to him, as they seem to the writer, sufficient ground for regarding the pre-coastal plain floor as wholly distinct, in nature and in origin, from the slopes connecting the supposed marine terraces.

Multiple projected profiles east-west, north-south, and north-

---

* For purposes of the present discussion I shall assume the essential correctness of the old view that there is, both in the Piedmont belt and in southern New England, a major upland peneplane level easily recognized in accordant summits of hilltops in each region. Should it be proved that this surface is terraced, or that it consists of remnants of peneplanes of many different dates as different writers have suggested, the particular thesis of the present volume will not thereby be affected.

west-southeast, across all of southern New England, and similar profiles thus far completed for a part only of the Piedmont, enable one to analyze more fully than did Barrell the character of the supposed terraces, and the intervening slopes. This study has tended to increase the writer's early doubt [11] as to the marine origin of these forms. The problem is of too wide a scope for discussion here, but among the grounds for doubt may be mentioned (1) the very faint inclination of the supposed ancient cliff lines, quite unlike any marine forms known to the writer, and sometimes requiring several miles for a descent of but a few hundred feet; (2) the great discrepancy between steepness of the supposed weathered sea cliffs and steepness of weathered hill slopes of similar age, the supposed sea cliffs often being far more faintly inclined, instead of steeper as one might expect from analogy with known marine forms; (3) the presence of the supposed sea cliffs back of outlying highlands at points which should have been protected from marine erosion; (4) their absence from Mount Desert, especially well situated to record them; [12] and (5) the presence of slopes similar to the supposed sea cliffs in form, but facing inland away from the ocean. It has seemed necessary to look to non-marine agencies for a satisfactory explanation of the terraces.

At the beginning of this study the writer was impressed with the fact that if the sedimentary cover in the Grand Canyon district had been removed at the period of the early monoclinal terraces described by Dutton [13] as descending eastward from the formerly high Nevada-land like a giant stairway, the crystalline floor of that region would then have presented features similar to the rudely terraced crystalline uplands of the east. Thus, on the basis of rough analogy, and without formulating any specific supporting tectonic theory, the writer has tentatively harbored the idea that the New England and Piedmont terraces might in part at least be the surface expression of deep-seated changes associated with broad uparching of the Schooley peneplane. [14] One would not ordinarily expect so vast an area of sedimentary rocks to be uplifted without some bending or breaking of the mass; and the absence of a sedimentary cover which would clearly reveal monoclinal warps in the New England-Piedmont areas should not blind us to the possibility that the warps may be there, even if detected with difficulty in the much dissected crystallines. So also in the region of closely folded sediments, where warping is equally difficult to detect, the difference in elevation between the Schooley and Kittatinny levels discussed in a later paragraph has tentatively been ascribed in part to monoclinal displacement of a single erosion surface.

Full discussion of the New England and Piedmont terraces will appear in a later paper. The subject is raised at this point merely to acknowledge the justice of a criticism offered by Professor Bailey Willis. This student of the tectonics of continent building correctly observed [15] that so long as the difference of slope between the upland peneplane and the crystalline floor under the coastal plain may be ascribed to warping or faulting, the validity of important arguments presented in the following pages must remain in doubt. This observation is all the more pertinent because the present writer has himself been willing to admit the possibility that other changes of slope on the uplands may be of tectonic origin; and it gains weight from the fact that Professor Willis has elaborated a theory of continental uplift which, if accepted, may adequately account for monoclinal surface warpings or step faulting associated with deep-seated shearing and giving terraces roughly parallel with the axis of uplift. It thus becomes incumbent upon the writer to state the grounds which cause him to regard the pre-coastal plain floor and the adjacent crystalline upland as two distinct peneplanes intersecting at a low angle, instead of two parts of a single warped or faulted peneplane.

The two-peneplane interpretation has already been set forth in the writings of Darton and Davis (for the Piedmont), Barrell (for New England), Shaw, Renner, and other students of the question. As a rule the only evidence for two separate peneplanes specifically noted by these writers is the difference in slope, although other grounds are sometimes implicit in the published discussions. Among my own reasons for accepting this interpretation are the following: (1) The change in slope which marks the transition from upland peneplane to pre-coastal plain floor is more sharply localized, more angular, than seems appropriate for a monoclinal warping of the upland surface. In form the " fall-line angle " thus accords better with the theory that it results from the intersection of two erosion planes than with the interpretation that it is due to faint monoclinal warping, whether or not associated with faulting. (2) The great linear extent of the fall-line angle seems to place it in a class by itself. No other feature on the upland surface which might reasonably be interpreted as the product of warping or step faulting has been traced across so vast an extent of territory. It continues around the southern end of the Appalachians and where studied by Shaw [16] in the Mississippi embayment is strikingly apparent. (3) The observed close approximation of the fall-line angle to the eroded inner margin of the Atlantic coastal plain is a necessary corollary of the theory of two intersecting peneplanes, but is difficult to explain on the theory of a single warped or faulted surface. Shaw [17] has properly stressed the improbable nature of the assumption that an axis of warping would for so great a distance coincide with the inner margin of an earlier-formed coastal plain wedge. On the other hand, I have found it equally difficult to imagine reasonable conditions under which erosion would strip the coastal plain cover from a previously warped combined mass (crystalline floor and coastal plain cover) to give the observed coincidence. Neither does it seem probable that deposition would lay down the inner edge of a coastal plain everywhere close to the axis of an earlier warp, especially when we have no reason to believe that such a warp would maintain a constant elevation with respect to sea level.

Faulting might, under proper conditions, determine the inner margin of the coastal plain; but one of the requisite conditions is significant displacement along the fault plane, and the fall line not only affords no evidence of such displacement, but does not appear to be genetically associated with any fault. (4) The indefinite southeastward descent of the relatively steep pre-coastal plain floor clearly differentiates it from the strictly limited descent of the inter-terrace slopes of the upland. The first is a broad regional feature, the apparently continuous steep descent of which has with the aid of well records been measured over a zone scores of miles in breadth; the latter are purely local features, each descending slope soon being cut short by the next lower terrace, usually within a space of from one to 3 miles, and always within less than 5 miles from its beginning. As seen in projected profiles the local terraces are situated upon, and are only minor departures from, the two major intersecting planes (pre-coastal plain surface and upland surface). (5) The degree of preservation of the upland peneplane seems incompatible with the theory that it is as ancient as the crystalline floor underlying the Cretaceous beds of the coastal plain. This point has been emphasized by Shaw [18] and more recently by Ashley.[19] It seems only reasonable to suppose that any land surface of Cretaceous age, continuously exposed to erosion since Cretaceous time, must have been long ago completely destroyed. Hence we must regard the present upland peneplane as distinct from, and of later date than, the peneplane underlying the Cretaceous sediments.

The evidence on the points noted above will be presented fully elsewhere. In the present outline of the theory of regional superposition of Appalachian drainage it is sufficient to warn the reader that if he be not convinced of the existence of two distinct peneplanes of widely different age intersecting along the fall line, he must remain equally skeptical of certain important conclusions (though not necessarily of all the conclusions) set forth in the following pages.

Relations of Piedmont and New England upland. — It is obvious that if the main upland peneplane of the Piedmont belt be of different date from the New England upland peneplane, as Davis and certain others have supposed, the observed intersection of a steep pre-coastal plain floor with a more gently sloping Piedmont surface near Baltimore and elsewhere, would throw no light on conditions in New England. It would still be permissible to regard the pre-coastal plain floor of the south as merely the seaward border of the Schooley (or Kittatinny) peneplane, not yet denuded of its overlapping coastal plain cover.

The New England area must tell its own story. Here we might expect to find (1) the Schooley peneplane arching down to pass beneath the coastal plain deposits, as Davis believed to be the case even after he accepted Darton's view for the Piedmont area; or we might expect (2) a post-Schooley peneplane, similar to the Piedmont, to intersect the steeper pre-coastal plain (Schooley) surface. A great number of projected profiles constructed in connection with this study abundantly confirm Barrell's conclusion that the pre-coastal plain surface is more steeply inclined and, as shown above, presumably older than is the New England upland peneplane. Other profiles drawn to connect the New England upland with the top of Schooley Mountain demonstrate with equal clarity the identity of these two surfaces, thus proving the correctness of Davis' idea that the New England upland is in fact an extension of the Schooley peneplane. Thus neither expectation based on supposed conditions in the Piedmont area is realized. We are forced to conclude that, in the New England area at least, the pre-coastal plain floor is the long-preserved remnant of a peneplane older than the Schooley. While this conclusion does not necessarily apply to the eastern border of the Piedmont, there is reason to believe that there also the pre-coastal plain floor is part of a pre-Schooley erosion surface, and it is so represented in our diagrams.

The Kittatinny-Schooley question. — Can it be that the pre-coastal plain floor is the Kittatinny peneplane of Willis,[20] which some regard as a distinct erosion surface older than the Schooley, partially preserved in occasional abnormally high erosion remnants in the Appalachians of Pennsylvania and elsewhere? Possibly, but the weight of evidence seems to be against such an interpretation. Full discussion of this problem must await completion of projected profiles for critical areas; but it has always seemed to the writer that the moderate differences in altitude of ridge crests in the Pennnsylvania-New Jersey region, often cited in support of distinguishing between a Schooley and a Kittatinny peneplane, may more reasonably be interpreted as due in part to warping or faulting of a single surface, and in part to the persistence on such a surface of low monadnocks where noses of pitching anticlines or synclines gave unusually broad areas of resistant rock. This view, shared by other students of the question, receives substantial confirmation in a recent study [21] of Appalachian wind gaps and water gaps made by Professor Karl Ver Steeg of Wooster College as part of the general problem of Appalachian evolution here discussed.

REFERENCES

[1] William Morris Davis: "The Triassic Formation of Connecticut," *U. S. Geol. Surv.*, *18th Ann. Rep.*, Pt. 2, 1–192, 1898, see p. 165; "The Rivers and Valleys of Pennsylvania," *Nat. Geog. Mag.* Vol. 1, 183–253, 1889; Geographical Essays, 413–484, Boston, 1909, see pp. 471–473.

[2] R. S. Tarr: Personal communication to William Morris Davis.

[3] Joseph Barrell: "The Piedmont Terraces of the Northern Appalachians," *Am. Jour. Sci.*, 4th ser., Vol. 49, 227–258, 327–362, 407–428, 1920, see p. 240.

[4] *Ibid.*, see pp. 418, 421.

[5] *Ibid.*, see pp. 230, 328, 410, 416, 423, 424.

[6] *Ibid.*, see pp. 424, 425.

[7] Joseph Barrell: "Central Connecticut in the Geologic Past," *Conn. Geol. and Nat. Hist. Surv.*, *Bull. 23*, 1–44, 1915, see p. 26.

[8] N. H. Darton: "Outline of Cenozoic History of a Portion of the Middle Atlantic Slope," *Jour. Geol.*, Vol. 2, 568–587, 1894, see pp. 570, 571. "Artesian Well Prospects in the Atlantic Coastal Plain Region," *U. S. Geol. Surv.*, *Bull. 138*, 1–232, 1896, plates VII and XIII. N. H. Darton and A. Keith: "Washington Folio," *U. S. Geol. Surv.*, *Geologic Atlas*,

Folio 70, 1901, Economic Geology sheet. See also, William Morris Davis: "The Peneplain," *Am. Geol.*, Vol. 23, 207–239, 1899, especially p. 214 and Fig. 1.

9 Eugene Wesley Shaw: "Ages of Peneplains of the Appalachian Province," *Geol. Soc. Amer., Bull.* 29, 575–586, 1918.

10 Joseph Barrell: "Piedmont Terraces of the Northern Appalachians," *Amer. Jour. Sci.*, 4th ser., Vol. 49, 227–258, 327–362, 407–428, 1920.

11 See *Bull. Amer. Geol. Soc.*, Vol. 24, 691, 1913.

12 Erwin J. Raisz: "The Scenery of Mt. Desert Island: Its Origin and Development," *N. Y. Acad. Sci., Annals*, Vol. 31, 121–186, 1929, see pp. 138–139.

13 Clarence E. Dutton: "Tertiary History of the Grand Canyon District," *U. S. Geol. Surv., Mon.* 2, 264 pp., atlas, 1882, see p. 115.

14 Douglas Johnson: "Appalachian Studies I," *Bull. Geol. Soc. Amer.*, Vol. 40, 131–132, 1929 (abstract).

15 Bailey Willis: Personal communication.

16 Eugene Wesley Shaw: "The Pliocene History of Northern and Central Mississippi," *U. S. Geol. Surv.*, Prof. Paper 108, 125–162, 1918. Figs. 22 and 23.

17 Eugene Wesley Shaw: "Age of Peneplains of the Appalachian Province," *Geol. Soc. Amer., Bull.* 29, 575–586, 1918, see p. 583.

18 *Ibid.*, pp. 575–586.

19 George H. Ashley: "Age of the Appalachian Peneplains," *Geol. Soc. Amer., Bull.* 41, 695–700, 1930.

20 Bailey Willis: "The Northern Appalachians," *Nat. Geog. Mon.*, Vol. 1, 169–202, 1896, see p. 189.

21 Karl Ver Steeg: "Wind Gaps and Water Gaps of the Northern Appalachians, Their Characteristics and Significance," *N. Y. Acad. Sci., Annals*, Vol. 32, 87–300, 1930.

Editor's Comments
on Papers 8 Through 10

8   **HACK**
     Interpretation of Erosional Topography in Humid Temperate Regions

9   **BRETZ**
     Dynamic Equilibrium and the Ozark Land Forms

10   **SCHUMM and LICHTY**
     Time, Space, and Causality in Geomorphology

QUANTITATIVE AND ANALYTIC ASPECTS OF PLANATION

A new trend of thought concerning planation was set in motion by the statistical stream and slope studies of Horton (1945) and Strahler (1950). This led to the application of the concept of dynamic equilibrium and steady states in open systems to fluvial drainage and denudation.

J. C. Hack (Paper 8) applies some of these concepts to the explanation of present landscapes in terms of work being done by present streams, which are in equilibrium with their load, discharge, slope, and the like, in present areas of high relief. The implication is that the equilibrium which produced the high relief is achieved after a short period of adjustment and that both equilibrium and high relief are inseparable, with little chance for a broad planation surface to develop. These conclusions, drawn from field studies of streams and the rock types being cut (here in a humid temperate region), would leave little planation history to decipher.

J. H. Bretz (Paper 9) confronts Hack's ideas with a field study of an area in the Ozarks that does support a planation surface. Bretz makes a strong plea for regional field studies in preference simply to stream measurements and the application of mathematical forms. Bretz does not picture a Davisian peneplain as the "product largely of stream planation." Davis, himself, gave this considerable emphasis in his paper "Rock floors in Arid and Humid Climates" (Paper 14). Bretz asks of Hack, "What are we disagreeing about when we both agree that slope equilibrium is involved?" The Bretz answer is instructive.

This section ends with consideration of two theoretical papers. The first, by Chorley (1962; not reproduced herein) is a thoughtful discussion of the uses of open and closed systems as models of geomorphic activity, with particular reference to planation surfaces. He deals with the dilemma of time-dependent and time-independent aspects of geomorphic development. He deplores the closed systems into which geomorphologists sometimes seal themselves, on the one hand retreating into regional historic studies and on the other withdrawing into restricted empirical and theoretical studies based on process. In spite of a harsh attitude toward the work of W. M. Davis, which may or may not be justified, a potentially conciliatory and progressive attitude toward the solution of complex geomorphic problems is expressed. (This paper may be found in a forthcoming Benchmark volume on the history of geomorphology by Cuchlaine King.)

Chorley, in a recent exhaustive treatise on the life of W. M. Davis (Chorley et al., 1973), gives him full credit as a scientist, educator, and writer. This generally sympathetic treatment has led Claude Albritton to remark at the end of a review: "Members of the Davis Protective Society may now lay down their cudgels."

Paper 10 elaborates on and in a sense synthesizes the ideas of Hack, Chorley, Bretz, and others, evoking the realization that there should be no conflict between the laboratory and field worker since they have different objectives.

With this general overview in mind we will turn next to a consideration of pediplains.

REFERENCES

Chorley, R. J., 1962. Geomorphology and general systems theory. U.S. Geol. Surv. Prof. Paper 500-B, p. 1–10.

Chorley, R. J., Dumont, A. J., and Beckindale, R. P., 1973. The life and work of William Morris Davis. Methuen, London, 864 pp.

Horton, R. E., 1945. Erosional development of streams and their drainage basins; hydrophysical approach to quantitative geomorphology. Bull. Geol. Soc. Amer., 56, p. 275–370.

Strahler, A. N., 1950. Equilibrium theory of erosional slopes approached by frequency distribution. Amer. J. Sci., 248, p. 673–696.

8

Reprinted from *Amer. J. Sci.*, **258-A**, 80–97 (1960)

INTERPRETATION OF EROSIONAL TOPOGRAPHY
IN HUMID TEMPERATE REGIONS*

JOHN T. HACK

U. S. Geological Survey, Washington, D. C.

ABSTRACT. Since the period 1890 to 1900 the theory of the geographic cycle of erosion has dominated the science of geomorphology and strongly influenced the theoretical skeleton of geology as a whole. Some of the principal assumptions in the theory are unrealistic. The concepts of the graded stream and of lateral planation, although based on reality, are misapplied in an evolutionary development, and it is unlikely that a landscape could evolve as indicated by the theory of the geographic cycle.

The concept of dynamic equilibrium provides a more reasonable basis for the interpretation of topographic forms in an erosionally graded landscape. According to this concept every slope and every channel in an erosional system is adjusted to every other. When the topography is in equilibrium and erosional energy remains the same all elements of the topography are downwasting at the same rate. Differences in relief and form may be explained in terms of spatial relations rather than in terms of an evolutionary development through time. It is recognized however that erosional energy changes in space as well as time, and that topographic forms evolve as energy changes.

Large areas of erosionally graded topography in humid regions have been considered to be "maturely dissected peneplains." According to the equilibrium theory, this topography is what we should expect as the result of long continued erosion. Its explanation does not necessarily involve changes in base level. Pediments in humid regions and some terraces are also equilibrium forms and commonly occur on a lowland area at the border of an adjacent highland.

INTRODUCTION

The part of geologic theory that deals with the interpretation of landforms and the history of landscape development has been dominated for several generations by the ideas of William Morris Davis and his followers. Davis' theory of landscape evolution was first fully presented in his essay, "The Rivers and Valleys of Pennsylvania" (Davis, 1889).[1] The important concepts that he introduced include the geographic cycle, the peneplain, and the formation of mountains by a succession of interrupted erosion cycles. Davis' theories became immensely popular among geologists in Europe as well as in America, though there were dissenters, including, for example, Tarr (1898) and Shaler (1899). His theory of the evolution of mountains as topographic features through the mechanism of multiple erosion cycles was especially influential and came to have a great influence on the theoretical skeleton of the whole science of geology. Its impact is still felt. Many of our ideas relating to the history of mountains, the internal constitution of the earth and the origin of some ore deposits are closely related to this theory. The idea that mountain ranges are vertically uplifted after they have been folded was conceived in order to explain the widespread existence of dissected peneplains (Daly, 1926). Another example is the theory of origin of bauxite and of manganese ores and other residual concentrates in the Appalachian Highlands, that are thought by some to have formed on a Tertiary peneplain surface (Hewett, 1916; Stose and Miser, 1922, p. 52-55; Bridge, 1950, p. 196.)

---

* Publication authorized by the Director, U. S. Geological Survey.

[1] Davis' major papers dealing with the sculpture of landscapes by streams (Davis, 1889, 1890, 1896a, 1896b, 1899a, 1899b, 1902a, 1902b, 1903, 1905a, 1905b) as well as others of his papers were collected in one volume published as "Geographical Essays" (Davis, 1909) and reprinted in 1954 (Davis, 1954). The 1954 edition of "Geographical Essays" has the same page numbers as the 1909 edition.

In the last 20 years, however, Davis' ideas have become less popular and the small but ever-present number of geologists who were skeptical of his theories has increased. Though many geologists have been dissatisfied with it, the theory of the geographic cycle and its application to the study of landforms has not generally been replaced by any other concept. Several alternative theories have been proposed, including the theory of Penck (1924, 1953) which relates the form of slopes to changes in the rate of uplift relative to the rate of erosion, and the "pediplain" theory of L. C. King (1953), an elaboration and expansion of Penck's concept of slope retreat. Both of these theories, however, are also cyclic concepts and hold that the landscape develops in stages that are closely dependent on the rate of change of position of baselevel.

During the course of my work in the Central Appalachians which began in 1952, seeking a different approach to geomorphic problems, a conscious effort was made to abandon the cyclic theory as an explanation for landforms. Instead, the assumption was made that the landforms observed and mapped in the region could be explained on the basis of processes that are acting today through the study of the relations between phenomena as they are distributed in space. The concept of dynamic equilibrium forms a philosophical basis for this kind of analysis. The landscape and the processes molding it are considered a part of an open system in a steady state of balance in which every slope and every form is adjusted to every other. Changes in topographic form take place as equilibrium conditions change, but it is not necessary to assume that the kind of evolutionary changes envisaged by Davis ever occur. The consequences and results of this kind of analysis in most cases differ from conclusions arrived at through the use of the cyclic concepts of Davis.

On rereading some of the classic American literature in geomorphology I realized that G. K. Gilbert used essentially this approach and that I have followed a way of thinking inherited either directly from him or from some of his colleagues. Even though Davis and Gilbert were contemporaries and friends, Gilbert makes little use of and few references to the theory of the geographic cycle or any of its collateral ideas. This omission is so conspicuous that it is difficult to believe Gilbert ever wholeheartedly accepted the idea. It seems to me that Gilbert's famous paper, "Geology of the Henry Mountains" (Gilbert, 1877, p. 99-150) outlines a wholly satisfactory basis for the study of landscape that does not foreshadow the developments in geomorphology that followed in the next 50 years.

In the pages that follow some concepts inherent in the theory of the geographic cycle that seem to me unsound are briefly discussed. The alternative approach to landscape studies based on spatial relations in a system in equilibrium is briefly presented. Very few of the ideas are original and most of them have been published in one form or another in the works of other geologists. In addition I wish to acknowledge the considerable assistance obtained in stimulating discussions with my friends and colleagues, especially C. S. Denny, J. C. Goodlett, C. B. Hunt, L. B. Leopold, C. C. Nikiforoff, and M. G. Wolman with whom I have been associated at various times during the formulation of these ideas. The manuscript has been read and criticized by R. P. Sharp of the California Institute of Technology, Sheldon Judson of Princeton University,

C. C. Nikiforoff (formerly of the Department of Agriculture) as well as by some generous colleagues in the U. S. Geological Survey.

THE GEOGRAPHIC CYCLE AND THE PENEPLAIN CONCEPT

The theory of the geographic cycle rests on the assumption that there is a base level toward which every area erodes and to which the streams become graded. After an initial uplift or rise of a part of the earth's crust, erosion proceeds through successive stages of youth, maturity, and old age; from a time in which stream grades are irregular, through a time when they become smooth, to a stage of low relief when the entire landscape is reduced close to base level. An important stage in the cycle is reached when the slopes of the larger streams are so reduced that they are able to transport just the amount of debris supplied from upstream and no more. At this point the stream is said to be graded and the stage of maturity begins. The trunk streams, unable to erode their beds, shift laterally, forming floodplains and meanders. The debris-covered area on the valley floor expands laterally as the interstream divides are lowered. At the final stage of old age the landscape is one in which the streams meander across broad plains covered by a sheet of waste. The divide areas are graded to the streams and are covered by a waste mantle transported by creep. They rise only slightly above the shallow valleys (Davis 1909, p. 254-272; 1899, p. 485-499). Davis envisaged that there must be interruptions of the ideal cycle of erosion and in fact that an ideal cycle is rarely completed. Alternate periods of uplift and stability of base level result in successive incomplete cycles during which the uplifted peneplains are dissected and new ones form along the streams. Hilly areas in which the hilltops rise to roughly the same height above the streams, like the Piedmont region of the central Atlantic States, are peneplains that have been uplifted and dissected by stream erosion until a stage of maturity has been reached (Davis, 1909, p. 272-274; 1899, p. 499-501).

The concept of planation.—The concept of *planation* or *lateral planation* is lucidly presented in the Henry Mountain report (Gilbert, 1877, p. 127). Gilbert described the process of planation in connection with the formation of smoothly graded, gravel-covered surfaces, cut on soft Mesozoic rocks at the foot of the Henry Mountains. Surfaces like these have come to be known as *pediments* and have been restudied in the Henry Mountains by Hunt, Averitt, and Miller (1953). It was recognized by Gilbert that the planation in this region occurs on soft rocks, such as weak sandstones and shales where slopes or declivities are small in comparison with the trachyte mountains in which the streams originate. Gilbert thought that lateral shifting of the streams is dependent on the fact that the bed load transported by the stream is more resistant than the rock through which it flows so that the stream cuts laterally against the soft bank. Where one of these streams cuts again through hard rock lateral planation ceases (Gilbert, 1877, p. 130) and canyons are formed. The process is dependent on the geology of the drainage basin of the laterally shifting stream, and on a contrast in rock resistance and slope between the upper and lower parts of the basin.

Davis applied the erosion cycle concept to the idea of lateral planation. In his theory a stream always has a tendency to erode laterally against its banks.

131

As the landscape passes through the evolutionary stages of the erosion cycle, first the larger streams and later the tributaries approach the base level of erosion. As they do so their ability to cut downward diminishes. They migrate laterally eroding the valley walls, producing a floodplain or surface of planation.

It is interesting to note the contrast between the planation observed and described by Gilbert and the planation envisaged by Davis. Gilbert's explanation of lateral planation involves a dynamic equilibrium of forces existing at the present time in actual drainage basins and the relation of these forces to the rocks. Davis' theory on the other hand assumes that lateral planation occurs in any drainage basin with the passage of time, regardless of its geology. In fitting the concept of planation into the framework of the geographical cycle Davis attempted to rationalize relations between things that change through time and hence cannot be observed or measured. In the transfer from a scheme of ideas that involves space to a scheme that involves time Davis ignored the spatial relations cited by Gilbert that make the concept valid.

Surfaces of planation are produced by streams under certain circumstances, but there is no reason to believe that such surfaces enlarge through time as relief is lowered, merely as a consequence of a reduction in slope. On the contrary it is likely that as gradation proceeds, the efficiency of the stream system in removing the waste of its drainage basin may increase.

The graded stream.—One of the key ideas in the theory of the geographic cycle is the concept of the graded stream. The word "grade" was used by Gilbert (1877, p. 112) in discussing the stable slope of the stream channel in the same sense that an engineer uses it to describe the slope of a railroad or highway. Davis borrowed the term at Gilbert's suggestion and used it in a more special sense to designate a certain stage in the evolution of stream profiles when the stream's ability to transport the load supplied to it from above is just balanced by the load that it has to carry (Davis, 1909, p. 392; 1902, p. 89).

This concept of grade has probably received more discussion among geomorphologists than any other aspect of the geographic cycle (for example Kesseli, 1941, Mackin, 1948, Woodford, 1951, Rubey, 1952, Leopold and Maddock, 1953, Wolman, 1955). As suggested by Kesseli, the concept as outlined by Davis seems rather elusive, so that it is difficult to identify a graded stream in nature. Mackin, however, in his study of the graded stream, clarifies some of the ideas and suggestions of Davis. The examples of graded streams cited by him are migrating laterally, depositing on the floodplain an amount of material equal to what they erode by lateral cutting. The graded stream is not actively cutting vertically downward and its longitudinal profile is being changed only very slowly as the relief or other conditions in the drainage basin change. Since it is cutting laterally, the channel of the graded stream is bordered by a floodplain underlain by thin river deposits and by terraces whose composition is entirely material carried from upstream and different from the underlying rock (Mackin, 1948, p. 472).

Leopold and Maddock (1953) considered the graded stream in relation to the hydraulic geometry of the channel. Their study of stream gaging and cross section data indicates just as consistent a pattern in the relationships between the variables width, depth, velocity, and sediment load in ungraded as in graded streams. They conclude that Mackin's concept of grade cannot be demonstrated by consideration of stream gaging data and they use the term "quasi-equilibrium" in reference to the equilibrium in stream channels observed by them in all the streams studied. They recognize that this equilibrium is distinct from the equilibrium implied by Davis and Mackin in the concept of the graded stream.

In Davis' concept of grade, high velocity of flow and a high capacity are associated with a steep channel slope. As slope diminishes during the evolution of the landscape through the erosion cycle, velocity diminishes as well as the capacity to transport debris (Davis, 1909, p. 397-398; 1902, p. 95-96). This idea may have seemed reasonable to Davis because he shared with many others the belief that mountain streams with steep channels have higher velocities of flow and therefore greater capacity than do large streams with lower slopes in lowland areas. This observation is not necessarily true. Actual measurements in many natural streams demonstrate that for equivalent frequencies of discharge average velocities tend to increase downstream rather than decrease (Leopold, 1953). Studies of some Appalachian streams, furthermore, indicate that the size of material a stream has on its bed and banks is not related directly to slope, but is related also to discharge and other variables in such a way that in many streams the competence (or size of material that a stream can transport) increases downstream as slope diminishes (Hack, 1958). These facts make it appear doubtful that streams reach a balanced condition through any evolutionary sequence involving a gradual reduction in slope. Probably the balance that exists in most streams is the quasi-equilibrium described by Leopold and Maddock (1953, p. 51). This is a balance among at least seven variables. It is so complex and there are so many alternative adjustments possible that equilibrium can be achieved under many conditions and is arrived at very quickly, almost immediately, in the development of a valley. The uniform, or regular concave-upward longitudinal profile that is characteristic of many streams and has been called "the profile of equilibrium" results not from the attainment of a certain stage in the evolution of a valley, but merely from the regular change downstream in some of the many variables involved in channel equilibrium. Most important of these is probably discharge that increases downstream as a consequence of a regular enlargement of the drainage area.

The streams cited by Mackin (1948) as examples of graded streams represent special cases that are exceptional rather than general. Like Gilbert's streams in the Henry Mountains, such streams head in hard rock areas of high slope and altitude. Their lower courses are in soft rocks and as a consequence have a low slope relative to the increased discharge. They migrate laterally and have a diminishing competence only because of the geologic pattern of the terrain they traverse. They represent a class of streams intermediate between those whose competence with respect to the load derived from upstream is increasing in a downstream direction or remains the same, and those whose competence decreases downstream so abruptly that they aggrade actively enough to build fans. They are no more in a state of equilibrium or disequilibrium than a mountain torrent that is engaged in cutting a gorge. The torrent also has a bed load and lag deposit or floodplain along the bank composed of material too coarse for the stream to move in the ordinary flood, but in this case the lag deposit is locally derived by washing or sliding down the adjacent slope or by plucking from the bed.

The stage of old age and the maturely dissected peneplain.—In the concept of the geographic cycle the appearance of the land surface in the stage of old age is dependent on the process of lateral planation. The ideal surface is a plain partly graded by planation and covered by a veneer of waste. Divide areas with convex upward slopes exist, but are relatively smaller in area than in earlier stages. Such graded surfaces, as stated above, do not now exist in nature. The extensive plainlands of the earth are either depositional surfaces like alluvial plains, deltas, drift plains, and coastal plains, or if they are erosion surfaces in humid areas, they are hilly with rounded divides and steep-walled valleys that have generally come to be described as "maturely dissected peneplains." Exceptions are pediments and terraces, that in humid regions occupy relatively small areas. Excellent examples of maturely dissected landscapes in America are found in the Piedmont region of eastern North America, or in the Central United States where the so-called Ozark peneplain has been "dissected and uplifted." Large areas of the Canadian shield have been said to be a dissected peneplain whose drainage has been disrupted by glaciation. The great limestone valley of the Appalachians, similarly, has been said to be a dissected peneplain as is the plateau area to the west of the Appalachians. Thus land surfaces that are worn down to the stage of old age, as conceived by Davis, are virtually nonexistent; on the other hand former old age surfaces that have been dissected to Davis' stage of maturity are ubiquitous in the older terrains of the earth, especially in humid regions. Indeed this kind of topography is so universal it suggests that the end product or end surface toward which erosion proceeds resembles the "maturely dissected" surface rather than the "old age" surface or peneplain. Such an end surface is one whose forms are graded for the efficient removal of waste rather than one on which the waste products accumulate and stagnate.

THE PRINCIPLE OF DYNAMIC EQUILIBRIUM IN LANDSCAPE INTERPRETATION

An alternative approach to landscape interpretation is through the application of the principle of dynamic equilibrium to spatial relations within the drainage system. It is assumed that within a single erosional system all elements of the topography are mutually adjusted so that they are downwasting at the same rate. The forms and processes are in a steady state of balance and may be considered as time independent. Differences and characteristics of form are therefore explainable in terms of spatial relations in which geologic patterns are the primary consideration rather than in terms of a particular theoretical evolutionary development such as Davis envisaged.

The principle of dynamic equilibrium was applied to the study of landforms both by Gilbert (1877, p. 123) and by Davis (1909, p. 257-261, 389-400; 1899, p. 488-491; 1902, p. 86-98). Recently Strahler has outlined the principle in more modern terms as it might be applied to landscapes (Strahler,

1950, p. 676). The concept requires a state of balance between opposing forces such that they operate at equal rates and their effects cancel each other to produce a steady state, in which energy is continually entering and leaving the system. The opposing forces might be of various kinds. For example, an alluvial fan would be in dynamic equilibrium if the debris shed from the mountain behind it were deposited on the fan at exactly the same rate as it was removed by erosion from the surface of the fan itself. Similarly a slope would be in equilibrium if the material washed down the face and removed from its summit were exactly balanced by erosion at the foot.

In the erosion cycle concept of Davis, equilibrium is achieved in some part of the drainage system when there is a balance between the waste supplied to a stream from the headwaters and the ability of the stream to move it, or in other words, when the slope of the channel is reduced just enough so that the stream can transport the material from above with the available discharge. As argued on page 9 this kind of equilibrium probably is achieved in a stream almost immediately and is not related to a particular stage in its evolution. Davis' concept would imply that some parts of a drainage system would be in equilibrium whereas at the same time other parts would not, and that the condition of equilibrium is in time gradually extended from the downstream portion to the entire drainage system.

Rather than a concept of balance between the load of a stream and the ability of the stream to move it, it is more useful in the analysis of topographic forms to consider the equilibrium of a particular landscape to involve a balance between the processes of erosion and the resistance of the rocks as they are uplifted or tilted by diastrophism. This concept is similar to Penck's concept of exogenous and endogenous forces (1924, 1953). Suppose that an area is undergoing uplift at a constant rate. If the rate of uplift is relatively rapid, the relief must be high because a greater potential energy is required in order to provide enough erosional energy to balance the uplift. The topography is in a steady state and will remain unchanged in form as long as the rates of uplift and erosion are unchanged and as long as similar rocks are exposed at the surface. If the relative rates of erosion and uplift change, however, then the state of balance or equilibrium constant must change. The topography then undergoes an evolution from one form to another. Such an evolution might occur if diastrophic forces ceased to exert their influence, in which case the relief would gradually lower; it might occur if diastrophic forces became more active, in which case the relief would increase; or it might occur if rocks of different resistance became exposed to erosion. Nevertheless as long as diastrophic forces operate gradually enough so that a balance can be maintained by erosive processes, then the topography will remain in a state of balance even though it may be evolving from one form to another. If, however, sudden diastrophic movements occur, relict landforms may be preserved in the topography until a new steady state is achieved.

The area in which a given state of balance exists and that may be considered a single dynamic system may be conceived as very small or very large. In the Appalachian region, it may be that large areas are essentially in the same state of balance. In the West, however, in an active diastrophic belt, a single dynamic system may constitute only a small area such as a single mountain range or a small part of a mountain range. Furthermore, because of sudden dislocations of the crust relict forms may be preserved in the landscape that reflect equilibrium conditions that no longer exist.

The crust of the earth is of course not isotropic and within a single erosional system, no matter how small, there is a considerable variation in the composition and structure of the crust. These variations are reflected by variations in the topography. Consider, for example, an area composed partly of quartzite and partly of shale. To comminute and transport quartzite at the same rate as shale, greater energy is required; and since the rates of removal of the two must be the same in order to preserve the balance of energy, greater relief and steeper slopes are required in the quartzite area. Similarly geometric forms differ on different rock types. An area that is underlain by mica schist or other igneous or metamorphic rock subject to rapid chemical decay, has more rounded divides than an area underlain by qaurtzite, if both are in equilibrium in the same dynamic system, for the schist is comminuted by weathering to silt and clay particles that are rapidly removed from hill tops on low slopes. On the other hand to remove quartzite from a divide at the same rate, steeper slopes and sharper ridges are required because the rock must be moved in the form of larger fragments.

The analysis of topography in terms of spatial or time-independent relations provides a workable basis for the interpretation of landscape. This kind of analysis is uniformitarian in its approach, for it attempts to explain landscapes in terms of processes and rates that are in existence today and therefore observable. It recognizes that processes and rates change both in space and time, and, by clarifying the relation between forms and processes, it provides a means by which the changes can be analyzed.

THE RELATION OF SOIL TO TOPOGRAPHY

Cyclic concepts of soil evolution have developed in a manner parallel to the erosion cycle concept of geomorphology. The idea of a cyclic evolution of soil through a stage of maturity to senility in which the profile becomes intensified and thickened through time is dependent on the concept of a topography that is stable, such as might exist on a peneplain or on a remnant of a dissected peneplain. Naturally enough, this idea lends support to the cyclic concept of landscape evolution.

An alternative theory of soil evolution based on dynamic equilibrium has been forcefully presented by C. C. Nikiforoff (1942, 1949, 1955, and 1959, p. 188) and parallels the concept of equilibrium in landscape evolution. As explained by Nikiforoff, nearly all soils achieve a state of dynamic equilibrium if they are exposed to the surface for a sufficient time. Factors in the equilibrium include .climate, slope, rate of erosion, composition of the parent material, vegetation, and others. The horizons of the soil become diversified and owe their existence to an equilibrium among processes that tend to accumulate certain substances at certain depths, and those that tend to remove them. Take the clayey "B" horizon as an example:

The cyclic viewpoint holds that the clay in the "B" horizon accumulates through leaching of the "A" horizon above it. The concentration of clay increases through time, though at a slower and slower rate until further accumulation is impossible. At this point the soil is mature and remains in this state until removed by erosion.

In terms of dynamic equilibrium, on the other hand, the amount of clay present in any horizon of the soil is the result of a balance at that place between the rate of clay accumulation and clay removal. These rates differ in different horizons and subhorizons and the balance between them determines the amount of clay present. Similar balance between rates of removal and of accumulation and the interactions between them determine the composition of all the horizons.

From the point of view of landscape interpretation this view of the soil has many advantages. It permits the lowering of the hilltop by erosion at a more or less constant rate, at the same time maintaining the equilibrium of the soil. In fact the soil profile is dependent on erosion as one of the factors in the equilibrium.

Compare a hilltop underlain by pure limestone or by clayey or silty limestone with a hilltop nearby underlain by cherty limestone containing massive beds of chert (fig. 1). The hilltop underlain by silty limestone does not ac-

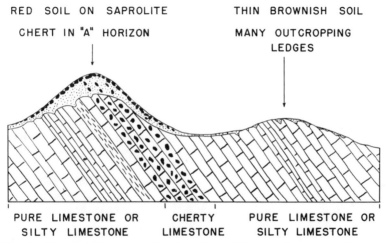

Fig. 1. Hypothetical cross section through two hills in a limestone region showing the relation of the bedrock and topography to the residual soils.

cumulate weathering products because they are removed by erosion as soon as they are freed from the rock by solution of the carbonate that binds them. On the other hill, underlain by cherty rocks, though solution goes on at the same rate, a debris of chert fragments is produced in addition to fine silt. These are too coarse to be removed by sheet erosion on slopes as gentle as those of the first hill. The chert accumulates forming a protective armor that prevents finer insoluble residues, also present, from being eroded. Accumulation of this deposit continues until the slopes are steep enough for the chert to be removed. The topography is now in equilibrium. One hill is covered with saprolite or residuum and the other is not, but the saprolite-covered hill has steeper slopes and rises higher above the streams than the other (Hack, 1958b). It might be said that the residual material is "stored" on top of the hill as its covering armor is comminuted to sizes that can be removed by creep and wash. During the period of "storage" the material becomes oxidized and reddened and a profile develops. The time of storage may be very long, even thousands of years, and may be long enough for the soil to have survived major changes in climate and to owe some of its characteristics to irreversible reactions that took place in the past (Nikiforoff, 1955, p. 48).

EXAMPLES OF EROSIONALLY GRADED OR EQUILIBRIUM TOPOGRAPHY

As an area is graded by erosional processes the differences in the bedrock from one place to another cause a differentiation of the forms on them. Landscapes that develop on intricate and actively rising fault blocks may bear a closer relation to major structural features than to the underlying rock, but in a landscape like that of the Appalachian region in which large areas are mutually adjusted, the diversity of form is largely the result of differential erosion of rocks that yield to weathering in different ways. Such topography may be referred to as erosionally graded.

Ridge and ravine topography.—Many of the erosionally graded landscapes in humid temperate regions belong to the almost ubiquitous type that is commonly known as the "maturely dissected peneplain". Preferring a term that has no genetic connotation a more descriptive one such as *ridge and ravine* topography is suggested. This term refers to the monotonous network of branching valleys and intervening low ridges that make up the landscape of large areas. This topography may be concisely explained in terms of dynamic equilibrium in the words used by Gilbert (1909) in his discussion of rounded hilltops. He conceived that the important elements of the topography could be divided into two domains. The first, a domain of stream sculpture represented by channels in which the slopes are concave upward, because, as he says (Gilbert, 1909, p. 344), the transporting power of a stream per unit of volume increases with the volume; the transporting power also increases with the slope; and a stream automatically adjusts slope to volume in such a way as to equalize its work of transportation in different parts. The other domain is that of creep, represented by the slopes between the channels. In this domain slopes are mostly convex. Gilbert states (1909, p. 345) that:

> This is because the force impelling movement of material is gravity which depends for its effectiveness on slope. On a mature or adjusted profile the slope is everywhere just sufficient to produce the proper velocity. It is greatest where the velocity is greatest and therefore increases progressively with distance from the summit.

The forms of well-graded ridge and ravine landscapes vary within wide limits. Typical examples, both in areas of high relief, have been described by Strahler (1950) and by Hack and Goodlett (in press, 1960). Somewhat gentler topography of the same type is widespread in the Piedmont Province. An example of such an area is shown in figure 2. This is in a drainage basin tributary to the Patapsco River in Carroll County, Maryland. The bedrock is phyllite that is cut by veins of quartz. The interstream divides are convex upward

SCALE IN MILES
CONTOUR INTERVAL 20'FEET

Fig. 2. Topographic map of area in the Piedmont Province of Maryland, showing typical ridge and ravine landscape (1950, U. S. Geol. Survey Winfield Quadrangle, Maryland).

and if measured on a coordinate system in which the origin, or zero point, is the top of the hill or ridge, they have the form of a parabolic curve like the one shown in fig. 3. They intersect the stream bottoms in steep slopes and

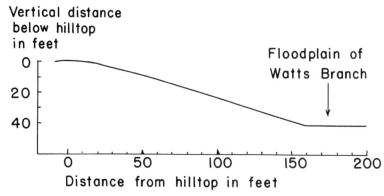

Fig. 3.   Profile of hill slope near Rockville, Maryland, on fine-grained mica schist.

sharp angles, though in places the foot of the slope is concave upward, especially in slopes that intersect the floodplain of a stream at a point opposite the channel. The regularity of the landscape and the rather uniform height of the hills owe their origin to the regularity of the drainage pattern that has developed over long periods, by the erosion of rocks of uniform texture and structure.

Differences in form from one area to another, including the relief, form of the stream profiile, valley cross sections, width of floodplain, shape of hill tops and other form elements are explainable in terms of differences in the bedrock and the manner in which it breaks up into different components as it is handled on the slopes and in the streams.

Pediments.—Where differences in rock resistance in graded landscapes are slight or confined to narrow or small areas, the differences in topography are small. Where, however, two large areas, one of resistant rock and the other of much softer rock, are juxtaposed, the differences are not only pronounced but there is a zone of transitional forms on the less resistant rock. In both areas ridge and ravine landscapes are developed, but as the more resistant area has greater relief, steeper slopes, and sharper divides, debris is shed from the higher to the lower area. This kind of situation is a common one in the Appalachian Highlands where many valley or lowland areas are underlain by limestone and shale and are bordered by ridges or series of ridges underlain by sandstone and quartzite. The transitional forms are broadly fan-shaped gravel-covered and dissected surfaces cut by streams on bedrock that closely resemble typical pediments in many western areas. They are called pediments because of this resemblance, but it is recognized that similar surfaces may be produced by different processes.

An example of a pediment area in the headwaters of the South Fork, Shenandoah River, Augusta and Rockingham Counties, Va., may clarify the equilibrium relations involved. As shown in figure 4, Dry River, Briery Branch,

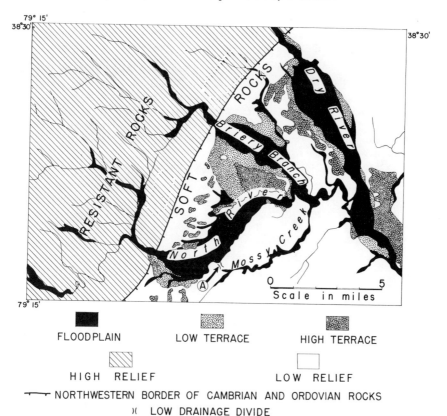

Fig. 4. Simplified map showing bottomlands and terraces on the northwest side of the Shenandoah Valley, Va., to illustrate the formation of pediments by lateral planation and piracy.

and the North River have their headwaters in an area of resistant rock of Silurian. Devonian and Mississippian age consisting mostly of sandstone, quartzite, and shale. The relief in this area averages 1,500 to 2,000 feet and ridges rise to altitudes as high as 4,400 feet. The soft rock area into which the three streams flow is underlain by Cambrian and Ordovician carbonate rocks and shale and has an average relief of only a few hundred feet. Most of it is less than 1800 feet in altitude. On entering the carbonate rocks each of the three streams is bordered by a broad floodplain and terraces composed of cobbles and sand derived from the mountains upstream. The low hilly plain at the foot of the mountains is an extensive complex of dissected terrace remnants that resembles, and in fact is similar to the dissected pediments common at the foot of mountain ranges in the western United States.

Mossy Creek, a tributary of the North River, does not head in the sandstone mountains, but in the limestone area. It has a flatter gradient than the North River even though a smaller stream and at locality A (fig. 4) where the two streams are separated by a low divide, Mossy Creek is about 60 feet below the elevation of the North River floodplain. In 1949 during a severe flood, water actually spilled over the divide into Mossy Creek. This is therefore an example of a stream piracy in progress.

In the resistant rock area the relief is high because the bedrock is removed largely through mechanical processes. Slopes are steep, divides sharp and there are many rock slides. Stream slopes throughout are adjusted for any given drainage area and discharge to transport rock fragments of large size. In the soft rock, or carbonate area, on the other hand, chemical weathering is more important and although the surface is being lowered at the same rate as in the resistant rock area the graded slopes are much gentler and the relief lower. Where the North River leaves the resistant rock area it moves a large load of sandstone cobbles on its bed. They are carried out onto the soft rock area where the channels are adjusted for the transportation of much finer debris. As a consequence the stream shifts laterally and deposits cobbles on the banks, forming a floodplain. Being more resistant to weathering than the carbonate rocks on which they have been deposited, the cobbles persist in the landscape for long periods and as the river continues to erode they form terraces and eventually cap divides.

Mossy Creek will continue to erode its channel and since it is not required to move a load of cobbles, momentarily cuts faster than the North River. Eventually piracy will occur and Mossy Creek Valley will be aggraded by cobbles brought in by the North River. The North River floodplain below locality A will then be abandoned and will become a dissected terrace. Captures similar to this one have already occurred at other places. Note for example in figure 4 the low terraces that connect Briery Branch with the North River.

In this explanation it has been assumed that the resistant and soft rock areas are in the same erosional system and that the average rate of erosion in the two parts is the same. The pediment area exists because cobbles are shed from one part to the other, thus introducing and maintaining a belt of resistant cobbles at the margin of the soft rock area. The pediment area is of course itself in equilibrium in the same system. Its size and the amount of relief are determined by the rate at which cobbles are carried into the soft rock area (a function of size of drainage basin) and by the rate at which they are weathered and broken into pebbles that can be moved out in the streams draining the soft rock area. Hunt, Averitt, and Miller (1953, p. 189) applied the same kind of explanation to the pediments in the Henry Mountains. In that area, however, because the climate is semi-arid the processes are not quite the same, for there is a loss of discharge involved as the streams leave the more humid mountain area.

In the Valley and Ridge Province of the Appalachian Highlands there are many extensive surfaces produced by the processes just described. Many valleys are floored by shale and bordered by relatively high sandstone ridges. The gravels shed by the ridges are too coarse to be carried off by the master stream flowing in shale, and so are stored in floodplains, terraces, and dissected terraces, that because of their resistance to erosion form high benches on either side of the master streams. Eventually the cobbles and gravels in these benches

142

are reduced by weathering and reworking in the laterally shifting streams to sizes that can be carried off down the main valley. Such extensive pediment-like landscapes have long been mistaken for a former gravel-covered broad valley stage or peneplain that is now dissected. Actually such gravelly surfaces testify to the contrast in resistance between the rocks of the mountains and the rocks of the valley and they are part of the equilibrium between the two.

Terraces.—Some stream terraces may have their origin as equilibrium forms. Mapping of surficial deposits in the Shenandoah Valley, Va., indicated that terraces are most common in soft rock areas along streams that originate in hard rock areas. This coincidence suggests that the terraces are preserved essentially because they contain components more resistant than the underlying rock. Terraces composed of chert cobbles are common in areas of cherty limestone. They are not common, however, in areas of homogeneous rocks of any kind that do not provide a possibility for a contrast in resistance between the stream load and the rock through which it moves.

THE RETREATING ESCARPMENT AND PARALLEL RETREAT OF SLOPES

According to cyclic theories elaborated by Penck (1953) and King (1953) the evolution of topographic forms involves parallel retreat of slopes. As the hill or mountain slope retreats away from the main stream a pediment or network of interconnected pediments forms at the foot of the retreating slope and is extended as erosion continues. The Badlands area of South Dakota provides an example (Smith, 1958). Such forms are, however, far from universal. The retreating escarpment appears to be characteristic of gently dipping stratified rocks in which some beds are more resistant to erosion than others. Pediments and foot slopes may or may not be associated with them. The Highland Rim of Tennessee is an example of an escarpment that may be retreating but is without pediments at its foot. The Badlands of South Dakota, on the other hand, have in front of them a broad network of miniature pediments.

Retreating escarpments and associated pediments appear to be especially characteristic of dry climates, a relationship pointed out by Frye (1959). They are certainly less common features, however, in the Appalachian Highlands and in the Piedmont of the eastern United States where not only is the climate humid but horizontally bedded rocks are rare. In these areas the escarpments are fixed in space by geologic contacts. There may, however, be exceptions. The Blue Ridge escarpment of Virginia and the Carolinas appears to divide rock areas that are in some places identical. This scarp may indicate a condition of disequilibrium between two areas (Davis, 1903; White, 1950; Dietrich, 1958).

EVOLUTION OF TOPOGRAPHIC FORMS THROUGH TIME

The theory of dynamic equilibrium explains topographic forms and the differences between them in a manner that may be said to be independent of time. The theory is concerned with the relations between rocks and processes as they exist in space. The forms can change only as the energy applied to the system changes. It is obvious, however, that erosional energy changes through time and hence forms must change. It is of interest, therefore, to speculate on the effect of a gradual reduction in relief of a well-graded landscape such as we assume occurs through long periods of geologic time as diastrophic forces cease to exert their influence and an isostatic balance is approached.

In a typical ridge and ravine landscape the general character of the topography is probably maintained as the relief is lowered. There is no reason to believe that the efficiency of the forms for the shedding of waste becomes any less. The forms in which the waste is removed may change, however, and the rate of removal may diminish. In an area of high relief the waste may be largely in the form of boulders and cobbles that are removed mechanically. As relief is lowered in the same area, perhaps chemical weathering becomes relatively more important. In a high relief area, the divides are sharp and slopes steep. As relief is lowered in the same area the slopes in interstream areas become more rounded, and the divides more blunt.

Speculating on the evolution of pediment landscapes that occur in soft-rock areas adjacent to hard-rock areas it is evident that if relief becomes lower the difference in the energy potential between the two areas will become less marked. It can be expected therefore that the pediment areas will diminish in size and may eventually disappear.

THE APPLICATION OF GEOMORPHIC CONCEPTS TO
GENERAL GEOLOGIC PROBLEMS

The theory of the geographic cycle has been widely used by geologists. Its abandonment must result eventually in changes in many of our concepts. Though it is not my purpose here to elaborate fully such changes, some examples are cited in order to illustrate the extent to which a change in theory may affect geologic problems.

Theories of ore genesis, particularly of deposits classed as supergene, are affected by abandonment of the cyclic concept of landscape evolution. The manganese deposits of the Appalachians associated with lower Cambrian carbonate rocks are a good example. These deposits generally occur in thick residuum preserved beneath quartzite gravels shed from adjacent highland areas. They have been interpreted as of Tertiary age, formed in the Harrisburg cycle of erosion (Hewett, 1916, p. 43-47). By application of the theory of the equilibrium landscape they may be interpreted not as relics of a Tertiary weathered mantle preserved beneath younger gravels, but as deposits that are forming at the present time, or under conditions like the present: They form beneath the gravelly mantle, and are preserved because the gravel covers them, and protects both the ore minerals and the residuum around them from erosion (Hack, 1959).

Some of the greatest changes in concept required relate to the concept of the dissected peneplain. Because we have accepted for many years the idea that a ridge and ravine landscape is formed by the dissection of a peneplain we have also accepted the idea that many highland areas like the Appalachians eroded in steps or cycles and that the orogenies that deformed the rocks of such highland belts were followed by long periods of vertical uplift of a cyclic nature involving repeated changes in the rates of deformation. Having abandoned the peneplain we must reexamine the history of such areas and apply areal studies of erosional process and form to the interpretation of their past history. In the Appalachian Highlands, for example, the general outlines of the present drainage may be inherited in part from conditions that existed as early as Permian or Triassic time. The present landscape may have formed through one continuous period of dying orogeny or isostatic adjustment. Differences in relief and form in different areas are explainable partly by the reaction of various erosive processes on a complex bedrock, and partly by what is probably a long history of complicated diastrophic movements.

Cyclic theories of landscape origin are close relatives of the theory of periodic diastrophism which holds that orogenies have generally occurred in geologic time in brief episodes of world wide extent. This theory is questioned and critically discussed by Gilluly (1949) who shows that the evidence of the sedimentary rock column supports the idea that diastrophism has not been periodic but was almost continuous through time, though the form and location of diastrophic movements has continually changed. This concept of continuity of diastrophic processes is, of course, discordant with cyclic geomorphic theories, but is in harmony with the equilibrium concept outlined here.

REFERENCES CITED

Bridge, Josiah, 1950, Bauxite deposits in the southeastern United States, *in* Snyder, F. G., ed., Symposium on mineral resources of the southeastern United States: Knoxville, Tenn., Tennessee Univ. Press, p. 170-201.

Daly, R. A., 1926, Our mobile earth: New York, Charles Scribner's and Sons, 342 p.

Davis, W. M., 1889, The rivers and valleys of Pennsylvania: Natl. Geog. Mag., v. 1, p. 183-253.

————, 1890, The rivers of northern New Jersey, with notes on the classification of rivers in general: Natl. Geog. Mag., v. 2, p. 81-110.

————, 1896a, Plains of marine and subaerial denudation: Geol. Soc. America Bull., v. 7, p. 377-398.

————, 1896b, The Seine, the Meuse, and the Moselle: Natl. Geog. Mag., v. 7, p. 189-202, 228-238.

————, 1899, The geographic cycle: Geog. Jour., v. 14, p. 481-504.

————, 1899b, The peneplain: Am. Geologist, v. 23, p. 207-239.

————, 1902a, River terraces in New England: Harvard College, Mus. Comp. Zoology Bull., v. 38, p. 281-346.

————, 1902b, Base level, grade, and peneplain: Jour. Geology, v. 10, p. 77-111.

————, 1903, The mountain ranges of the Great Basin: Harvard College, Mus. Comp. Zoology Bull., v. 42, p. 129-177.

————, 1903, The stream contest along the Blue Ridge: Geol. Soc. Philadelphia, Bull., v. 3, p. 213-244.

————, 1905a, The geographical cycle in an arid climate: Jour. Geology, v. 13, p. 381-407.

Davis, W. M., 1905b, Complications of the geographical cycle: Internat. Geog. Cong., 8th Rept., p. 150-163.

————, 1909, Geographical assays: Boston, Ginn and Co., 777 p.

————, 1926, Biographical memoir of Grove Karl Gilbert, 1843-1918: Natl. Acad. Sci., 5th Mem., v. 21, 303 p.

————, 1954, Geographical Essays: Dover Publications, Inc., 777 p.

Dietrich, R. V., 1958, Origin of the Blue Ridge escarpment directly southwest of Roanoke, Virginia: Virginia Acad. Sci. Jour., v. 9, New Series, p. 233-246.

Frye, John C., 1959, Climate and Lester King's "Uniformitarian Nature of Hillslopes": Jour. Geology, v. 67, p. 111-113.

Gilbert, G. K., 1877, Geology of the Henry Mountains (Utah): Washington, D. C., U. S. Geog. and Geol. Survey of the Rocky Mts. Region, U. S. Govt. Printing Office, 160 p.

————, 1909, The convexity of hill tops: Jour. Geology, v. 17, p. 344-350.

Gilluly, James, 1949, Distribution of mountain building in geologic time: Geol. Soc. America Bull., v. 60, no. 4, p. 561-590.

Hack, John T., 1958a, Studies of longitudinal stream profiles in Virginia and Maryland: U. S. Geol. Survey Prof. Paper 294B, p. 45-97.
————, 1958b, Geomorphic significance of residual and alluvial deposits in the Shenandoah Valley, Virginia (abs.) : Virginia Jour. Sci., v. 9, p. 425.
————, 1959, The relation of manganese to surficial deposits in the Shenandoah Valley, Virginia (abs.) : Washington Acad. Sci. Jour. Proc., v. 49, p. 93.
Hack, J. T., and Goodlett, J. C., 1960, Geomorphology and forest ecology of a mountain region in the Central Appalachians: U. S. Geol. Survey Prof. Paper 347 in press.
Hewett, D. F., 1916, Some manganese mines in Virginia and Maryland: U. S. Geol. Survey Bull. 640-C, p. 37-71.
Hunt, C. B., Averitt, Paul, and Miller, R. L., 1953, Geology and Geography of the Henry Mountains region, Utah: U. S. Geol. Survey Prof. Paper 228, 234 p.
Kesseli, J. E., 1941, The concept of the graded river: Jour. Geology, v. 49, no. 6, p. 561-588.
King, L. C., 1953, Canons of landscape evolution: Geol. Soc. America, Bull., v. 64, no. 7, p. 721-752.
Leopold, L. B., 1953, Downstream change of velocity in rivers: AM. JOUR. SCI., v. 251, no. 8, p. 606-624.
Leopold, L. B., and Maddock, Thos., Jr., 1953, The hydraulic geometry of stream channels and some physiographic implications: U. S. Geol. Survey Prof. Paper 252, 57 p.
Mackin, J. H., 1948, Concept of the graded river: Geol. Soc. America Bull., v. 59, no. 5, p. 463-511.
Nikiforoff, C. C., 1942, Fundamental formula of soil formation: AM. JOUR. SCI., v. 240, no. 12, p. 847-866.
————, 1949, Weathering and soil evolution: Soil Sci., v. 67, p. 219-230.
————, 1955, Harpan soils of the Coastal Plain of southern Maryland: U. S. Geol. Survey Prof. Paper 267-B, p. 45-63.
————, 1959, Reappraisal of the soil: Science, v. 129, no. 3343, p. 186-196.
Penck, Walther, 1953, Morphological analysis of landforms (translation by Hella Czeck and K. C. Boswell) : New York, St. Martin's Press, 429 p.
————, 1924, Die morphologische Analyse. Ein Kapitel der physikalischen Geologie: Stuttgart, Geog. Abh. 2 Reihe, heft 2, 283 p.
Rubey, W. W., 1952, Geology and mineral resources of the Hardin and Brussels quadrangles (in Illinois) : U. S. Geol. Survey Prof. Paper 218, 179 p.
Shaler, N. S., 1899, Spacing of rivers with reference to hypothesis of base-leveling: Geol. Soc. America Bull., v. 10, p. 263-276.
Smith, K. G., 1958, Erosional processes and landforms of Badlands National Monument, South Dakota: Geol. Soc. America Bull., v. 69, no. 8, p. 975-1008.
Stose, G. W., and Miser, H. D., 1922, Manganese deposits of western Virginia: Virginia Geol. Survey Bull. 23, 206 p.
Strahler, A. N., 1950, Equilibrium theory of erosional slopes approached by frequency distribution analysis: AM. JOUR. SCI., v. 248, no. 10, p. 673-696; no. 11, p. 800-814.
Tarr, R. S., 1898, The peneplain: Am. Geologist, v. 21, p. 341-370.
White, W. A., 1950, The Blue Ridge front—a fault scarp: Geol. Soc. America Bull., v. 61, no. 12, pt. 1, p. 1309-1346.
Wolman, M. G., 1955, The natural channel of Brandywine Creek, Pennsylvania: U. S. Geol. Survey Prof. Paper 271, 56 p.
Woodford, A. O., 1951, Stream gradients and the Monterey sea valley: Geol. Soc. America Bull., v. 62, no. 7, p. 799-851.

9

Reprinted from *Amer. J. Sci., ***260**, 427–438 (June 1962)

DYNAMIC EQUILIBRIUM AND THE OZARK LAND FORMS

J HARLEN BRETZ

Department of Geology, The University of Chicago, Chicago, Illinois

ABSTRACT. Significant geomorphic features and their relationships in the Ozarks record events and sequences that are impossible by the noncyclic theory of dynamic equilibrium. Only the Davisian concept of cycles of erosion separated by recurrent uplift of the Ozark dome will account for the region's geomorphic history.

The concept of "dynamic equilibrium" (Hack, 1957, 1959, 1960a,b) denies that peneplanation ever occurs under stream attack, however long continued. It considers the end product of erosion under a humid climate to be the "ridge and valley" topography that Davis (1909) called mature. Broad, flattish interfluve uplands with accordant summits, supposedly peneplain remnants, are said to be the consequence of quasi-uniform stream valley spacing. Dynamic equilibrium is "a steady state of balance in which every slope and every form [in an erosional system] is adjusted to every other". When "erosional energy remains the same, all elements of the topography are downwasting at the same rate". It is a noncyclic sequence that can not reduce the region beyond a condition resembling Davisian maturity. The slopes become gentler until equilibrium is reached. Such slopes are dependent on the rocks and on the environment, not on time or any evolutionary development.

Neither the Davis sequence nor that espoused by Hack can be demonstrated from observations through time nor proved by continuously operating models or laboratory experiments. The series of changes envisaged in both hypotheses can, of course, be dealt with by mathematical models in which one accepts postulates built on observed measurements of streams at work and assumptions of variables that have vanished. Arguments thus obtained for either sequence will still remain arguments.

But there are relationships among existing land forms that provide definite statements regarding the past events that have culminated in those forms. Field evidence of this kind affords a far safer approach to satisfactory conclusions than does extrapolation from stream measurements or deductions from mathematical assumptions. Field relations are the final court of appeal.

The field for the present study is the Ozark highlands of southern Missouri and northern Arkansas. Details, which would demonstrate accuracy and completeness of the observer's study, are reserved for a larger work. Here we shall only outline those field facts whose relationships bear directly on the history of the Ozark land forms.[1] This study holds that these forms can not be explained by the noncyclic concept. They are interpretable only by the Davis concept of multiple erosion cycles caused by uplifts multiple in time.

Commonly called the Ozark Plateau, these highlands are structurally and topographically an imperfect dome of Paleozoic sedimentary rocks about 250

[1] Critical reading of an early draft of this paper by T. R. Beveridge, R. P. Sharp, and A. C. Trowbridge is gratefully acknowledged. This draft was also submitted to J. T. Hack who commented helpfully on some misinterpretations of his statements and noted certain differences of interpretation of the Ozark land forms. Discussion of these differences appears in the present article.

miles in maximum diameter and surrounded by lowlands. The irregularly radial drainage goes to Missouri River on the north and northwest, to Mississippi River on the east, and to the Arkansas River system on the south and southeast. The long radial interfluves are so strikingly and broadly flattish on top that about two-thirds of the region's cities and larger towns and a large fraction of its arable land are located on them, and the traveler along their railroads and main highways is seldom aware of the many deep, steep-walled river valleys. Because the structural dip is slightly greater than longitudinal gradients of both the valleys and their separating interfluves, these landscape features, with few exceptions, cross from older to younger rocks in their descent off the dome. Interrupted escarpments, rudely tangential to the dome, face inward. In most places, interfluve summits smoothly truncate structures, including faults, folds, and minor domings subsidiary to the main uplift.[2]

The surrounding lowlands are underlaid by Mississippian, Pennsylvanian, and younger formations, whereas the summit areas are largely Cambrian and Ordovician formations. Tracts of residual Pennsylvanian and Mississippian rock on the summit and other high areas indicate that the dome structure has been denuded of some hundreds of feet of these Paleozoic formations.

Previous studies (Hershey, 1895, 1901; Marbut, 1896; Fenneman, 1909, 1938) have interpreted this denudation as carried to the penultimate result of a Davisian cycle of erosion, a peneplain. Hack refers to the "so-called Ozark peneplain". By either the Davis or the noncyclic concept (whose land forms approximate only the first half of a Davisian cycle), the erosional reduction that made this broad flat summit can only be the result of slope wash into the headwater creeks of the radial river systems. This conclusion must be accepted whether the dome has been rising continuously or has had successive pauses in its uplift.

One of the best showings of these summit flats of the Ozark topographic dome is on the 15 minute Edgar Springs and Raymondville, Missouri, Quadrangles. Here it is more than a mile wide in places and for a length of about ten miles it ranges in altitude through only 80 feet. Ravine heads encroaching from either side have streams descending 100 feet or more per mile. They are fed from broad shallow drainage grooves on the summit with gradients of rarely more than 20 feet per mile. The flat on which the town of Licking stands has a slope of 10 feet to the mile.

This summit flat is underlain by both Roubidoux sandstone and Jefferson City dolomite with no hint in the flat of any different response to the erosion that has scalped it of upper Ordovician, Mississippian, and Pennsylvanian rocks.

A hill of dolomite three miles south of Licking and nearly a mile long stands more than 100 feet above the surrounding flat, and three smaller hills nearby rise 75 feet or so above. A few miles south of Raymondville is a scattering of similar isolated hills up to 80 feet above the flat, which partially surrounds them and carries the town.

---

[2] Exception to the above characteristics is a small section near the eastern end of the uplands, the St. Francois Mountains of Precambrian rocks.

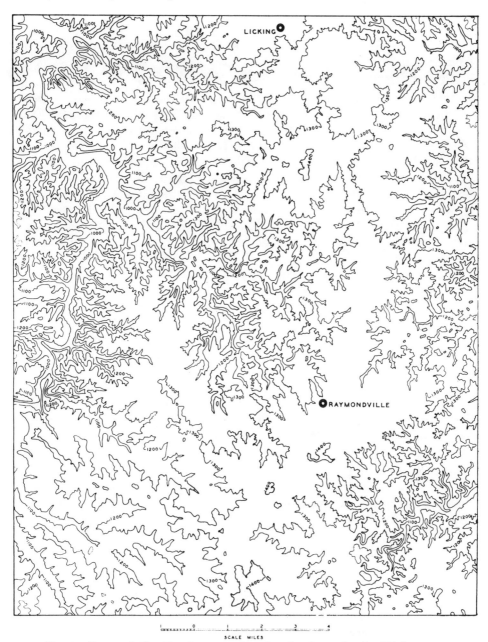

SCALE MILES

Fig. 1. Raymondville 15 minute quadrangle, showing only the 100-foot contours. Most monadnock elevations on this portion of the Ozark peneplain are caught by the 1400-foot contour. All remnants of the peneplain lie above 1200 and most of their area is enclosed by the 1300-foot contour. Many "island" and "peninsula" residuals among mar- gining ravines indicate that this district was a broad and gently sloped peneplain interfluve between drainage of south-flowing Current River and north-flowing Big Piney River.

The interfluve summit is continued for 12 miles farther north at and above 1200 feet A.T. (Edgar Springs Quadrangle) and there possesses the same diagnostic characters of monadnocks and of once greater widths. Thence northward, it is attenuated in places but identifiable as a continuous feature across portions of Rolla, Salem, Linn, Redbird, Bland, Gerald, and Hermann Quadrangles to terminate in the summit of Missouri River bluffs 100 miles from Licking. For the latter half of this distance the summit flat is on Pennsylvanian rock. Total descent is 325 feet. The river bluffs of Hermann are 450 feet high, a fact whose significance the proponents of dynamic equilibrium should ponder.

Chemical weathering may be considered to have reduced the dolomite to solubles and clay, available for rainwash removal under present conditions—although these hills have survived. But the flat on sandstone is not so easily explained.

The Edgar Springs-Licking-Raymondville flat is only part of the main divide of the dome. For more than a hundred miles this divide varies but little in altitude and shows no consistent differences as it crosses three formations of dolomite and sandstone. The isolated hills near the two towns named are duplicated in many other places, some of Jefferson City dolomite and some of Mississippian limestone. None of these hills fits the requirements of the dynamic equilibrium procedures, namely, markedly more resistant rock or "sudden diastrophic movements". Relict hills of the Ozark main divide and of dozens of other interfluves in the province are truly monadnocks standing above the trace of a dissected peneplain. The shallow drainage grooves on virtually all of the broad flat uplands of the province are relicts also, orphans of the Ozark cycle of erosion.

A considerable remnant of the Ozark peneplain, with a broad monadnock on it (Lanes Prairie, Vienna, and Redbird Quadrangles) lies halfway down the northern slope of the dome. It is bounded on the west by the 400-foot-deep, canyonlike Gasconade River valley. On the east, the upland grades rather gently into headwater creeks of Bourbeuse River. The entire drainage of the Prairie goes to these creeks. Not a drop reaches the Gasconade except from long, narrow, west-pointing fingers of the Prairie surface that record an earlier westward extension and carry the Prairie's altitudes. Vigorous attack on this margin by a swarm of ravines born of the Gasconade's deep trench indicates the complete failure of the dynamic equilibrium theory to explain the upland flat and its strongly contrasted east and west margins.

If the Pennsylvanian rocks surfacing Lanes Prairie were markedly more resistant to erosion than the subjacent dolomites, one might see in this an explanation for the broad upland flat and the marginal shallow valleys of the Bourbeuse drainage. However, the Pennsylvanian strata are shales and sandstones, considerably weaker than the dolomites, and their existence on that summit is explicable only by the view that, like the stream gravel on the Ozark peneplain remnants about Owensville, a few miles distant, they record a former base-leveled lowland. If the slopes of the Bourbeuse headwater creek valleys are in dynamic equilibrium, they are quite out of equilibrium with both Bourbeuse and Gasconade Rivers whose deep valleys semi-isolate the upland.

Hack says (1960b) that "parallelism between . . . profiles of . . . channelways and . . . intervening ridge crests argues against the existence of relict [peneplain trace] forms". Such parallelism is completely lacking between profiles of the broad summit interfluve flats and the Ozark's major radial rivers. For example, the interfluve along the east side of the Big Piney-Gasconade River, a spur of the main divide projecting northward from near Dunn, Cabool, and Sterling and including the Licking-Raymondville and Lanes Prairie summit flats, descends quite uniformly 600 feet across Ordovician and Pennsylvanian formations in its 125 miles of length and terminates in 400-foot bluffs

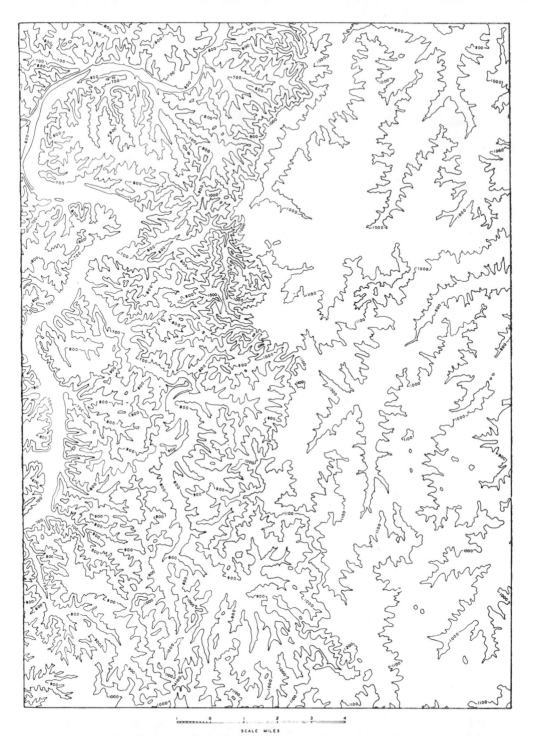

SCALE MILES

Fig. 2. Lanes Prairie and environs. Only the 100-foot contours are shown; taken from the Vienna, Rolla, Redbird and Meramec Spring quadrangle maps. The Springfield peneplain remnant is largely within the 1100-foot contour and its highest part rises to 1200 on Meramec Spring and Rolla Quadrangles. The monadnock has gentle slopes north-ward down to the Ozark peneplain remnant at ca. 1100 ft A. T. On the Meramec Spring Quadrangle however, the descent is 100 ft in half a mile. A prominent strath in Gasconade River valley lies at ca. 800 ft A. T. in the northwestern part of the area shown.

of Missouri Valley while the closely parallel Big Piney-Gasconade River descends about 1000 feet in that same distance.

The St. Francois Mountains have been resurrected from a deep burial under Paleozoic formations, although many of their Precambrian valleys still are floored with nearly flat-lying sedimentaries. The region is replete with narrows and "shut-ins", but a few streams escape without encountering Precambrian rock. The Big River system, draining northward, offers the best showing of this exceptional relation and affords striking demonstrations that the profiles of older valley bottom remnants are traceable downvalley and out into the plateau to become, as the valleys deepen, the flattish interfluve summits already noted as Ozark peneplain remnants. One needs only to start with the broad upland tracts on the Fletcher Quadrangle map and trace them southward up Big River drainage onto either the Farmington or Potosi Quadrangles to see the cogency of this argument. For the oldland uplands have a gradient sufficiently lower than that of the streams themselves to bring their rising profiles down, within confines of Precambrian hills, to benchlands but little incised by the headwater streams. Indeed, one Big River tributary heads on a broad, low, gently graded divide on Paleozoic rock (Farmington Plain), the southern slope of which drains to St. Francis River and thus crosses the mountainous area in the opposite direction to that taken by the Big River system.

According to the noncyclic theory, as we read it, stream rejuvenation strictu sensu is virtually impossible. Only structurally determined or alluvial terraces can be expected. Bluff-bounded river valleys, which possess straths containing meanders entrenched in bedrock and associated with erosionally determined, flat-topped divides with much gentler longitudinal profiles, are unthinkable in noncyclic erosion.

In the Ozarks, however, most major river valleys are of this forbidden type. Their remarkably tortuous patterns of entrenched meanders are featured in many geological studies, textbooks, and compendia. The White River valley in southwestern Missouri and northwestern Arkansas, well shown on six $7\frac{1}{2}$ minute quadrangle maps with 20-foot contour intervals (Lampe, Table Rock Dam, Hollister, Branson, Forsyth and Mincy, Stone and Taney Counties, Missouri) is worthy of study because, through the twistings of the meanders, the river flows toward all points of the compass, structures in the one formation involved are ignored, and the valley actually has two tiers of bedrock strath terraces, 250 and 350 feet above the river or 300 and 400 feet below the flat uplands shown on the $7\frac{1}{2}$ minute quadrangles (Reeds Spring, Spokane, Highlandville, Selmore, and Chadwick) farther north. It is explicable only if interpreted as a record of a sufficiently low gradient attained by valley deepening toward a base level that subsequently has been lowered twice.

The Gasconade and Osage Rivers' entrenched meanders on the north slope of the dome are noteworthy because they are developed across four formations of Cambrian and lower Ordovician age with no "planar" or "prismatic" structures that could have caused or even influenced the meander development. These and other irregularly shaped and spaced river courses have been correctly interpreted by Davis (1909), Rich (1914), and Tarr (1924).

Except for the St. Francois Mountains section of the Ozark Plateau, there are no spectacular fater or wind gaps in the province, nor any major piracies on record. Escarpment-making edges of the inclined formations have certainly migrated down dip during the dome's earlier history to reach the positions they occupy today, but only a few rivers show any control during such scarp retreat in determining any part of their courses.

There is enough departure from such relationships in Ozark rivers to show that the locations of most present crossings were not so determined. Indeed, two streams in the west-central Ozarks, Turnbo Creek-James River and Pomme de Terre River, cross the Eureka Springs escarpment and, farther downstream, *cross back again.* This is the most prominent escarpment in the Ozark province. The edges of the more resistant formations have become the asymmetrical ridges of today's topography since the streams established their present transverse courses. This becomes clear when the escarpment summits are examined. They are beveled off to levels consistent with summit levels of neighboring radial interfluves that carry the profile of the Ozark peneplain down off the main divide. Also, they carry comparable monadnock hills. Escarpment summits are relicts of the same peneplain, and the present stream valley crossings date back only to the post-Ozark cycle of erosion. Such a sequence in relations of streams to escarpments is very unlikely under noncyclic dynamic equilibrium. Dominance of such relations would be an impossible coincidence.

The northwestern flank of the Ozark dome in the drainage area of Osage River has the lowest relief of any part of the region here considered. Its east-flowing rivers (Marmaton, South Grand, Osage, and Little Osage) actively meander on wide floodplains strewn with oxbow lakes and swamps. Their larger creeks similarly meander and have deferred junctions, which strongly suggest that the wide bottom flats and very low stream gradients are the result of extensive valley alluviation. Except where meandering has undercut, the bluffs are gently sloped. Not all summits fall into striking accordance because the region is one of Pennsylvanian rocks, chiefly sandstone and shales, with scattered hills of varying heights determined by lingering sandstone masses. On east-west belts of $7\frac{1}{2}$ minute quadrangle maps, the generalized hilltop level rises eastward, downstream, while the floodplain altitudes decrease.

This downstream increasing relief is more pronounced as the rivers, combined into the Osage trunk, pass from the area of Pennsylvanian rocks to Mississippian limestone and thence to lower Ordovician and Cambrian rocks as the valley enters a local dome on the larger feature's northwestern flank. Here the Ozark peneplain trace is readily identified in the uplands of the Osage-Missouri divide and, from other field evidence, can with confidence be extended westward, upstream, in the lower summit altitudes of the Pennsylvanian hills.

This two-fold reason for the eastward increasing depth of the Osage valley has long been known (Tarr, 1924) and correctly interpreted. The Ozark peneplain here is warped down westward, or up eastward, or both. Little was stripped off, during the Ozark cycle, from the area drained by the four named rivers. The Osage eastward drainage was established before the warping, and the deeper valleys farther down the river are post-Ozark.

153

It has also long been known (Marbut, 1907) that this deeper eastern stretch of the Osage valley has a much dissected but prominent strath held up by several formations along its length. What is new, from the present study, is that this strath has no detectible gradient. Thus the profile of the bordering hilltops rises eastward, the strath lies nearly horizontal, and the river profile descends to the east. Thus, also, the strath profile, when traced upstream, finally coincides with that of the present floodplain (Taberville and Monegaw Springs Quadrangles, St. Clair County, Missouri) or is shallowly buried by a later alluviation.

So interpreted, rivers farther upstream are still flowing essentially at the Osage strath's base level, and the lack of gradient for the strath in the locally domed tract is the result of a slight back-tilting that occurred after the pause which gave the strath its origin. What alluviation has occurred is ascribed to that same slight tilting.

If a proponent of noncyclic erosion limited his observations to the low Pennsylvanian tract and was unaware that water-worn chert gravel is strewn over a few hilltop flats that correlate with both Springfield and Ozark peneplain remnants in adjacent regions, he could find a fair case for his theory. But the uplift downstream has clearly occurred twice and the Osage strath there is overlooked on the north by an escarpmentlike upland with steep, ravined slopes toward the river valley and strikingly gentle slopes northward away from it. The river's entrenchment into the strath follows a meander pattern, with some very tight S curves that can not be dated as earlier than the old valley bottom (with stream gravels) now recorded in summits of the peninsular projections inside the river loops. Such dating is because restoration of the strath produces only a uniform floor between the old valley sides. There are no surviving meander peninsulas on it.

Throughout the country once reduced to the Ozark peneplain, there are fragmentary surfaces of the slopes off that peneplain's divides and into its valleys. The Osage valley across the local dome has one of the most convincing relics of such slopes, well shown on Proctor Creek, Irontown Ferry (Bollinger Creek), and Purvis (Sunrise Beach) Quadrangles, Morgan County. It is now only a ridge descending southeastward from the escarpmentlike interfluve that bounds the north side of Osage valley. The interfluve summit at this place carries a hill, 1165 feet A.T., and nearly 100 feet higher than elsewhere along its neighboring crest. From this hill, the ridge projects out into the wilderness of the ravine-dissected Osage valley slope while holding 1000 feet A.T. altitudes and carrying half a dozen 1100-foot elevations elongated with its axial line. The hill nearest the distal end of this ridge is 1147 feet A.T., almost as high as the monadnock noted but is eight miles or so out into the valley. The ridge summit, dropping to about 1000 feet altitude, continues for another five miles, still well above all other divides among the south-draining creeks. Thence its summit descends until the ridge form loses character in the dissected Osage strath.

The first eight miles constitute the trace of the Ozark peneplain with a few "baby monadnocks", all that has been left in the retreat of the escarpment face under ravine attack. The next five miles represent the side slope of the

Osage valley of peneplain time and indicate a depth of about 60 feet. Thence the ridge crest ranges between 900 and 940 feet A.T. for three miles or so and there is taken to be a remnant of the peneplain valley floor. Beyond that, the profile drops to the level of the dissected strath and loses individuality among the minor interfluves of post-strath dissection.

Although precision in figures is, of course, impossible in such a situation, the valley-slope ridge is surely a remnant of the escarpment face when the strath was the valley bottom, and the hill at the northern high end of the ridge is only one of several residual elevations sitting on the otherwise evenly beveled escarpmentlike summit. Time is of the essence in producing these relations and changing base level has been a vital factor. Noncyclic theory cannot account for the features of this ridge.

The writer has described (1953) red clay fills in Ozark caves, now partially removed by present-day cave floor streams: unctuous deposits that lack silt, sand, gravel, and flowstone. They are interpreted as laid down when the phreatic circulation that made the caves had come to complete stagnancy because of erosional reduction of the region's relief during an epeirogenic stillstand. Obviously, this concept requires postulation of a long time when ground water stagnation succeeded a phreatic circulation and preceded a vadose circulation that resulted from renewed uplift and stream rejuvenation. This interval marks an Ozark peneplanation. Clay fractions of the peneplain's soil supplied the largely unlaminated fillings.

This long pause and the following complete change in ground water circulation in Ozark cave histories is inexplicable by the dynamic equilibrium sequences of geomorphic development.

Another significant feature bearing directly on Ozark geomorphic history is the occurrence of bedded stream gravel on some interfluve summits. Of more than a dozen known cases, the deposits near Owensville (Bland and Gerald Quadrangles, Osage and Gasconade Counties) on the Bourbeuse-Gasconade divide are most striking. Here the broad, flattish upland on which the city stands is an integral part of the dissected plain whose lowering niveau is readily recognized across diverse formations all the way (70 miles) north from the main divide of the dome. Six miles distant, Gasconade River is down in a cliff-margined valley 600 feet lower.

Pennsylvanian sandstones and shales constitute the summit rock for the upland flat, and numerous pits for fire clay have exposed a patchy cover of rounded, sorted, and stratified chert gravel up to 12 feet thick. This chert is definitely from Ordovician rock and must have come off higher portions of the dome to the south after scalping had exposed the Ordovician; it never was collected from the shales and sandstones. Nor could it have been transported across the intervening valleys of Bourbeuse and Meramec Rivers unless they were then completely obliterated by alluvial fill, a condition for whcih there is no accessory evidence. These river valleys, by the noncyclic concept, have held their present courses since the beginning of Ozark denudation, and the interfluve chert gravel blanket on the youngest rock of the dome thus is inexplicable. If the valleys began on an Ozark peneplain, an explanation for the topographic location of the gravel is readily available.

These gravels can not record a pediment veneer. The last escarpment to migrate down off the dome and cross the site of this upland was the Mississippian Burlington. It now is about 25 miles distant in the summits of the bluffs along the north side of Missouri River. Burlington-Devonian contact there is but little lower than the lowest gravel at Owensville. Pennsylvanian shales at Owensville lie on Ordovician Jefferson City dolomite, a record of the scarp's retreat before Pennsylvanian time, whereas the gravels lie above the shales.

The Owensville gravel is here interpreted as an alluvial fan deposited on the lower slope of the dome when post-Ozark uplift had begun in central parts, rejuvenating the peneplain streams there, but had not yet spread radially to the periphery.

The various gravel patches on interfluves are not all of the same age and genesis. To illustrate this, one other area should be noted. It is near Grover (Eureka Quadrangle, St. Louis County) a few miles from the western limits of the city of St. Louis. Several deposits are known here, distributed through a vertical range of 90 feet. From character and field relations they can hardly be parts of one original deposit. The lowest, largest, and thickest (30 feet, Rubey, 1952) contains much cobbly debris and some boulders. Some of these large fragments are of quartzite, strongly marked with Liesegang bands, that apparently could have come only from the Proterozoic Baraboo Ranges of south-central Wisconsin. Altitude here is 300 feet above the Missouri floodplain less than five miles away and the Mississippi floodplain a little more than 20 miles distant.

Trace of the Ozark peneplain is fairly definite in hilltops of the vicinity although no erosional flat tops them. The highest gravel caps a low elevation above that trace, the large deposit falls below.

All who have studied this largest river deposit agree that it is not Pleistocene gravel. The only logical alternative is to consider it as a late Tertiary record of an ancestral main stream from the north.

Sedimentological studies will eventually give us a better idea of the character of the stream, but simple qualitative observations seem at present to mean that the highest deposit near Grover is pre-peneplain in age and the lowest records an early Mississippi River undergoing rejuvenation. At any rate, there is clearly more than a noncyclic episode of "dynamic equilibrium" on record at Grover.

The planned brevity of statement in this paper forbids the detail by which the writer identifies two peneplains and two later straths in the Ozarks. The earlier peneplain was largely destroyed during the Ozark cycle and its remnants have been suffering further reduction during all subsequent time. Its record is best seen in summits of the Eureka Springs escarpment within the drainage of White River and in the numerous monadnocks on the Ozark peneplain. It has a few monadnocks of its own and one known instance of stratified stream gravel. For the record, the writer correlates the older with the Dodgeville peneplain of Wisconsin, Iowa, and Illinois and the younger Ozark peneplain with the Lancaster peneplain of the same states.

What is a peneplain? If, under dynamic equilibrium, the overall rate of erosion exceeds the rate of uplift, the end result in interfluve surfaces must be exceedingly gentle slopes, far beyond Davis' maturity, worn down well toward base level for those slopes. Isn't that a peneplain? If uplifted later, and its streams rejuvenated, it could become the Ozarks of today.

Hack appears to interpret a Davisian peneplain as the product largely of widespread lateral stream planation. But most of these Ozark relict uplands clearly were divides on an oldland surface, were not valley bottoms. They were graded to adjacent wide valleys, were "in equilibrium" with them but were never crossed by streams. Gravel on some of them, however, puts such summits in a stream planation category.

The writer does not find the type of peneplain made by excessive lateral planation at base level in the Ozarks. Surviving land forms show that peneplain valleys were not exceedingly broad features separated by relatively limited interfluve tracts. The surviving Ozark peneplain surfaces largely lay above the bottoms of their valley ways. Perhaps these "so-called" peneplaned surfaces would fit the dynamic equilibrium concept. Nonetheless, they are the end products of an earlier cycle and there is in the record a succession of partial cycles of reduction toward later, lower base levels. These are the bedrock straths.

Landscape in equilibrium but of greatly reduced relief endangers the noncyclic argument. Such reduction demands time for the earlier "ridge and ravine" character to disappear. It was sequential development toward, if not to, the topography most geomorphologists call a pene (L.poene = almost) plain. Fenneman (1936) has given us, facetiously, the graded series: a peneplain, almost a peneplain, and almost a poor peneplain. Doesn't the end product of the dynamic concept belong to that series? Would anyone question that all three of Fenneman's grades possess dynamic equilibrium of slopes? What, then, are we disagreeing about?

In consistency, Hack must interpret the upland erosional plains of the Ozarks as products of the *presently operating drainage* and must account for the steep valley slopes that bound them by appeal to the factors of rock and/or environment, neither of which obviously is the correct explanation. We are disagreeing on the occurrence of cyclic erosion in the Ozarks.

However impressive may be a mathematical treatment of possibilities and probabilities among past events, the geomorphic processes involved in the Ozarks have left forms whose existence and interrelations are facts. These include: (1) the relict shallow valleys on the summit flats; (2) the irregular dissection of the main divide to leave wide, isolated summit flats interspersed with much dissected stretches carrying the same crest accordances and composed of the same rocks; (3) the nonparallelism of divide and stream valley longitudinal profiles; (4) the relict hills that rise above the even crest lines of the divides; (5) the continuity of profile of benchlands in valley heads with interfluve summits farther downstream; (6) the failure of rock and structure contrasts to appear in the upland topographies or in the strath surfaces; (7) the failure of escarpments to influence river courses; (8) the beveling of escarpment crests to conform to neighboring radial summit flats; (9) the entrenched meanders inherited from strath surfaces whose origin *does* "necessarily involve changes in baselevel"; (10) the unctuous clay fill in many Ozark caves, and (11) the bedded stream gravel on uplands.

Whatever theoretical learnings the investigator may have, he cannot gainsay these field facts and relationships. As earlier noted, they are the court of last appeal.

REFERENCES

Bretz, J H., in preparation, Geomorphic history of the Ozarks of Missouri.
———— 1953, Genetic relations of caves to peneplains and big springs in the Ozarks: Am. Jour. Sci., v. 251, p. 1-24.
Davis, W. M., 1909, Geographical essays: Boston, Ginn & Co., 777 p.
Fenneman, N. M., 1909, Physiography of the St. Louis area: Illinois Geol. Survey Bull. 12, 83 p.
———— 1936, Cyclic and non-cyclic aspects of erosion: Science, v. 83, no. 2144, p. 87.
———— 1938, Physiography of eastern United States: New York, McGraw-Hill, 714 p.
Hack, J. T., 1957, Studies of longitudinal stream profiles in Virginia and Maryland: U. S. Geol. Survey Prof. Paper 294-B, p. iv, 45-97.
———— 1959, Intrenched meanders of the North Fork of the Shenandoah River, Virginia: U. S. Geol. Survey Prof. Paper 354-A, 10 p.
———— 1960a, Interpretation of erosional topography in humid temperate regions: Am. Jour. Sci., v. 258A, p. 80-97.
———— 1960b, Geomorphology and forest ecology of a mountain region in the Central Appalachians: U. S. Geol. Survey Prof. Paper 347, 96 p.
Hershey, O. H., 1895, River valleys of the Ozark plateau: Am. Geologist, v. 16, p. 338-357.
———— 1901, Peneplains of the Ozark highland: Am. Geologist, v. 27, p. 25-41.
Marbut, C. F., 1907 (1908), Geology of Morgan County, Missouri: Missouri Bur. Geology and Mines (2), 7, 97 p.
———— 1896, Physical features of Missouri: Missouri Geol. Survey, v. 10, p. 11-100.
Rich, J. L., 1914, Certain types of stream valleys and their meaning: Jour. Geology, v. 22, p. 469-497.
Rubey, W. W., 1952, Geology and mineral resources of the Hardin and Brussels quadrangles [in Illinois]: U. S. Geol. Survey Prof. Paper 218, 179 p.
Tarr, W. L., 1924, Intrenched and incised meanders of some streams on the northern slope of the Ozark plateau in Missouri: Jour. Geology, v. 32, p. 583-600.

10

Reprinted from *Amer. J. Sci.*, **263**, 110–119 (Feb. 1965)

TIME, SPACE, AND CAUSALITY IN GEOMORPHOLOGY*

S. A. SCHUMM and R. W. LICHTY

U. S. Geological Survey, Denver, Colorado

ABSTRACT. The distinction between cause and effect in the development of landforms is a function of time and space (area) because the factors that determine the character of landforms can be either dependent or independent variables as the limits of time and space change. During moderately long periods of time, for example, river channel morphology is dependent on the geologic and climatic environment, but during a shorter span of time, channel morphology is an independent variable influencing the hydraulics of the channel.

During a long period of time a drainage system or its components can be considered as an open system which is progressively losing potential energy and mass (erosion cycle), but over shorter spans of time self-regulation is important, and components of the system may be graded or in dynamic equilibrium. During an even shorter time span a steady state may exist. Therefore, depending on the temporal and spacial dimensions of the system under consideration, landforms can be considered as either a stage in a cycle of erosion or as a system in dynamic equilibrium.

INTRODUCTION

Current emphasis on the operation of erosion processes and their effects on landforms (Strahler, 1950, 1952) not only has opened the way to new avenues of research but also introduces the possibility of misunderstanding the role of time in geomorphic systems. As Von Bertalanffy (1952, p. 109) put it, "In physical systems events are, in general, determined by the momentary conditions only. For example, for a falling body, it does not matter how it has arrived at its momentary position, for a chemical reaction it does not matter in what way the reacting compounds were produced. The past is, so to speak, effaced in physical systems. In contrast to this, organisms appear to be historical beings". From this point of view, although landforms are physical systems and can be studied for the information they afford during the present moment of geologic time, they are also analogous to organisms because they are systems influenced by history. Therefore, a study of process must attempt to relate causality to the evolution of the system.

It is the purpose of this discussion to demonstrate the importance of both time and space (area) to the study of geomorphic systems. We believe that distinctions between cause and effect in the molding of landforms depend on the span of time involved and on the size of the geomorphic system under consideration. Indeed, as the dimensions of time and space change, cause-effect relationships may be obscured or even reversed, and the system itself may be described differently.

ACKNOWLEDGMENTS

The writers wish to thank several colleagues who reviewed this paper. The comments and criticism of the following were most helpful: M. Morisawa, Antioch College; A. N. Strahler, Columbia University; R. J. Chorley, University of Cambridge; and John Hack, R. F. Hadley, and H. E. Malde of the U. S. Geological Survey. Malde not only made numerous suggestions for improvement of the paper, but he also suggested the format for tables 1 and 2.

* Publication authorized by the Director, U. S. Geological Survey.

TIME, SPACE, AND THE FLUVIAL CYCLE OF EROSION

The description of the changes occurring in a landscape with time, according to the cycle of erosion as propounded by Davis, is encountered less frequently in current geomorphic writings. In the study of geomorphic processes earth scientists are applying themselves to modern problems, and the spatial-temporal range of their research is considerably curtailed. This is necessary if the knowledge of processes is to be developed; however, even in this work the historical aspect of landscape evolution or the time dimension should not be neglected. The neglect of time leads to confusion and needless controversy. For example, recent papers by Hack (1960) and Chorley (1962) may startle some geomorphologists by the rejection of the time dimension, which is a major concern of the geologist. The discussion that follows is an attempt to show that what Hack and Chorley suggest need not be a break with tradition but is simply a method of considering the landscape within narrow temporal limits.

Hack (1960, p. 85) suggests that many elements of the landscape are in dynamic equilibrium with the processes acting upon them; that is, "The forms and processes are in a steady state of balance and may be considered as time independent". He compares this condition with that of a soil undergoing erosion at the surface at the same rate as the lower boundaries of the soil horizons move downward into the regolith (Nikiforoff, 1959). Hack (1960, p. 94) continues his argument as follows: "The theory of dynamic equilibrium explains topographic forms and the differences between them in a manner that may be said to be independent of time. The theory is concerned with the relations between rocks and processes as they exist in space. The forms can change only as the energy applied to the system changes".

This concept is synonymous to that of the physical system described by Von Bertalanffy in which, "The past is, so to speak, effaced". Nevertheless, after excluding time from his system, Hack considers it in the following qualification, "It is obvious, however, that erosional energy changes through time and hence forms must change". A change of erosional energy can be initiated by many factors, of which diastrophism or climate change are the most obvious. In addition, with the passage of time erosional modification of the landforms themselves will affect erosional energy. Therefore, it appears impossible to exclude time and history from a consideration of landforms except during the study of purely empirical relations among variables, which may or may not reflect causality.

Chorley likewise feels that freedom from the historical approach is desirable, because research efforts are then directed toward a study of the rate and manner of operation of erosional processes, the empirical relations that exist between a landscape and its components, and the relations between the erosion processes and the landform. Chorley (1962, p. 3) illustrates the difficulty of reconciling the two approaches, the Davisian cycle of erosion and dynamic equilibrium, as follows: "In the former, the useful concept of dynamic equilibrium or grade rests most uncomfortably; in the latter . . . the progressive loss of a component of potential energy due to relief reduction imposes an unwelcome historical parameter".

To resolve the controversy resulting from these two viewpoints it may be necessary to think only in terms of large and small areas or of long and short spans of time. A choice must be made whether only components of a landscape are to be considered or whether the system is to be considered as a whole. Also, a choice must be made as to whether the relations between landforms and modern erosion processes are to be considered or whether the origin and subsequent erosional history of the system is to be considered. In table 1 an attempt is made, using a hypothetical drainage basin as an example, to demonstrate that the concepts of cyclic erosion with time and timeless dynamic equilibrium are not mutually exclusive.

The variables listed in table 1 are arranged in a hierarchy we believe approximates the increasing degrees of dependence of the variables considered. For example, time, initial relief, geology, and climate are obviously the dominant independent variables that influence the cycle of erosion. Vegetational type and density depend on lithology and climate. As time passes the relief of the drainage system or mass remaining above base level is determined by the factors above it in the table, and it, in turn, strongly influences the runoff and sediment yield per unit area within the drainage basin. The runoff and sediment yield within the system establish the characteristic drainage network morphology (drainage density, channel shape, gradient, and pattern) and hillslope morphology (angle of inclination and profile form) within the constraints of relief, climate, lithology, and time. The morphologic variables, in

Table 1

The status of drainage basin variables during time spans
of decreasing duration

Drainage basin variables	Status of variables during designated time spans		
	Cyclic	Graded	Steady
1. Time	Independent	Not relevant	Not relevant
2. Initial relief	Independent	Not relevant	Not relevant
3. Geology (lithology, structure)	Independent	Independent	Independent
4. Climate	Independent	Independent	Independent
5. Vegetation (type and density)	Dependent	Independent	Independent
6. Relief or volume of system above base level	Dependent	Independent	Independent
7. Hydrology (runoff and sediment yield per unit area within system)	Dependent	Independent	Independent
8. Drainage network morphology	Dependent	Dependent	Independent
9. Hillslope morphology	Dependent	Dependent	Independent
10. Hydrology (discharge of water and sediment from system)	Dependent	Dependent	Dependent

turn, strongly influence the volumes of runoff and sediment yield which leave the system as water and sediment discharge.

Among the variables listed on table 1, every cause appears to be an effect and every effect a cause (Mackin, 1963. p. 149); therefore, it is necessary to set limits to the system that is considered. Obviously neither the causes of geology, climate, and initial relief nor the effects of water and sediment discharge concern us here.

The three major divisions of table 1 are time spans which are termed cyclic, graded, and steady. The absolute length of these time spans is not important. Rather, the significant concept is that the system and its variables may be considered in relation to time spans of different duration.

Cyclic time, of course, represents a long span of time. It might better be referred to as geologic time, but in order to keep the terminology of the table consistent, cyclic is used because it refers to a time span encompassing an erosion cycle. Cyclic time would extend from the present back in time to the beginning of an erosion cycle.

Consider a landscape that has been tectonically stable for a long time. A certain potential energy exists in the system because of relief, and energy enters the system through the agency of climate. Over the long span of cyclic time a continual removal of material (that is, expenditure of potential energy) occurs and the characteristics of the system change. A fluvial system when viewed from this perspective is an open system undergoing continued change, and there are no specific or constant relations between the dependent and independent variables as they change with time (fig. 1a).

During this time span only time, geology, initial relief, and climate are independent variables. Time itself is perhaps the most important independent variable of a cyclic time span. It is simply the passage of time since the beginning of the erosion cycle, but it determines the accomplishments of the erosional agents and, therefore, the progressive changes in the morphology of the system. Vegetational type and density are largely dependent on climate and

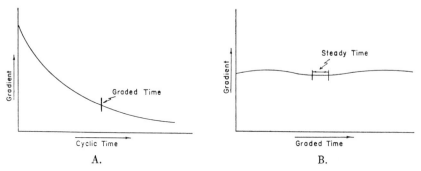

Fig. 1. Diagrams illustrating the time spans of table 1. Channel gradient is used as the dependent variable in these examples.

a. Progressive reduction of channel gradient during cyclic time. During **graded time,** a small fraction of cyclic time, the gradient remains relatively **constant.**

b. Fluctuations of gradient above and below a mean during graded time. Gradient is constant during the brief span of steady time.

lithology, but they significantly influence the hydrology and erosional history of a drainage basin. If all the independent variables are constant except time, then as time passes the average relief and mass, volume of material remaining within the drainage system, will decrease. As the relief or mass of the system changes so will the other dependent morphologic and hydrologic variables.

With regard to space or the area considered, it is possible to consider an entire drainage system or any of its component parts during a cyclic time span. For example, the reduction of an entire drainage system or only the decrease in gradient of a single stream may be considered (fig. 1a) during cyclic time.

The graded time span (table 1) refers to a short span of cyclic time during which a graded condition or dynamic equilibrium exists. That is, the landforms have reached a dynamic equilibrium with respect to processes acting on them. When viewed from this perspective one sees a continual adjustment between elements of the system, for events occur in which negative feedback (self-regulation) dominates. In other words the progressive change during cyclic time is seen to be, during a shorter span of time, a series of fluctuations about or approaches to a steady state (fig. 1b). This time division is analogous to the "period of years" used by Mackin (1948, p. 470) in his definition of a graded stream by which he rules out seasonal and other short-term fluctuations, as well as the slow changes that accompany the erosion cycle.

As an erosion cycle progresses, more and more of the landscape may approach dynamic equilibrium. That is, the proportion of graded landforms may increase, and it seems likely that temporary graded conditions become more frequent as time goes on. However, it is apparent that during this time span the graded condition can apply only to components of the drainage basin. The entire system cannot be graded because of the progressive reduction of relief or volume of the system above base level, which occurs through export of sediment from the system. A hillslope profile or river reach, however, may be graded. Therefore, unlike cyclic time when no restriction was placed on space or area considered, graded time is restricted to components of the systems or to smaller areas.

During a graded time span, the status of some of the variables listed on table 1 changes. For example, time has been eliminated as an independent variable, for although the system as a whole may be undergoing a progressive change of very small magnitude, some of the components of the system will show no progressive change (that is, graded streams and hillslopes). Initial relief also has no significance because the landform components are considered with respect to their climatic, hydrologic, and geologic environment (Hack, 1960), and intial relief with time has been designated as not relevant on table 1.

In addition, some of the variables that are dependent during a long period of progressive erosion become independent during the shorter span of graded time. The newly independent hydrologic variables, runoff and sediment yield, are especially important because during a graded time span they take on a statistical significance and define the specific character of the drainage channels and hillslopes, whereas during a cyclic time span there is a progressive change in these morphologic variables.

163

The geomorphic variables of hillslope and drainage network morphology of graded time may be considered as "time-independent" in the meaning of Hack (1963, written communication). That is, relict features may not be present, and the landforms may be explained with regard to the independent variables without regard to time.

During a steady time span (table 1) a true steady state may exist in contrast to the dynamic equilibria of graded time (fig. 1b). These brief periods of time are referred to as a steady time span because in hydraulics steady flow occurs when none of the variables involved at a section change with time. The landforms, during this time span, are truly time independent because they do not change, and time and initial relief have again been eliminated as independent variables. During this time span only water and sediment discharge from the system are dependent variables.

Obviously the steady state condition is not applicable to the entire drainage basin. Although an entire drainage basin cannot be considered to be in a steady state over even the shortest time span, yet certain components of the basin may be. For example, a stream over short reaches may export as much water and sediment as introduced into the reach, yet the river as a whole is reducing its gradient in the headwaters (cyclic erosion). In addition, the entire drainage basin may be losing relief as hillslopes are lowered (cyclic erosion); however, segments of the hillslopes may remain at the same angle of inclination and act as slopes of transportation (steady state), or they may retreat parallel, maintaining their form (dynamic equilibrium), but the volume of the drainage basin is being reduced nevertheless. Thus over short periods of time and in small areas the steady state may be maintained. Over large areas progressive reduction of the system occurs, and this is true over long periods of time.

The preceding discussion and the relations presented in table 1 and figure 1 have the sole object of demonstrating that, depending on the time span involved, time may be either an extremely important independent variable or of relatively little significance to a study of landforms.

FLUVIAL MORPHOLOGY AND HYDRAULICS

In this section a specific example of river channel morphology and hydraulics will be cited to illustrate how, as time spans are shortened, there is a shift from dependence to independence among the variables and how, during the shortest span of time, an apparent reversal of cause and effect may occur.

In table 2 an attempt has been made to illustrate the effect of time span on the interrelations between dependent and independent variables of a river system. A similar table has been presented by Kennedy and Brooks (in press) to compare independent and dependent variables in the flume and field situations. As in the preceding discussion, it is the time span or duration of a time period that is important, but Pleistocene and Recent geomorphic history is such that in many areas it is possible to divide the time involved into three spans, termed geologic, modern, and present, each shorter and more recent than the span that precedes it.

TABLE 2

The status of river variables during time spans of decreasing duration

River Variables	Status of variables during designated time spans		
	Geologic	Modern	Present
1. Time	Independent	Not relevant	Not relevant
2. Geology (lithology and structure)	Independent	Independent	Independent
3. Climate	Independent	Independent	Independent
4. Vegetation (type and density)	Dependent	Independent	Independent
5. Relief	Dependent	Independent	Independent
6. Paleohydrology (long-term discharge of water and sediment)	Dependent	Independent	Independent
7. Valley dimension (width, depth, and slope)	Dependent	Independent	Independent
8. Mean discharge of water and sediment	Indeterminate	Independent	Independent
9. Channel morphology (width, depth, slope, shape, and pattern)	Indeterminate	Dependent	Independent
10. Observed discharge of water and sediment	Indeterminate	Indeterminate	Dependent
11. Observed flow characteristics (depth, velocity, turbulence, et cetera)	Indeterminate	Indeterminate	Dependent

It is almost impossible to assign temporal boundaries to these time spans because their duration will vary with each example considered. Nevertheless, geologic time in this sense begins during the Pleistocene Epoch, because it is during the higher discharges of the glacial stages that the width and depth of many valleys were established (Dury, 1962). Since the Pleistocene there undoubtedly have been some modifications of these valleys, but their major characteristics were determined both by the higher discharge of the Pleistocene and the post-Pleistocene adjustments to a changed hydrologic regimen. For this reason geologic time probably should end at perhaps 5000 or 10,000 years ago. However, to keep table 2 consistent with table 1 geologic time is defined as beginning 1,000,000 years ago and extending to the present.

During geologic time (table 2), time, geology, and climate are independent variables. Initial relief is not included because geologic time does not extend back to the origin of the system. Vegetation and relief are considered to be dependent variables, as is the paleohydrology of the system, which controlled the dimensions of the valley during geologic time. For geologic time we can know little or nothing about the dependent variables in the hierarchy below valley dimensions (table 2), and these variables are classed as indeterminate.

In the time span termed modern (table 2) (arbitrarily defined as the last 1000 years) the number of independent variables increases, and some previously indeterminate variables become measurable. For example, valley dimensions become independent during modern time because they were defined by the paleohydrology of geologic time and inherited from geologic time. The mean discharge of water and sediment during modern time is also considered an independent variable, because it determines the morphology of the modern channel. Only channel morphology is dependent during modern time. Modern time is 1000 years in duration; therefore, the observed discharge and flow characteristics, which can be measured only during a brief span of time, are indeterminate.

During the short span of present time (defined as 1 year or less), channel morphology assumes an independent status because it has been inherited from modern time. The present or observed discharge of water and sediment and flow characteristics can be measured at any moment during present time, and these variables are no longer indeterminate.

It is during the brief span of present time that the possibility of an apparent reversal of cause and effect may occur, due to feedback from the dependent to the independent variables. For example, a major flood during this brief span of time might so alter the flow characteristics that a modification of channel dimensions and shape could occur. Just as water depth and velocity can be adjusted in a flume to modify sediment transport, so there is a feedback from flow velocity to sediment discharge and channel morphology. That is, as discharge momentarily increases, sediment that was previously stationary on the channel floor may be set in motion. The resulting scour, albeit minor, will influence channel depth, gradient, and shape. Thus, short term changes in velocity can cause modification of some of the independent variables.

These modifications are usually brief and temporary, and the mean values of channel dimensions and sediment discharge are not permanently affected. Nevertheless, a temporary reversal of cause and effect can occur, which, when documented quantitatively, may be a source of confusion in the interpretation of geomorphic processes. This is best demonstrated by comparing the conflicting conclusions that could result from studying fluvial processes in the hydraulic laboratory and in a natural stream. The measured quantity of sediment transported in a flume is dependent on the velocity and depth of the flowing water and on flume shape and slope. An increase in sediment transport will result from an increase in the slope of the flume or an increase in discharge. In a natural stream, however, over longer periods of time, it is apparent that mean water and sediment discharge are independent variables, which determine the morphologic characteristics of the stream and, therefore, the flow characteristics (table 2, modern time). Furthermore, over very long periods of time (geologic) the independent variables of geology, relief, and climate determine the discharge of water and sediment with all other morphologic and hydraulic variables dependent. Both Mackin (1963) and Kennedy and Brooks (in press) used this identical example to illustrate the need to consider how time spans are relevant to the explanation of fluvial phenomena.

Kennedy and Brooks (in press) state it thusly (Q and Q_s are water and sediment discharge),

> Streams are seldom if ever in a steady state (because of finite time required to change bed forms and depth) and transitory adjustments are accomplished by storage of water and sediment. Water storage is relatively short (hours and days) and occurs simply by the increasing of river stage or overbank flooding; sediment storage (+ or −) occurs by deposition or scour. Thus for the short term, Q_s may be considered a dependent variable, with departures of the sediment inflow from the equilibrium transport rate being absorbed in temporary storage (for months or years). But in the long term the river must assume a profile and other characteristics for which on the average the inflow of water and sediment equals the outflow; consequently for this case (called a "graded" stream by geologists), Q and Q_s are . . . independent variables.

Table 2 is more than a scholarly exercise in sorting variables, for although, as our knowledge increases, it will require modification, it can be of immediate use in the consideration of problems of fluvial morphology. For example, assuming we have arranged the variables in correct order, if flow characteristics are dependent variables for a modern river then currents or the helicoidal flow measured at river bends should not be the cause of meanders. In other words, sinuosity of the river (the ratio of valley slope to channel gradient) influences the flow character not the converse. Specifically, helicoidal flow exists because of a meander, which in turn may exist partly because of past conditions of flow (paleohydrology).

As another example, in a set of data collected for Great Plains rivers it was found that a highly significant correlation exists between valley slope and stream gradient. At first, this appears trivial for slope is correlated with slope. However, if table 2 is correct, the slope of modern valleys is an independent variable dependent on the paleohydrology of geologic time, and the existing, modern channel slope is a dependent variable.

Variations in the sinuosity of the Great Plains rivers have been explained (Schumm, 1963) by the decrease in post-Pleistocene discharge and a change in the amount and type of sediment transported by the rivers. Depending on the changes in sediment load and discharge, the present stream may require a slope identical with the valley slope (sinuosity is 1) or much less than the valley slope (sinuosity is greater than 1 but usually less than 2.5). The valley floor, therefore, is the surface upon which the present river flows, and depending on changes in sediment and discharge, the river may flow at the slope of the valley or at a slope considerably less. Valley slope, as indicated on table 1, is an independent variable exerting a control on stream gradient.

CONCLUSIONS

The distinction of cause and effect among geomorphic variables varies with the size of a landscape and with time. Landscapes can be considered either as a whole or in terms of their components, or they can be considered either as a result of past events or as a result of modern erosive agents. Depending on one's viewpoint the landform is one stage in a cycle of erosion or a feature in dynamic equilibrium with the forces operative. These views are not mutually exclusive. It is just that the more specific we become the shorter is the time span with which we deal and the smaller is the space we can consider. Conversely when dealing with geologic time we generalize. The steady state concept can fit into the cycle of erosion when it is realized that steady states can be maintained only for fractions of the total time involved.

The time span considered also influences causality, as the sets of independent and dependent variables of tables 1 and 2 show. If the variables were not considered with respect to the time span involved, in many cases it would be difficult to determine which variables are independent. Mackin's (1963) and Kennedy and Brooks' (in press) suggestions forestall any arguments between workers in the laboratory and workers in the field. In the same manner the disparate points of view of the historically oriented geomorphologist and the student of process can be reconciled.

REFERENCES

Chorley, R. J., 1962, Geomorphology and general systems theory: U. S. Geol. Survey Prof. Paper 500-B, 10 p.

Dury, G. H., 1962, Results of seismic exploration of meandering valleys: Am. Jour. Sci., v. 260, p. 691-706.

Hack, J. T., 1960, Interpretation of erosional topography in humid temperate regions: Am. Jour. Sci., v. 258-A, Bradley v., p. 80-97.

Kennedy, J. F., and Brooks, N. H., in press, Laboratory study of an alluvial stream at constant discharge: Federal Interagency Sedimentation Conf. Proc., Jackson, Miss., 1963, in press.

Mackin, J. H., 1948, Concept of the graded river: Geol. Soc. America Bull., v. 59, p. 463-511.

———— 1963, Rational and empirical methods of investigation in geology, *in* Albritton, C. C., Jr., ed., The fabric of geology: Reading, Mass., Addison-Wesley Publishing Co., p. 135-163.

Nikiforoff, C. C., 1959, Reappraisal of the soil: Science, v. 129, p. 186-196.

Schumm, S. A., 1963, Sinuosity of alluvial rivers on the Great Plains: Geol. Soc. America Bull., v. 74, p. 1089-1100.

Strahler, A. N., 1950, Equilibrium theory of erosional slopes approached by frequency distribution analysis: Am. Jour. Sci., v. 248, p. 673-696, p. 800-814.

———— 1952, Dynamic basis of geomorphology: Geol. Soc. America Bull., v. 63, p. 923-938.

Von Bertalanffy, Ludwig, 1952, Problems of life: London, Watts and Co., 216 p.

Part II

PEDIPLAINS

The papers in this section deal with problems related to the recognition and mode of formation of pediments and their coalescent equivalents, pediplains. These are compared to peneplains.

Although Gilbert, Dutton, and others (see general introduction) used the name and in some cases recognized the erosional nature of surfaces called pediments, it was left to W. J. McGee (Paper 11) to realize and describe their wide extent in the Sonoran Desert of the United States, a region formerly considered as an example of mountains buried in their own debris. He also offered an explanation for these widespread erosion surfaces.

Although pediments and peneplains were described at about the same time, almost 20 years elapsed before such workers as Lawson and Bryan further pursued the idea of pediments in the semiarid Southwest. Here the controversy over the processes involved began. It might be dubbed the great "wash or wear" controversy. Some workers ascribed the development of pediments to unconcentrated overall washing away of weathered material; others saw their development as resulting from the concentrated channel wear of laterally swinging streams.

During this controversy, the need for quantitative work on slope measurements and on the nature and quantity of rock removal was demonstrated. Some of this work is included here.

The idea of the coalescence of adjoining pediments to form pediplains affords a new conceptual method of general planation, but it also engenders problems. One is that surfaces heretofore described as peneplains may have some of the attributes of pediplains. Conversely, there is the possibility of pediplains "softening" into peneplains. L. C. King's conviction that all planation surfaces are pediplains and subject to intercontinental correlations is an interesting but perhaps simplistic view which has produced much reaction.

This section ends with consideration of climatic change subsequent to the formation of any given pediplain or peneplain. The basic problem is: How can we develop criteria to determine whether a particular planation surface is presently exposed to the same climatic conditions under which it was produced? Some progress has been made in this direction and will be considered further in the section on etchplains.

Editor's Comments
on Papers 11 Through 14

RECOGNITION OF PEDIMENTS AND POSSIBLE MODES OF ORIGIN

Horse hooves ringing against bare rock called McGee's attention to the essentially erosional nature of the Sonoran plains near the backdrop mountains of Arizona (Paper 11). He gives a careful description of the surface and its relation to the mountains. He also reasons that such a broad erosion surface must have been produced by a widely operating agent. Sheetfloods, observed in the region, were assigned this role. Others suggest that sheetfloods may be an effect rather than the cause of planation. McGee also recognizes a prepediment peneplain at high altitude, which suggests a possible climatic change from humid to semiarid conditions.

In Paper 12 Lawson (1915) considers surface development in an arid region where drainage goes to the rising base level of an enclosed and alluviating basin. The result of stream and wash activity is the retreat of mountain slopes and the production of a suballuvial bench, which is exposed, if at all, only near a narrow working edge near the mountain front. Continued action results eventually in a "panfan" consisting mainly of a gradually thickening alluvial wedge overlying a convex upward suballuvial bench. In some respects this is considered analogous to a peneplain.

Bryan, in Paper 13, extends Lawson's studies to an area of exterior drainage in the Papago country of Arizona. The important conceptual change is the extension mountainward of a concave, upward, planation surface controlled by a stationary or falling base level. Bryan calls the resulting surface a *mountain pediment*. Bryan's glossary (here omitted) defines a mountain pediment as a

> plain which lies at the foot of a mountain in an arid region or in headwater basins within a mountain mass. The name is applied because the plain appears to be a pediment upon which the mountain stands. A mountain pediment is formed by the erosion and deposition of streams usually of the ephemeral type and is usually covered by a veneer of gravel in transit. . . . It simulates the form of an alluvial slope.

Adjacent mountain slopes are formed by wash, but pediments are cut by ephemeral streams.

Davis takes the firm position in Paper 14 that rock floors in arid and humid climates are homologous and are made by virtually the same geologic agents working at different intensities and under different constraints. The actual homologues are functions of the degree of cover of the rock floor and the available energy of the running water.

This paper contains the germ of the basic difference between peneplains and pediplains. The implications are that (1) both types of surface are homologous, (2) a climatic change at some time during the cycle could transform one developing type of surface into the other, or (3) one type of surface could be changed into the other near the end of a cycle.

In a 1933 paper (not reproduced herein), C. H. Crickmay disagreed with Davis on the relative work done by streams and by valley slide processes such as creep and slope wash. Crickmay championed the unlimited lateral corrasion of streams, which would result in the eventual removal of lateral divides and the production of a surface called a *panplain*. Like Davis, he could point to no active example of his concept, but stated that pediments closely approximated incipient panplains. Crickmay here assumes that pediments (and pediplains) are entirely produced by the lateral corrasion of stream channels.

REFERENCE

Crickmay, C. H., 1933. The later stages in the cycle of erosion. Geol. Mag., 70, p. 337–347.

11

Reprinted from *Bull. Geol. Soc. America,* **8,** 87–92, 96–111 (1897)

SHEETFLOOD EROSION

BY W J MC GEE

(Presented before the Society August 22, 1896)

CONTENTS

DEFINITIONS.

Commonly, running water gathers into streams and corrades channels; exceptionally, running water spreads into sheets of limited or unlimited width, and, by a combination of erosion and deposition, produces plains.

Pure water flowing over a smooth indestructible surface does not move as a uniform film: if the surface is broad the sheet differentiates into parallel streams of greater depth and relatively rapid flow, separated by shallower bands of relatively sluggish flow; and at the same time both streams and intervening bands differentiate into series of transverse waves which move forward more rapidly than the body of the differentiated sheet. The tendency of flowing water to divide into streams is well known; it is undoubtedly by reason of this tendency that running waters commonly flow in streams which cut channels and eventually fashion most of the lands of the earth. The tendency of flowing water to break up into transverse waves is not so well known, though beautifully exemplified in sluices for irrigation or for bringing logs and lumber down from mountains (where it is the succession of waves which prevents the sluice from clogging and equalizes the movement of the floating load); it is of subordinate importance geologically.

Under certain conditions, sand-laden water flowing over an erodable plain tends at first to divide into parallel streams like those of pure water on an indestructible surface, yet, since the streams formed in this way at once begin to scour and overload themselves and thus check their own flow, this tendency is soon counteracted and the water is distributed again; so that the ultimate tendency is toward movement in a more or less uniform film or sheet. This tendency is well known to laymen in those regions in which it prevails, but it seems not to be generally recognized among geologists. Colloquially a moving water-body of this type is sometimes known as a " wash ; " but since the term is commonly applied primarily to the product and only secondarily to the agency, and since it is usually restricted to limited, though broad, channels (*e. g.*, in San Francisco wash), it seems desirable to use some other designation for the water-body ; and the term *sheetflood* has come into use in notes and in conversation.

Thus there are in nature two strongly contrasted types of moving water bodies, namely, (1) streams, and (2) sheetfloods. The first type is characterized by a tendency toward concentration in narrow and relatively deep bands which quickly cut channels for themselves ; the second is characterized by a tendency to spread widely in relatively shallow sheets. The second is logically coordinate with the first as a geologic agent. In a general way streams prevail in humid regions, sheetfloods in arid regions, though streams occur locally in arid lands, while it seems probable that sheetfloods occur under certain conditions in nearly all lands, howsoever humid.

Sheetflood Work in the Sonoran District.

FEATURES OF THE DISTRICT.

Location.—Sheetflooding is characteristic of the broad expanse of plain and mountain in southwestern Arizona and western Sonora (Mexico), stretching from the Sierra Madre to the gulf of California, and lying between Gila and Yaki rivers. In physical characteristics and geologic history this territory forms part of Powell's great province of " Basin

Ranges,"* constituting a considerable part of that portion of the province characterized by " Open Basins ; " but the district is so distinctive that, for purposes of description at least, it may be set apart as the *Sonoran district*. It corresponds approximately with the region known among the Mexican inhabitants as Papagueria, or land of the Papago Indians.

The Sonoran district slopes from the high Sierra of Mexico and the Mogollon escarpment at its northern extremity to tide level in the gulf of California. To the casual observer traversing its expanse it seems a region of mountains, for rugged buttes, mesas, and sierras are always in sight and usually dominate the landscape ; but more careful observation shows that it is primarily a plains region, since fully four-fifths of its area consists of plains, hardly one-fifth of mountains, and the elongated sierras and scattered buttes are nearly always flanked by vast tracts of lowland. The mountains range up to 6,000 or 6,500 feet above tide in the interior (imposing Baboquivari, southwest of Tucson, measures 6,798 feet) and still greater altitudes toward the Sierra Madre ; the intermontane plains rise gently from sealevel to 3,500 or 4,000 feet,† while the buttes and sierras may rise anywhere from 100 to 4,000 feet above the circumscribing plain. The mountains are notably rugged, abounding in lofty precipices, vertical and sometimes overhanging cliffs, knife-edge aretes, and leaning (apparently or actually) picachos, and they rise with remarkable abruptness from the smooth plains, so that, to the casual observer, they seem an archipelago of rocky peaks rising from an ocean of desert sands. Though monotonously smooth, partly by contrast with the rugged mountains, the plains slope gently away from the mountains and merge in flat-bottomed valleys leading directly or deviously toward the sea, so that the entire surface, with the exception perhaps of a few playas near the coast, drains seaward. In the valleys occupied by washes (rivers during the wet seasons, sand wastes during most of the year) the grades may be as low as 20 or even 15 feet to the mile ; over the average plain the slope is 50 or 75 feet to the mile, and on the great apron-like foot-slopes pushing among the aretes and lateral spurs of the larger sierras the grade may be 100, 200, even 300 or more feet to the mile.

Climate.—The Sonoran district is excessively hot and arid. Yuma, noted as the hottest station in the United States, is in its northwestern corner, and while records are lacking, it is probable that Caborca and Hermosillo are considerably hotter ; the rainfall in the Sierra may reach 15 or 20 inches,

---

* National Geographic Monographs, vol. i, 1895, pp. 95–98.

† Captain D. D. Gaillard, of the International Boundary Commission, writes : " The sacred peak of the Papago, Baboquivari, is 6,798 feet above mean sea level of San Diego bay. The lowest part of the valley on the east of this peak (called the " Sasabe flat ") is about 3,200 feet above the same datum plane, while the lowest part of the valley on the west of the peak (called the " Moreno flat ") is only about 2,300 feet above the plane of reference " (official letter dated February 6 1896).

but the meager records in even the most fertile valleys in the foothills seldom rise above 10 inches, while the average over the interior probably falls below 5, and may be no more than 2 or 3 inches during the year. There are two nominally wet seasons, occurring respectively about mid-winter and midsummer. The midwinter precipitation is generally the heavier and the more widely distributed, but both in summer and in winter the greater part of the rainfall occurs in local storms. Snow falls on the Mogollon and the Sierra, remaining half the year on a few of the highest points; to this fact the perennial character of some of the northern tributaries of the Gila and the stronger branches of the Yaki may be ascribed. During the occasional drizzles of the wet seasons the scanty moisture is chiefly absorbed by the sun-baked earth, so that floods do not ensue, though the spring-fed streams may rise by reason of the diminished evaporation; during the moderate storms that occur here and there from year to year torrents are produced which rush tumultuously down the slopes and become potent geologic agents, and during the great storms occurring from decade to decade or from century to century whole plains are flooded; yet so dry are air and earth that the deluge is absorbed within a few miles or scores of miles. Thus, although the entire surface slopes seaward, no living water reaches the sea between the Colorado and the Yaki, 700 miles away.

By reason of heat and aridity the Sonoran district is desert or subdesert throughout; the vegetation is too scant, stunted, and scattered to protect the surface from storms; the meager flora forms little or no humus, and thus there is no soil and little of that chemic action initiated by vegetal growth and decay. Through a combination of biotic conditions of great significance the vegetal life and the sedentary animal life are concentrated in scattered colonies with bare earth between, and the colonies collectively form but a small fraction of the total area. Thus the region is one in which physical agency operates directly, with little aid or obstruction from the biotic agency always present and often predominant in humid districts.

Since the waters of the Sonoran district never reach the sea the territory is complete in itself as a geologic province; the storm waters gather detritus in the mountains and transport it into the valleys, but their agency is limited to shifting the rock matter from one point to another in the same vicinity, and thus degradation and aggradation go hand in hand, and gradation is completed within the district.

Topography.—At first sight the Sonoran district appears to be one of half-buried mountains, with broad alluvial plains rising far up their flanks, and so strong is this impression on one fresh from humid lands that he finds it difficult to trust his senses when he perceives that much of the valley-plain area is not alluvium, but planed rock similar to or identical with that constituting the mountains. To the student of geomorphy this is the striking characteristic of the Sonoran region—the mountains rise from plains, but both mountain and plain (in large part) are carved out of the same rocks. The valley interiors and the lower lowlands are, indeed, built of torrent-laid debris, yet most of the valley area carries but a veneer of alluvium so thin that it may be shifted by a single great storm. Classed by surface, one-fifth of the area of the Sonoran district, outside of the Sierra and its foothills, is mountain, four-fifths plain; but of the plain something like one-half, or two-fifths of the entire area, is planed rock, leaving only a like fraction of thick alluvium. This relation seems hardly credible. During the first expedition of the Bureau of American Ethnology (in 1894) it was noted with surprise that the horseshoes beat on planed granite or schist or other hard rocks in traversing plains 3 or 5 miles from mountains rising sharply from the same plains without intervening foothills; it was only after observing this phenomenon on both sides of different ranges and all around several buttes that the relation was generalized, and then the generalization seemed so far inconsistent with facts in other districts that it was stated only with caution even in conversation. During the 1895 expedition a skilled student of geomorphy (Mr Willard D. Johnson) was added to the party, partly in the hope that observation in this direction might be extended and verified or corrected. Even then the observations and inferences seemed hardly worthy of trust until the shores of the gulf of California in Seriland were examined and the superb section from Punta Ygnacio to Puerta Infierno was found to exhibit clearly the inferred relation. "A quarter of this 15-mile exposure is the current-built point, another quarter cuts butte or range of igneous rock or ancient granite, while the remaining half traverses typical intermontane plain in cliffs of 20 to 50 feet, and fully 5 out of the $7\frac{1}{2}$ miles of the low cliff reveal the substratum of planed granite beneath a torrential veneer, while there is more of alluvium-free granite than of graniteless alluvium."[*] Sierra de Tonuco, lying a few miles outside the northeastern corner of Seriland,[†] is a typical mountain mass of the Sonoran district; it is a deeply furrowed block of semi-metamorphic limestone resting on an inclined table of slightly schistoid granite; its crest is perhaps 2,500 feet above the plain, and its upper three-fifths is limestone; its base appears to be heavily burdened with taluses of limestone and granitic debris, but occasional arroyas cut through these aprons and show that they are not of great depth. These rubble-cumbered slopes pass within a fraction of a mile into a lowland plain so

[*] National Geographic Magazine, vol. vii, 1896, pp. 127, 128.
[†] Map of Seriland, op cit., pl. xiv.

smooth that, except where outlying granite buttes rise from its expanse, it may be traversed by wagons in any direction; yet for 5 miles from the mountain base on northeast, north, west, south, and southeast the wheels grind over granite half the time, while on the east the alluvial veneer appears barely to conceal the granite over an area larger than that occupied by the sierra. In the northeastern portion of the Seriland map over a dozen buttes are shown rising sharply from the subdesert plain, and though these lie almost in the delta of Rio Bacuache, analogy with similar buttes a few miles further northeastward, which were carefully examined, indicates that they are not scattered island summits, as their appearance suggests. but merely knobs rising from a baseleveled plain of granite traversed somewhere by an ancient valley a few hundred feet deep. Again. 5 miles southeast of Puerta Infierno the tide-carved coast cuts a typical granitic butte a hundred feet high and as many yards across, rising sharply from the inclined foot-slope of Sierra Seri, yet the rugged-faced knob is seen to surmount a granite pediment nearly half a mile across in the line of section.

It is to be remembered that the surface and structure of the district are known only through two expeditions, each traversing it twice ; but so far as the observations go they indicate that the vast plains diversified by scattered sierras and buttes do not represent an alluvium-buried mountain land so much as a planation level with a few monadnocks still surviving.

Geology.—The structural geology of the Sonoran district is too little known to warrant detailed description. In a general way the rocks appear to be (1) ancient granites, (2) partially metamorphosed limestones, shales. etcetera, and (3) moderately old igneous sheets with associated tuff beds.

[*Editor's Note:* Some geological details not pertinent to the main topic are deleted.]

STREAM EROSION IN THE DISTRICT.

Character of streams.—The current maps of the Sonoran district commonly represent a considerable number of rivers gathering from many tributaries in the mountains and flowing northward into the Gila or southward and westward into the Altar and Sonora or directly into the gulf. Better maps based on actual surveys, such as those of the General Land Office in the United States, show multitudes of small streams flowing down the mountain slopes and either ending on the plain or uniting in rivers which wander toward some principal waterway and end blindly; but the general maps express hypotheses based on the behavior of streams in humid lands, and even the best maps represent the sand washes produced by great storms in lieu of permanent water. In the more elevated eastern part of the district, and during the rainy seasons, especially that of winter, when the storm water in the mountains is supplemented by melting snows, most of the sand washes are, indeed, converted into streams, which, although shallow, are rapid and even torrential, and carry vast volumes of sediment-charged water down the slopes; but throughout most of the territory the sand washes are hardly wetted during the normal rainy season, and are transformed into torrents only during great storms or cloud-bursts occurring at intervals of years or decades. So the map representing sand washes as rivers is misleading unless its purely conventional character is clearly understood. Although the aggregate stream length represented in such a map is considerable, perhaps several times greater than the aggregate length of the streams actually flowing on a given date during the wettest season, yet during the dry season the aggregate length of actual water lines is reduced to a minute fraction of the aggregate length of sand wash, and nearly all of the channels are dry. So the stream of living water is an exceptional, indeed an exceedingly rare, feature in the Sonoran district, and is hardly known save in its more elevated eastern portion. *

Sources of streams.—The sources of stream water are, as usual, three— (1) melting snow on the high Sierra and the Mogollon, (2) ground-water appearing as seepage or springs in the deeper valleys, and (3) the product of rains. The first of these sources might be neglected, save that it contributes (*a*) directly to the longer rivers, San Pedro and Santa Cruz in Arizona, and Altar, Magdalena, Bacuache, Sonora, San Miguel, and the main and minor branches of the Yaki in Sonora, and (*b*) indirectly through ground-water to these and other streams during a part of the year.

The ground-water is of considerable importance as a source of streams, partly since its flow is moderately steady, and its tendency is thus to maintain the stream as such and to enable it to continue corrasion and transportation, albeit feebly, throughout the year. In the mountain gorges the ground-water commonly emerges as permanent or temporary springs, while in the alluvium-lined valleys it simply seeps through the sand, generally below the surface, though sometimes in a slender streamlet winding through the broad sand wash. Among the rocks of the mountains the ground-water movement is conditioned as in humid lands, but in the broader valleys it is conditioned by a variety of factors, including the conformation of the alluvium-lined basin with its various arms and interruptions, as well as by the rate of evaporation, etcetera. Thus the alluvial mass and the adjacent hard rocks are wetted or saturated during the rainy season, and the water percolates down the slopes along lines generally corresponding with those of surface drainage; and wherever the surface drainageway is exceptionally deep, there the ground-water most frequently emerges to flow for rods or miles, or until evaporated or reabsorbed by the sands. So it happens that water may often be found by digging in the dry sands of a wash; that a nominally permanent stream may appear, disappear, and reappear half a dozen times in the course of a day's journey down a single storm-fashioned wash, and that the rippling streamlets lengthen during the night and in cool or cloudy weather and shrink during the day and the heat, when evaporation culminates.

It is by reason of this relation between land surface and ground-water surface, in conjunction with the characteristic migration of divides, that a considerable part of the Sonoran district is habitable by man, for it is in the narrow gorges produced by the retrogressive erosion trenching the

---

* The bounding rivers, Gila and Yaki, are maintained by tributaries from without the district.

ranges that the ground-water reaches or approaches the surface so that settlements can be maintained.

The chief source of stream water is the sporadic storm, especially the thunder-gust or cloud-burst, which fills old channels and gouges out new ones, though the flow may last but a few minutes and seldom continues more than a few hours.   In the sierras the slower drizzles produce stream floods which sometimes find their way out on the valley-plains, though the drizzle on the plain commonly does no more than wet the surface or produce feeble sheetfloods; and on the broader plains only a relatively small part of even the heaviest rainfall ever collects in streams.

In brief, the streams of the district are strikingly short and small in proportion to the area, and only less strinkingly few and feeble in proportion to the scant precipitation.

Streamways and stream-work.—In the sierras the permanent streams are slender threads of water slipping over ledges, now gathering in tinajas,* and again disappearing in fissures or gravel pockets at the bottoms of rugged barrancas; and the barrancas dividing narrow aretes are exceptionally parallel and close laid, while it is the combes or amphitheaters in which the barrancas head, in conjunction with the peaks in which the aretes join the crests, that produce the characteristic sierra profile.   Most of the multitudinous barrancas are supplied only by storm torrents, and these usually end about the base of the sierra, the margin of the valley-plain; it is only the deeper and longer barrancas that send arroyas or permanent channels far enough over the plain to unite with other waterways in dendritic systems; and it is partly for the reason that most barrancas end at the plain that their remarkable parallelism is maintained. Outside the sierras the typical channel is at first a rugged or flat-bottomed barranca cut in the country rock; it soon diminishes in depth and increases in width and becomes lined with boulder beds; still further down stream it changes into a broad, steep-banked arroya cut in alluvium and burdened with gravel beds or sand sheets; and it finally ends in an alluvial fan, usually of imperceptible slopes miles in length and furlongs in width.   If the stream is permanent it is, in its low-water stage, but a thin ibbon of water rippling over the rocks of the upper course or the sands of the middle course.   The streamways are notable, first, for high grades, and, second, for the width of their channels, which may exceed that of the Ohio, or even that of the Mississippi, for a stream less than 50 miles long; but during most of the year nine-tenths of the few streamways are broad wastes of barren sand, the most forbidding lines of the desert, often littered with skeletons of famished stock.

---

* Natural bowls, or water-pockets; defined in Science, new series, vol. iii, 1896, p. 494.

The character of the channels expresses well the characteristics of stream erosion: In the sierras the storm torrents gather loosened rock masses (there is little disintegrated detritus and still less decomposed rock matter on the steep slopes), hurl them down the cliffs and hurry them through the barrancas, bursting them asunder and knocking loose other masses on the way; toward the base of the sierras the larger boulders lodge, to be removed and reduced during later storms, while the pebbles and finer debris are swept further. Then general or local conditions either spread the torrent into a sheetflood, or else maintain the stream character; and in the latter case cobbles and pebbles are laid down after much trituration, the sand is carried far, scouring the channels as it goes, and finally nothing coarser than silt is borne by the diminishing flood, which is constantly robbed by dry earth below and drier air above; the silt burdens the flood without giving much aid in corrasion, and it gradually expands either into a labyrinth of interlacing channels or into a sheetflood, when evaporation and absorption rapidly sap the strength of the torrent until it ceases to be. It is significant that despite the high declivity of the barrancas the freshet torrents are often surprisingly clear, evidently by reason of the dearth of comminuted and lixiviated detritus, so that the streams are often underloaded and thereby enfeebled as erosive agents.

The streamwork in the district is notable partly in that it is exceptional in occurrence, partly in that it is reduced and rendered subordinate by the tendency of the streams to pass into sheetfloods with diminishing declivity; for most of the barranca torrents are transformed at once on reaching the valley-plain, while the sand-lined channel of the typical arroya is but a sheetflood of limited width. So strong indeed is the tendency toward transformation that it is only in the few streams of permanent supply and in the valley-plains oversupplied by exceptionally extensive drainage basins that definite channels are maintained.

SHEETFLOODING IN THE DISTRICT.

Character of sheetfloods.—In distribution the streams are mainly confined to the sierras, including the Sierra Madre and the higher foot-ranges as well as the lower outlying ranges and masses and the isolated mesas and buttes, and are local and exceptional in the valley-plains, while half the vast valley-plain area is the area of sheetflooding. Although there is a general increase in precipitation with altitude throughout the district, the cloudburst and drizzle usually affect both sierra and valley-plain; and in such cases the plain is flooded by the direct rainfall as well as by drainage from the sierra. The character of the flooding is known from direct observation, from indirect observation on flotsam, and from consistent lay testimony.

During the 1894 expedition a moderate local rain occurred while the party were at a Papago rancheria near Rancho de Bosque, some 15 miles north of the international boundary at Nogales; the rainfall was perhaps one-fifth of an inch, sufficient to moisten the dry ground and saturate clothing despite the concurrent evaporation, and was probably greater in the adjacent foothills of Santa Rita range. The road was sensibly level, having only the 20-feet-to-the-mile grade of Santa Cruz valley; it ran across the much stronger slope from the range toward the river, and an arroya embouched from low terraciform foothills not more than 200 or 300 yards up the slope. Thus the arroya opened not on a perceptible fan but on a sensibly uniform plain of sand and silt with occasional pebbles sloping perhaps a 150 feet to the mile. The shower passed in a few minutes and the sun reappeared, rapidly drying the ground to whiteness. Within half an hour a roar was heard in the foothills, rapidly increasing in volume; the teamster was startled, and set out along the road up the valley at best speed; but before he had gone 100 yards the flood was about him. The water was thick with mud, slimy with foam, and loaded with twigs, dead leaflets and other flotsam; it was seen up and down the road several hundred yards in either direction or fully half a mile in all, covering the entire surface on both sides of the road, save a few islands protected by exceptionally large mesquite clumps at their upper ends. The torrent advanced at race-horse speed at first, but, slowing rapidly, died out in irregular lobes not more than a quarter of a mile below the road; yet, though so broad and tumultuous, it was nowhere more than about 18 inches and generally only 8 to 12 inches in depth, the diminution in depth in the direction of flow being less rapid than the diminution in velocity. The front of the flood was commonly a low, lobate wall of water 6 to 12 inches high, sloping backward where the flow was obstructed by shrubbery, but in the open curling over and breaking in a belt of foam like the surf on a beach; and it was evident that most of the water first touching the earth as the wave advanced was immediately absorbed and as quickly replaced by the on-coming torrent rushing over previously wetted ground. Within the flood, transverse waves arose constantly, forming breakers with such frequency as to churn the mud-laden torrent into mud-tinted foam; and even when breakers were not formed it was evident that the viscid mass rolled rather than slid down the diminishing slope, with diminishing vigor despite the constant renewal from the rear. Such were the conspicuous features of the sheetflood—a thick film of muddy slime rolling viscously over a gently-sloping plain; and this film was a transformed stream still roaring through a rugged barranca only a few miles away. A special feature soon caught attention : On looking across the flood it was seen from the movement of waves and flotsam that the rate of flow was generally uniform, a little more sluggish about the mesquite clumps, a little swifter over the interspaces ; but now and then a part of the sheet (usually between and below mesquite clumps or slight elevations by which the current was made to converge) began to move more rapidly, when almost immediately the flotsam would shoot forward at twice or thrice the ordinary rate, the flood surface would sink toward the upper end and swell toward the lower part of the rush line, while the roar would rise above the rustling tumult of the more sluggish waters ; within 50 feet or 50 yards the swelling rush would be churned into foam and rise several inches above the general level of the flood, and then the waters would diverge and slacken and quickly mingle with the general sheet ; sometimes the crest of a delta or fan would show through or above the water at the lower end of the rush, and would push up stream through growth chiefly at its upper end ; but in any event the whole process of the gathering and respreading of the waters commonly lasted but a few seconds, or perhaps a minute or two, and left but a faint trace in unusual rippling of the flood. So common were these rushes that two or three or even half a dozen might be within the field of vision at the same time—some just starting, some dying away. For perhaps five minutes the sheet-flood maintained its vigor, and even seemed to augment in volume ; the next five minutes it held its own in the interior, though the advance of the frontal wave slackened and at length ceased ; then the torrent began to disappear at the margin, the flow grew feeble in the interior, the water shrank and vanished from the margin up the slope nearly as rapidly as it had advanced, and in half an hour from the advent of the flood the ground was again whitening in the sun, save in a few depressions where muddy puddles still lingered.

The after effects of the flood were not conspicuous, though significant. The most striking effect was the accumulation of flotsam, chiefly twigs and branches, against the upper sides of clumps of shrubbery, ant-hills, ground-squirrel mounds, and other elevations ; and from these the extent of the flood could be traced even beyond the limits of vision, showing that its width was at least one and perhaps two or three miles, and that it nearly blended with other similar floods emerging from neighboring barrancas in the mountains and arroyos in the foot-slopes. A less striking effect was the accumulation of a nearly continuous film of sediment, chiefly fine sand or silt, hardly distinguishable after drying from the general surface deposit, with which it undoubtedly soon blended. This film was usually an inch or less in thickness, though sometimes it lined depressions to depths of several inches. A highly significant effect was found on examining the track of one of the more violent rushes within the flood: At the upper end this was a gully reaching two feet in depth and one or two yards in width, newly gouged in the gravelly and sandy silt of the plain; at the lower end it was an elongated delta or fan, composed chiefly of sand but containing occasional pebbles (which were not borne by the sheetflood, and must accordingly have been washed out of the gully). A score of other gullies and deltas were seen, some well developed, some nearly obliterated, and it seemed clear that the well developed examples were only those produced toward the end of the freshet, the earlier examples having been masked or obliterated by the later flooding. These distinctive marks of the sheetflooding were distributed as extensively as the flotsam heaps, and in like manner were found on the plains below neighboring arroyos

On traversing a characteristic torrential apron stretching southward from the southern end of Baboquivari range toward the great arroyo known as Rio Seco, in northern Sonora, the track of a still more extensive sheetflood was crossed, its traces (apparently some months old) being the characteristic accumulation of flotsam or drift lodged against shrubbery and elevations, and the equally characteristic gullies terminating in fans. The route lay almost directly across the slope, three to five miles from the base of the mountain; and the sheetflood-plain was so smooth that, with a little care in avoiding the occasional gullies, the wagon passed over it at a rapid trot, save in crossing a single sand-lined arroyo eight or ten feet deep and twenty yards across, the torrent marks indicating that the entire plain had been flooded for a width of nearly or quite ten miles to a depth exceeding a foot and not exceeding a yard, save in the central arroyo and one on either side (down which the drainage from the barrancas had evidently flowed streamwise after the force of the torrent was spent). This plain is a typical torrential apron of the Sonoran district; its slope, five miles from the mountain base, is perhaps 150 feet to the mile, increasing to 200 feet near the mountains and perhaps 300 in the reentrants between projecting spurs, while toward Rio Seco the inclination diminishes to 100 feet to the mile or less where the surface passes by a low crenulate escarpment into the broad wash of the pseudo-river, itself sloping probably 40 or 50 feet to the mile. The deposits of the plain as revealed in gullies and the marginal and central arroyos are gravelly and even bouldery sands near the mountain, grading into silty sands toward Rio Seco. These deposits have the customary air of great depth, yet, as shown in the walls of the barranca (the main head of Rio Seco) passing the frontier post of Sasabe and the lesser barranca near the Indian village of Poso Verde, they are usually but a yard or two in thickness for several miles from the mountains and rest on an eroded surface of non-decomposed mountain rocks.

Later the 1894 expedition passed up the broad valley extending north-westward from Hermosillo between two outlying ranges for about 100 miles. The greater part of this valley is a single torrential plain tilting up laterally into the bounding ranges and rising gradually northwest-ward from an altitude of perhaps 100 feet to over 2,000, where it seems to drown the mountains, save a few peaks rising sharply from its gentle surface; the regularity of the valley and its apparent lining being inter-rupted in four or five places by sharp-cut drainage-ways which have retrogressed through the westerly bounding range. About midlength of the valley (south of the tanque known as Agua Nueva) the route crossed obliquely the trail of a sheetflood, marked by flotsam and gullies, several miles broad. In this case the nearest upslope mountains were 20 or 30 miles away and not more than 2,500 feet higher than the flood-marks, while the surface inclined for 50 miles directly to the lower reach of Rio Sonora, which is never wet save by local showers or the storm freshets descending from the Sierra Madre during midwinter or midsummer rains. Accordingly it seemed clear that the sheetflood had been confined to the plain—that the waters of a single storm had accumulated, rushed down the 1:25 slope for ten or more miles, and then disappeared through absorption and evaporation.

During the exceptionally humid autumn of 1895, the second expedi-tion experienced a single rain, or rather a succession of showers, not suf-ficient to produce either streams or sheetfloods, on the gentle slopes of Altar valley about the settlement and entrenched buttes of San Rafael de Alamito, where camp was pitched for several days. Yet this plain was marked for an area of at least 100 square miles by bunches of flotsam and driftwood lodged against the sparse shrubbery. The area was alto-gether out of reach of possible floods in Altar wash, to which, indeed, the marks did not extend, and the slope and direction of flow were nearly at right angles to its line. Except the geographically insignifi-cant buttes, there were no mountains within a dozen miles in which the torrent might have gathered; and when the party ascended the 30-mile slope toward the Altar-Sonora divide the torrent-marks were found to diminish and gradually disappear in such manner as to demonstrate that the waters had gathered mainly, if not wholly, on the plain itself, and then rushed down toward, but apparently not quite to, the sand wash of the Altar. Where the flotsam was accumulated most abundantly the slope was probably 50 feet to the mile or less; and the fan-ended gullies were few and small, becoming more conspicuous further up the incline where the flotsam was less abundant.

The 1895 expedition skirted the western and southern bases of Sierra de Tonuco, just outside the area mapped as Seriland; and along the southwestern margin, near Rancho de Tonuco, the steep mountain side passed within a fraction of a mile into fairly smooth plain, sloping perhaps 150 feet to the mile. Numerous water-cut gullies and fans of exceptional size were found. Most of these were apparently some years old; and the deltas and fans contained quantities of angular and subangular boulders, sometimes reaching a foot in diameter. The deeper scorings here revealed a sheet of fragmental debris, evidently drift from the neighboring sierra, rarely reaching five feet in thickness, composed of little-worn gravel with occasional boulders embedded in a matrix of loam or silt; this rested on a sharply-eroded surface of granite or black marble not at all decomposed. The mantle was variable in thickness, averaging probably less than a yard over considerable areas, and frequently disappearing, leaving the rock to form the surface. About the rancho the alluvium thickened locally, and the ground water circulating on the subjacent rock surface was tapped by a well 40 or 50 feet deep; but even here, in the line of natural flowage from the deepest barranca of the sierra toward Encinas desert, where the slope was least, the usual gullies terminating in low fans abounded, some having been evidently produced by the latest storm. On the steeper slopes adjacent a number of low circular or elliptical mounds, apparently made up chiefly of angular boulders, were noted; and comparison of the evidently older with the manifestly young indicated that these were remnants of ancient gully-fans modified, rounded, and relatively raised above the mean surface by subsequent erosion, though in a few cases the boulder mounds were half buried by finer silt deposited about their flanks.

The 1894 expedition traversed Baboquivari or Moreno valley (west of Baboquivari range), crossing the course of " Fresnal creek " as shown on the excellent official map of Pima county, Arizona; and though the line of the " creek " was not traceable (the wash having been obliterated in consequence of local failure of rains during recent seasons), a considerable flotsam-marked area was found in the eastern part of the valley, below the Papago Indian villages collectively known as Fresnal. Here, too, in the shadow of the high and remarkably rugged granite range and its dominant peak, boulder mounds similar to those adjacent to Sierra de Tonuco were observed, and in two or three instances low buttes, together with the more considerable eminence known as Fresnal hill, were noted as apparently the product of continued circumdenudation, initiated about accumulations produced in the manner inferred in the Tonuco region. In this valley, too, it was observed that sometimes gullies of considerable dimensions are produced well out on the plain; in one instance the waters of a storm (apparently a local thunder-gust) had scooped out a basin 25 or 30 feet wide, 10 or 12 feet deep, and 250 feet long, then dammed its lower extremity with silty debris, and, finally, as the flow slackened, lined it with impervious silt which held water for months, and so located a Papago settlement; and cases were reported in which basins or tinajas of this character, refilled from season to season for some years, drew about themselves agricultural rancherias of Papago Indians, who built houses and planted fields around the banks in order that they might enjoy the priceless gift of their most potent deity.

In southwestern Arizona the sheetflood is well known to the Indians and to those Mexican and American rancheros who chance to be favorably situated, and they are well and expensively known to railway corporations, who sometimes have five miles or more of track washed out by a single storm perhaps sweeping over a smooth plain without a single waterway before, and with only a few new-made gullies after, the catastrophe. In essential features the local lay testimony is everywhere alike; there is a storm with exceptional precipitation, the water simply floods the surface in a muddy torrent or " wash " stretching as far as the eye can reach, the ground is swept of loose debris, and even of surface sands, while flotsam and sand heaps are piled up here and there; and in a few minutes, or perhaps a few hours in the lower valleys, the flood slackens and almost immediately disappears.

Such is the character of the sheetflood, as determined from direct observation, lay testimony, and the evidence of effects; and were it needful this evidence might be many times multiplied.

Conditions requisite for sheetflooding.—The main (perhaps the sole) source of the sheetflood is storm-water, comprising that shed from the mountains and that falling directly on the intermontane plains; and since the mountains are low and form only a small fraction of the surface, it seems clear that the chief source is the storm of the plains.

The first requisite for typical sheetflooding, then, is precipitation so rapid as to exceed immediate absorption by the dry earth and immediate evaporation in the drier air (for usually, in the Sonoran district, the precipitation horizon is some yards or hundreds of yards above ground, and the lower strata of the air are so hot and dry as to take up much of the falling and a part of the fallen moisture); and this involves several attendant conditions: One condition is that the temperature shall be high and the capacity for aqueous vapor great, in order that a large quantity of water may be produced when precipitation occurs; and this condition is amply met in the highly heated Sonoran district. An attendant condition is that the precipitation shall be rapid; and this also is met by the subtropical climate and wind-disturbing topography of the district. A third condition is that the soil shall be readily pervious only in limited degree or to limited depths; and this condition is met on the lightly veneered baselevels adjacent to the mountains, where the mantle only is porous and the under rocks sound and hard; it is not met in the deeper central portions of the valleys, where the permeable sands are of considerable depth. There are also other conditions which need not be noted in detail.

The second requisite is that there shall be abundant detritus, whereby the moving water may be readily loaded to the full limit of its capacity; and this requisite also involves several conditions: An important condition is absence of sward or turf to hold the earth-particles in union; this condition is fulfilled by the bare sands and naked rocks making up nine-tenths or more of the Sonoran surface. It is probable that another condition resides in chemical inertia of the mechanically comminuted rock-matter in the dry, coupled with some chemic activity promoting miscibility when wetted; and this condition is found in the friable sands and silts of the region, which form a tenaceous mud on saturation and a viscid slime on flooding. A third condition is dimensional heterogeneity in the debris, so that every part of the sheetflood may be loaded to its full capacity, whether its movement be swift or slow; and this condition is fully met on the upper plains, where silt, sand, gravel, and boulders in all sizes and shapes are intermingled, though it is less perfectly met in the valley interiors, where the materials are more completely assorted. Other conditions exist, but they are apparently of minor importance.

A third requisite is that the slope of the surface shall be of a certain somewhat variable value (not yet determinate save empirically as say 75 to 200 or possibly 300 feet per mile). The optimum or most efficient slope is evidently conditioned by thickness of the detritus mantle, which would appear to be considerable on the higher slopes, less on the lower slopes, and so great as to be an obstruction toward the valley interiors; by coarseness of the detritus, which always is greatest on the higher slopes; by dimensional heterogeneity, which in like manner culminates on the higher slopes; by porosity and friability, and by various other conditions. With slopes above the limit of efficiency, sheetflooding does not occur; the detritus is simply swept away, and the under-loaded storm-waters gather into streams, which carve channels. When the slope is below the limit of efficiency, the mechanism becomes clogged, the declivity and consequent velocity do not permit the incipient stream to overload itself quickly, and there is a tendency to assume the habit of streaming rather than that of sheetflooding. The various conditions of slope requisite for sheetflooding are strikingly met in the Sonoran district; two-fifths or more of its area consists of vast torrential aprons lightly veneered with detritus resting on baseleveled rocks, the inclination ranging between 75 and 250 feet per mile.

The final requisite for sheetflooding is that the volume of water, the mass of available debris for loading it, and the slope (and hence the velocity) shall be so interrelated and balanced that every part of the sheetflood may be loaded to its capacity, and that any temporary or sporadic increase in velocity may be quickly checked by overloading and consequent reduction of velocity. The conditions affecting this requisite are multifarious, probably beyond analysis, but it may be suggested that an important—perhaps the essential—condition is a progressive paralysis or weakening of the torrent due to the constant absorption and evaporation of its liquid element, the solid element remaining to burden immediately the falling or inflowing water. This condition is fully met in the parched air and burning sands of Papagueria.

On juxtaposing the requisite and qualifying conditions of sheetflooding, they are found in harmony with the distinctive characteristics of the greater part of the Sonoran district, and when the characteristics are compared with the conditions observed in special localities the harmony is rendered still more complete. It becomes evident that the sierras and buttes lie outside the domain of sheetflooding, since their slopes are too steep and their detritus too scanty ; that the mountain-born streams on reaching the torrential plains must become quickly diverted and attenuated into sheetfloods, provided their volume be sufficient; that the light shower falling on the baselevel plain is absorbed, while the heavy shower must spread into a moving film and the cloud-burst into an irresistible sheetflood, sweeping all before it; and that the final feeble flow, whether from distant barrancas or local seepage in depressions, must resume the habit of the stream, pushing down toward the sea until the waters are finally lost.*

Erosive work of sheetfloods.—It may be affirmed from observation, both direct and indirect, and from necessary inferences concerning land-forms

---

* Reference is due to the work of Mr Willard D. Johnson during the second expedition, and to his opinions concerning the somewh: t anomalous topographic features of the Sonoran district. Although sheetfloods were not witnessed during this expedition, Mr Johnson had opportunities for studying flotsam records, and in the course of his admirable topographic surveys he was much in contact with the features of the sand washes forming the fans or deltas of Rio Sonora, Rio Bacuache, and other principal drainageways. His conception of the Sonoran flood was commonly expressed by the term "interlacing drainage," consisting of a multitude of broad, shallow and swift streams, approximately parallel in course, constantly divaricating and reuniting in such manner as to leave numerous islands, which were from time to time invaded and swept away. This conception is undoubtedly accurate so far as the lower reaches of the waterways are concerned, and it doubtless applies also to what the writer would consider as the lower or peripheral zone of sheetflooding proper, where there is a tendency toward the resumption of stream habit. Unfortunately Mr Johnson's observations and conclusions are not yet published ; circumstances have prevented even the completion of his map, excepting the portion including Seriland ; but important results may be anticipated.

and structure, that the sheetflood is an efficient agency (1) in transportation, (2) in corrasion, and incidentally (3) in deposition. Its efficiency is enhanced in the Sonoran district by various conditions, comprising all those essential to the existence of this form of water body, and including notably the local tendency toward the production of disintegrated and comminuted detritus rather than residua over the surface of mountain and plain—vegetation is too meager to produce appreciable quantities of humus acid, water is too scanty to aid materially in chemic action, frost is too rare and superficial and the included water too minute to rend the rocks effectually. Its efficiency is limited by the rarity of rainfall in sufficient quantity to produce flowage in any form—over the greater part of Papagueria, where the mean annual precipitation is probably no more than three to five inches, many tracts of 1,000 square miles or over are missed by the midwinter and midsummer storms of one or more years, the freshet-making storm may be three or five years in coming, and the great mantle-moving torrents are apparently separated by decades or centuries.

The efficiency of the sheetflood in transportation was amply shown by the relatively trivial torrent at Rancho de Bosque, which was literally a thin mortar of mud rather than water, and the well-preserved traces of sheetflood work in other localities gave unmistakable indications that great quantities of detritus were collected and transported. Inferences as to the behavior of greater sheetfloods rendered necessary by analogy and by the distinctive topographic configuration indicate with strong probability that the volume of material transpor'ed by the debacle increases in a higher ratio than the volume of water, so that the entire detrital mantle may be saturated to the point of mobility and carried down the slope in a sort of mud-flow peculiar only in its magnitude and in the dimensional heterogeneity of the moving particles.

The efficiency of the sheetflood in corrasion is made manifest by consideration of its mechanism: The velocity of flow must be considerable, else the flood is absorbed and evaporated; the water must be laden to its full capacity with abrasive mat rial, else it runs clear and gathers into streams. This material, whether fine or coarse, is exceptionally hard and sharp rather than softened and rounded by decomposition, and the internal currents in the shallow sheet are such as constantly to batter and scour the subjacent surface as the mass half rolls and half slides across it. The inference from the character of the sheetflood is consonant with the necessary inference from the character of the baselevel surface. Over dozens or scores of square miles in carefully examined localities, hard rocks like those of the mountains, and with no sign of decomposition, are planed almost as smooth as the subsoil by the plowshare, with nothing either in configuration or in covering to indicate that streams have flowed over them, and extended consideration has yielded no other suggestion as to the eroding agent than that found also in analogy with the observed sheetflood. Moreover, these planed surfaces are not rare or exceptional; they occur under a definite law of distribution (with respect to sierras on the one hand and alluvium-lined valleys on the other) in all parts of the Sonoran region; their area may be estimated as two-fifths of the entire tract, or over 100,000 square miles. The efficiency of the sheetflood as a corrading agent is connected with its efficiency as an agent of transportation, for the rapid corrasion constantly furnishes material to be carried down the slopes, and this material in turn is available for cumulatively increasing the effective work of the agent.

The efficiency of the sheetflood in deposition would appear to be subordinate. In the miniature examples observed, a relatively considerable sheet of sediment was indeed laid down, yet it was no more than might be taken up and carried further down the slope by the next torrent, so that deposition in this case would seem to be little more than the mark of decadence; and in the various torrential plains on which the traces of past sheetfloods were found the conspicuous marks were those of degradation rather than aggradation. So far as the characteristics of the sheetflood are susceptible of analysis, too, it would seem that this form of flowage cannot maintain its distinctive attributes unless the detritus mantle is of limited thickness, so that deposition may be considered as essential and characteristic only to the extent of supplying abrasive materials for the next debacle within the sheetflood zone proper, while the external or peripheral deposition would appear rather to represent ordinary delta-building or stream-work. It is significant in this connection that many of the valley-plains (e. g., San Luis valley, east of Baboquivira range) are much more strongly diversified by waterways and their attendant bluffs in their medial portions than half way up the gentle slopes toward the bounding mountains.

When the functions of sheetfloods are combined, it is found that their tendency is to recede or retrogress from the valley interiors, as these are progressively clogged with sediments, toward and into the bounding ranges; and this retrogressive cutting is one of the most significant features of the erosive process, partly in that it tends to produce anomalous profiles, passing abruptly from steep mountain side to nearly flat plain, in lieu of the gently sloping concave profile characteristic of stream-work. When the torrent filling the mountain barranca embouches on the plain, the first effect of the slackening in flow is the discharge of the detritus in a fan, which of itself tends to spread the water; and this fan protects the subjacent rock from corrasion. The slackened torrent may, indeed, divide on the fan surface into a number of streamlets divaricating over the contiguous part of the plain, but these are subject to the same law as the main stream and are continually divided and subdivided as they move down the diminishing incline; and thus the ultimate effect is to distribute the water widely and plane the fan into gentle curves, blending with the adjacent baselevel and sharply discordant with the mountain slope. As the process continues, the fan of one episode, particularly if produced by an exceeding torrent, may resist the powers of succeeding freshets, and thus initiate a butte in the line of the barranca, like Fresnal hill; and in some cases the salients between adjacent barrancas are cut through by the divergent floods in such manner as to form outlying buttes or cusps in line with the aretes ribbing the sierra, like some of the isolated buttes rising from the eastern apron of Sierra Seri, and the granite picacho with its half-mile pediment in the Seriland section near Puerta Infierno. The immediate consequence of this retrogression is the cumulative sharpening of the inflection in profile marking the base of the mountain; and it is noteworthy that the smaller mountain remnants on the flatter plains are (other things equal) steeper and more rugged than the greater mountain masses, as illustrated by Sierra de Tonuco and the little buttes in the northeastern part of Seriland in comparison with the high Sierra.

In brief, it may be affirmed, with so much confidence as the conditions of observation during two expeditions warrant, that the general effect of sheetflooding in the Sonoran district is to carve baselevel plains, lightly veneered by the carving material, about the medial altitudes; that these plains tend ever to retrogress into the mountains, and thereby steepen their slopes and render them exceptionally rugged; and that the anomalous topography of the region is not susceptible of explanation by other agencies.

It would be of interest to consider the natural history of sheetflooding in the region, and to obtain thereby a distinctive geomorphic record which might be compared with the stratigraphic record in reading the geologic history of the southwestern portion of the continent; but the time for this study has not yet come.

Sheetflood Work in other Districts.

While the requisite conditions for sheetflooding are especially favorable in the Sonoran district, and while the effects of the process are proportionately conspicuous, it is not to be supposed that the process is confined to the district—indeed it may be suggested that the process will be found a main or minor one in various districts, particularly those whose climate and configuration approach those of Papagueria. It may be noted as probable, also, that even in the more humid provinces a process analogous to that of sheetflooding may exist; for wheresoever rain falls the waters gather into a moving film before rivulets and streams are formed, and this film must be a more or less active geologic agent. Finally, it may be noted that certain obscure phenomena of various waterways which, like the Susquehanna and other Piedmont rivers in parts of their course, tend to corrade their channels laterally rather than vertically, appear to be akin to those of sheetflooding.

[*Editor's Note:* All plates are omitted. None show the effects of sheetfloods *per se* on the pediment photographs. Similar surfaces are shown elsewhere in this volume (Papers 13 and 18).]

12

Originally published by the University of California Press; reprinted by permission of the Regents of the University of California from *Univ. Calif. Publ. Geol.,* **9**(3), 23–34 (1915)

THE EPIGENE PROFILES OF THE DESERT*

BY

ANDREW C. LAWSON

RAINLESS REGIONS

In a rainless region an uplifted land-mass, an orogenic block, is subject to degradation by:

1. The process of mechanical disintegration at the surface due to differential dilations and contractions under varying temperatures; (2) the gravitational transportation of the fragments thus detached from higher to lower levels; (3) the transportation by the wind of such of these detached fragments as may be small enough to be removed; (4) the corrasive action of wind-moved detritus and the removal of the products of such corrasion by the wind.

The relative efficacy of these processes in a purely *a priori* discussion depends altogether upon what assumptions are made. If, for

---

* Read before Geological Society of America at the Berkeley Meeting, Aug., 1915.

example, we assume that our orogenic block is composed of fine-grained incoherent rock and assume further the prevalence of powerful winds, then the last two of these processes might well be most important agencies of degradation. But if we assume that our orogenic block is composed of coherent, hard, elastic rock, then the products of mechanical disintegration would in general be of such a size as to be but slightly affected by the wind, and the first two processes would be almost the only agencies concerned in the degradation of the uplifted mass. It is the latter case, the effect of mechanical disintegration and gravitational transportation, that I desire here to discuss briefly as an introduction to the consideration of more complex conditions in the later part of the paper. For the purpose of simplifying the discussion I will further assume that the mountain is lithologically and structurally homogeneous.

Under the conditions assumed, our uplifted block may have initial slopes which are (1) less than the slope of repose for loose material, or, (2) greater than the slope of repose. The gentler slopes will remain unaltered throughout the persistence of the rainless climate. The surface will of course be at first subject to mechanical disintegration, but in the absence of any transporting agency the products of disintegration will encumber the firm rock and protect it from further attack. There can be no change of slope due to degradation. The steeper slopes, on the other hand, cannot retain the fragments shed by disintegration, and these lodge at the base of the slope in the form of talus. The slope of the talus thus becomes the limiting slope of all mountain facets evolved under such conditions. If we suppose a steeper slope to come into existence, as for example by repeated faulting, the disintegration of the surface and the direct action of gravity will give rise to a talus at its base. As the latter grows it maintains the constant slope of repose of loose material; but its upper edge encroaches upon higher and higher parts of the battered scarp. While this encroachment is in progress the batter of the scarp approaches the slope of the talus; and when the two coincide the combined slope of the battered scarp and talus becomes the final, unchanging mountain front. The talus ceases to grow and the rock slope above ceases to diminish, the loose fragments on both lying at the limiting angle at which gravity will move them.

During the encroachment of the talus upon the battered scarp the latter has been reduced in acclivity by more active recession in the

upper than in the lower part.[1] The recession has been progressive
with the growth of the talus, and the portion of the battered scarp
which becomes buried is therefore a parabolic curve, the volume of
the talus being assumed equal to the volume of the rock in place,
while the portion which remains above the talus maintains a straight
slope. The development of the final combined facet from the initial
scarp is shown in the diagram, figure 1.

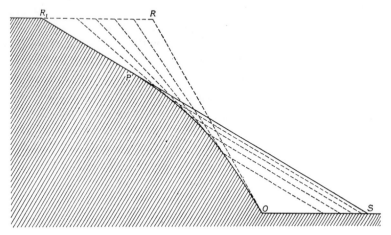

Figure 1.—Under rainless conditions a mountain face *OR* would be reduced
to *OPR₁*, the lower curved portion of which *OP* would be burried in talus *OSP*.
The combined rock and talus slope *SPR₁* is the unchanging profile under the
assumed conditions.

In the foregoing statement it is of course not implied that the
talus profile is found only in rainless climates; it may and does occur
under humid conditions, but here it is a temporary phase of the geomorphic evolution, whereas under rainless conditions it is a permanent
and characteristic feature of mountain fronts.[2]

ARID REGIONS

Elements of the profile.—The scant rains of the arid regions are
sufficient to differentiate sharply the epigene forms there displayed
from those deduced above as characteristic of the relief of a rainless

---

[1] This is due, of course, chiefly to the fact that the amount of disintegration
is a function of the extent of surface exposed, and that this is greatest at the
top of the scarp where it intersects the reverse slope or a gentler slope in the
same direction, but if the scarp be high it may also be due partly to larger
contrasts of temperature in higher altitudes.

[2] This was written before the appearance of Lane's abstract dealing with
the same profile in Bull. G. S. A., Vol. 26, No. 1, p. 75.

region. The rain wash co-operates with gravity and transportation is far more efficient. The rock slopes are generally much less than the angle of repose of loose material, the talus is replaced by broad alluvial fans, and for every enclosed basin there is the aggradational flat of the playa. For mountains of approximately homogeneous rock the epigene profile thus comprises three elements: (1) The rock slopes, having an angle of less than 35°; (2) the alluvial fan slope rarely exceeding 5°; and, (3) the flat of the playa.

In general these three elements of the profile are easily discriminated one from the other; but in certain cases the angle of the rock slope approaches that of the alluvial fan and the two may be confused. Similarly where the alluvial fan is very flat it may not easily be distinguished in its lower part from the playa. In many desert valleys the playa is located at one end, toward which the drainage flows, so that the fans from the two sides meet in the central part and there is no playa element in the profile for long stretches.

The proportion between the length of the rock slope and that of the fan slope varies greatly. The fans are composed of the detritus washed down from the rock and, as they grow, they rise about the mountain flanks and steadily diminish the extent of the rock slopes. Ultimately the upper edge of the fan reaches the crest of the mountain and there is no farther addition to its surface. A small alluvial embankment on the flank of a mountain ridge is, therefore, significant of an early stage of the degradational process; and a large embankment extending nearly or quite to the crest indicates a late stage. In the deserts of the Great Basin both extremes may be observed. At Genoa, for example, there is a bold, high mountain front of bare rock with a small embankment at its base, and in places none at all.[3] At Cima in Southern California the mountain ridge is almost completely buried in its own alluvial waste. From this variable quantitative relation of rock slope and alluvial slope it may be safely inferred that the mountains of the region, considered as features of the relief, are of diverse age. On the basis of this relation it is possible to classify the desert mountains chronologically. Some came into existence so long ago that their rock slopes have been almost entirely buried in their own waste; while others have had their origin so recently that the effect of aggradation about their flanks is relatively slight. Between these extremes a graded series may be recognized significant of a time-sequence. This criterion for the chronological

---

[3] Bull. Seism. Soc. Am., Vol. II, No. 3, Sept., 1912; U. S. G. S. Prof. Paper 73, p. 189, 1911.

classification of the salient features of the relief of the desert may attain a notable degree of exactitude if mountains lithologically similar be compared within the same climatic province.

The degree of acclivity of the epigene slopes of the desert is, on the other hand, no indication of the stage of advancement of the degradational cycle. Under humid conditions hard rocks that are susceptible of chemical decay, and so form soil, tend to acquire more and more gentle slopes as time goes on. A steep slope in such rocks is indicative, in general, of geomorphic youth, whereas a gentle slope is characteristic of old age. In the desert, however, hard rocks present persistently steep slopes throughout the entire period of their degradation. The epigene rock slopes appear to be just as steep in old residual mountains, almost buried in alluvium, as in youthful mountains with but a small embankment of detritus at their base; and it will be shown that they are in reality somewhat steeper.

If the mountain mass be heterogeneous the rock slope may not be uniform, and it may so far depart from the law governing the slopes of homogeneous material that it will assume temporarily the form of a vertical cliff. This is particularly well exemplified where a hard, resistant stratum lies at low angles upon soft beds, as in the case of the Vermilion Cliffs. Here the factors of heterogeneity and structure dominate the degradational process and the climatic control is relatively slight. The result is an escarpment which is not peculiar to the desert, but which may equally well be produced in humid climates, where the same structure and lithologic heterogeneity prevail, as is exemplified in the Niagara escarpment.

But even where the condition of approximate homogeneity obtains it is possible that other agencies than those noted may modify the profile. The wind may be both a transporting and a corrading agent, making for aggradation in one place and degradation in another. While this agency is recognized as of possible importance in regions where powerful winds prevail, particularly where mountains are largely composed of soft, fine-grained, incoherent rocks, it will not here be further considered. The wind, in regions of hard elastic rocks such as are commonly found in the ranges of the Great Basin, is an extremely inefficient agent in the evolution of the profiles of the relief. It is concerned chiefly with local transport of the silts and sands of the playas and with fitful whirling of sand over the fan slopes. The occasional sandstorms doubtless carry notable quantities of fine sand from one part of the desert to another and to regions beyond its connnes. But the wind thus employed plays almost no part in the direct reduction of the rock slopes, in the building up of the fans or in the filling of the playas, the processes which it is the purpose of this paper to discuss.

Degradation.—The encroachment of alluvial embankments upon the mountain slopes from which they are derived is a general phenomenon peculiar to the desert. To arrive at a proper appreciation of the process of encroachment we must first consider the degradation of the rock slope. The waste of mountains in arid regions is effected, as is well known, chiefly by the mechanical disintegration of the surface due to differential dilatation and contraction. The chemical decomposition of the rocks, due to bacterial action, which so largely contributes to the formation of soil in humid regions, is here relatively insignificant. The degradational process is one of transportation of the products of disintegration to lower levels. Corrasion by running water is quite a subordinate part of the process except in those ranges which are so high as to have a relatively abundant precipitation upon their summits. The streams thus fed corrade and deposit in accordance with the general laws so admirably elucidated by Gilbert.[4] But in most cases the quantity of detritus shed from the sides of the cañons increases faster than the enlargement of their carrying capacity due to the widening of their catchment area, so that early in their career they become habitually incapacitated in their lower stretches, and build up fans or cones of detritus which extend well up into the trunk cañon, forming accentuated features of the general alluvial embankment on the flanks of the mountain. If we confine our attention to the movement of detritus upon the slopes, rather than in the line of these exceptional streams, the rains which supply the transporting agency are characteristically of brief duration, local and violent. Practically all the reduction of the rock slopes and all contribution to the alluvial embankments, whether apexing in the line of perennial streams or not, is effected at these brief and infrequent periods of heavy downpour. When the rock slopes have once been reduced to an angle less than the angle of repose of loose material, the removal of the fragments upon their surface is conditioned by their size and shape. Should they be prevailingly rotund the fragments would move down a gentler slope than if they were irregularly prismatic. To simplify the discussion I shall assume that rock frag-

---

[4] Geology of Henry Mountains.

ments in general are irregularly prismatic in shape, which assumption agrees with observation.

For a given size of fragment the ease with which it may be moved diminishes with the decrease of the angle of slope at the place where it rests. Finally a minimum or limiting angle is reached below which the fragment cannot be moved. This limiting angle thus becomes a constant feature of that portion of the slope, say near the top of the slope, throughout all its future degradation under the assumed climatic conditions. For, if the angle of slope become lower, this and other similar sized fragments would encumber it and protect the underlying rock from further disintegration, and degradation would cease. But the fragments shed from a rock slope vary greatly in size, and those which determine the angle of slope are the spauls of maximum size, or perhaps better, the maximum order of size. All loose detritus shed from the rock slope of less order of size is readily carried down the slope in times of cloud-burst. Those of the maximum order represent the limit of size that can be transported. It may be presumed that rocks similar as to lithology and structure shed fragments of about the same maximum order of size, and, therefore, have slopes of about the same angle for similar parts of mountain fronts; whereas dissimilar rocks shed fragments the maximum order of size of which may differ greatly.

But the angle of slope for the upper part of a mountain front in homogeneous rock is not necessarily the angle at the lower part of the same declivity. It is clear from the bare character of the desert fronts that the sheets of water which wash down the detritus in times of cloud-burst are underladen; otherwise there would be a residue of debris left behind and the slope would be undergoing alluviation and not degradation. The fact that the flowing water is underladen and that the volume of water increases in an arithmetical ratio as it descends determines an acceleration of velocity toward the lower part of the rock slope. This in turn determines that spauls of maximum order of size may be moved on a lower angle of slope than at the top of the slope where the velocity is less. A lower angle is, therefore, developed at the lower part of the rock slope. This in turn tends to check the increase of velocity; but, in view of the fact that the slopes are bare and that the water which keeps them bare is underladen, it is evident that the acceleration of the velocity while lessened is not inhibited. Now in any desert range, as the rock slope shortens by the rise of alluvium on its flank, the contrast between the velocity and volume of water near the top of the rock slope and near the bottom steadily diminishes, so that the angle near the bottom of the shortened slope is greater than at earlier stages of longer slope; while the angle near the top of the slope remains constant. It follows from this that the general angle of the shortened slope, from top to bottom, is higher than for the earlier longer slope.

In this deduction we have the explanation of the observational facts: (1) That the hard-rock slopes of desert ranges which shed large spauls are steep, while those which shed small fragments have a low angle; (2) That ranges composed of hard rock, which are thus naturally steep, maintain their steepness as long as the rock slopes endure. We discover, moreover, that they become gradually steeper with age, the slight upward concavity approximating more and more a straight profile as the rock slope becomes shorter. There is, however, notwithstanding this straightening of the concavity of the profile, a persistent tendency to over-steeping at the very top of the rock slope, where the gravitative work of moving fragments has little aid from flowing water.

Alluviation.—Assuming the correctness of the hypothesis that the angle of the epigene rock slope in the desert is determined by the maximum order of size of the rock fragments shed from its surface by mechanical disintegration, and, therefore, after adjustment does not diminish throughout the entire period of its recession, we may apply it to the development of the mountain profiles. There is abundant evidence that, in the geological epoch of which the present forms a part, but of which the beginning is indefinite, the great alluvial fans which flank most of the mountains of the Great Basin have been steadily growing. For the greater part of this growth the slope of the surface has been approximately constant for every embankment. The increments of growth have, therefore, been of the nature of additions of uniformly thick layers of detritus to the top of the embankment. For each layer thus added to the embankment an equal volume, less the voids, has come from the mountain front. In general, therefore the average vertical cross-section of the embankment may be considered for the purpose of this discussion to have the same area as the average cross-section of rock removed from the mountain front, if we ignore wind action. The growth of the alluvial embankment proceeds, however, not only upward by the accretion of successive layers, but also horizontally by the increasing extent of the layers. Each layer added to it extends farther toward the crest of the range than the preceeding layer and so diminishes the height and breadth of the rock slope, the recession of which is at all stages limited by the upper edge of the embankment. The rock surface below this upper edge becomes fixed in position by burial.

Truncation.—It follows from this that the buried slope is a shelf or bench cut into the mountain, the inclination of which is somewhere between that of the still exposed rock slope and that of the alluvial embankment. This relation is shown in the diagram, figure 2. The upper limit of the buried bench, the upper edge of the alluvial embankment and the lower limit of the subaerial front are coincident at all stages of the general process. When the edge of the alluvium reaches the crest the subaerial front has disappeared. What is true

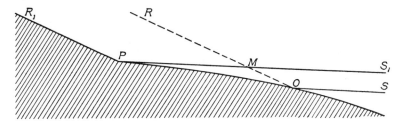

Figure 2.—If the surface of the alluvial embankment *OS* rises to *PS₁*, the rise is coincident with, and due to, the recession of the mountain front from *OR* to *PR₁*. The upper edge of the embankment migrates from *O* to *P*, and the rock-cut bench *OP* is evolved, but is buried as fast as it is cut.

of one side of the mountain ridge is true of the other, so that the rocky crest disappears by the meeting of the feather edges of the alluvial embankments, or rather by their coalescence with the unmoved products of the disintegration of the crest rock. Thus, if the mountain be lithologically and structurally homogeneous, it is reduced by truncation to a low rock ridge with symmetrical slopes wholly buried by alluvium, which on either side is wedge-shaped in cross-section, the edge of the wedge being at the crest and the butt out in the valley. The upper surface of this alluvium is a gentle slope of great stability for the prevailing climatic conditions.

But homogeneity is rarely realized in mountain masses and the heterogeneity varies the rate of the process in different parts and diversifies the ideal symmetry of the result. Owing to differential disintegration and recession, one part of the rock crest may disappear while another portion is salient, and the migration of the lowering rock crests may proceed unevenly. Similarly, resistant portions of the mass become shoulders and these are usually isolated in the rising flood of alluvium, as it creeps up the slope, and appear as island-like, conical rock hills protuberant above the slope. This diversity in plan and in profile is usually notable in the incomplete stages of the general process. At completion all the rock crest has disappeared, but the line of meeting of the opposing embankments may remain sinuous. Many of the island-like conical hills become eventually buried; but others that rise above the ultimate limit of the alluvial slope remain, of course, protuberant, till as local centers of recession and alluvial distribution they waste slowly away, and become mere patches of the bedrock obscured by the residual unmoved products of disintegration, nearly, but not quite flush with the general slope.

Under exceptional conditions the resultant profile may be asymmetric. For example, if the initial profile of the mountain be steep on one side and so gentle on the other that alluvial debris cannot be transported over it, then the reduction of the mass will proceed on the steep side only. When the rock surface on this side has been wholly buried by alluvium and the process of reduction has come to an end, the profile will be that of the fan slope; but on the other side it will be that of the initial slope, although the surface may be encumbered by the sedentary products of disintegration.

The Resulting Surface.—The result described has been attained in southern California; and various stages of the process may be observed throughout the Great Basin. It is a stage of geomorphic development at which the processes of degradation and aggradation, in so far as they are due to the agency of water, both almost cease. There are no exposed rock slopes to be degraded and the embankments can receive no farther increment. The slopes of the latter are close to the minimum angle for transportation and therefore are not themselves susceptible of vigorous attack. With the disappearance of the rock crest the process of degradation changes from one of relative rapidity to one of extreme slowness. The wind doubtless tends to modify the result, but in the Great Basin this tendency is extremely ineffective; and no important modifications have been observed to affect any stage of the general process or the result, except that locally sand derived from the playas may drift over the alluvial slopes in the form of dunes, which may thus be partially incorporated into the growing embankments.

The symmetry of profile which is attained in the normal process of concurrent degradation and aggradation in the desert is of course a symmetry of angles of slope. It does not extend to the length of the slope. This quantity is determined by the distance between the ultimate crest of the degraded buried range and the median line of the ultimate aggraded valley on either side. If these distances are unequal, then the playas in the valleys stand at different altitudes, the longer slope reaching out to the lower valley. It follows from this that difference of altitude of the playas on the two sides of a symmetrically sloped range introduces no element of instability into the profile. The playas represent the bottoms of distinct enclosed basins, but when the general condition above outlined has been attained there is no tendency for the lower basin to capture the higher, so long as the climatic conditions remain constant. Such capture may be effected in part by the migration of the crest of an initially asymmetric or heterogeneous range as an incident of the general process, but once the rock crest vanishes the tendency to migration ceases and capture is no longer possible. The surface thus evolved is, in its ideal completion, wholly one of aggradation, a vast alluvial fan surface to which for convenience in discussion I propose to give the name *panfan*.

The Panfan Stage of the Geomorphic Cycle of the Desert.—The panfan may be regarded an end stage in the process of geomorphic development in the same sense that the peneplain is an end stage of the general process of degradation in a humid climate. The peneplain closes the cycle of degradation and is a cut surface; the panfan closes a cycle of degradation and aggradation, is evolved by both cutting and filling, and is a built surface. Like the peneplain, the panfan is rarely observed in its ideal completion. The time required for its completion is so long that diastrophic or climatic changes usually interrupt the process, and even where it escapes those interruptions the final result is subject to destruction by the same changes. We may not expect, therefore, to find the panfan a prevalent feature of the relief of deserts which have been subject to such diastrophic and climatic vicissitudes as have from time to time overtaken the Great Basin or portions of it. Exceptionally, however, it does occur, as in parts of southern California; and in various stages of incompletion is one of the commonest features of the desert.

The recognition of the panfan as a phenomenon to be classified is, however, not so important as is the perception of the process[5] whereby

[5] This process has been fully recognized by Sidney Paige in his paper ''Rock-cut Surfaces in the Desert Ranges,'' Journal of Geology, vol. 20, 1912, and he has the honor of first summarily stating it. The conclusions set forth in the present paper may be considered as an amplification of his observations, although they were arrived at independently, and are, therefore, a corroboration and not merely a restatement of his views.

it is evolved, or the conception of it as a *quasi* limiting stage of the degradation cycle of the desert. Appreciation of the process enables us to understand certain features of the desert relief which are otherwise unintelligible; and without the concept of the desert cycle, as here defined, it is impossible to read the geomorphic history of such regions.

From the foregoing discussion it is apparent that bold relief in the desert is subdued partly by degradation and partly by aggradation in reciprocal relation, and that the degradational process produces two rock surfaces, one of which vanishes at the close of the cycle while the other then attains its maximum extent. The former may be referred to as the subaerial front and the latter as the suballuvial bench. Neither of these rock surfaces is in evidence at the panfan stage of the cycle. The subaerial front has been reduced to nothing and the suballuvial bench is, as at all stages, buried. If, however, there be introduced at some part-way stage a diastrophic movement which not only interrupts the normal course of the cycle, but also promotes the destruction of the alluvial embankment, then we may by resurrection have the suballuvial bench revealed as an element in the visible profile. Such stripping of the fans might be effected by faulting or doming which would increase the angle of the surface slope. A general elevation of the region without local faulting or doming might produce the same result, if precipitation upon the summit region were thereby increased to the extent of supplying streams competent to dissect the fans.

Such diastrophic or climatic interruption of the normal course of the desert cycle, resulting in the stripping of the upper parts of the embankments, affords the probable explanation of certain broad rock terraces which are found on the flanks of some of the desert mountains of California, Nevada, Arizona, New Mexico and Mexico.

Such rock terraces have been described by McGee,[6] who ascribed their origin to sheet flood erosion, and by Paige,[7] who dissents from McGee's hypothesis of sheet flood erosion and correctly interprets them as resurrected surfaces due to the stripping of the alluvium which once rested upon them.

[*Editor's Note:* The remaining pages elaborate principles that were discussed earlier.]

---

[6] Bull. Geol. Soc. Am., vol. 8, pp. 87–112, 1897.
[7] *Loc. cit.*

13

Reprinted from *U.S. Geol. Surv. Bull.*, **730**, 19–20, 36–38, 52–65 (1922)

EROSION AND SEDIMENTATION IN THE PAPAGO COUNTRY, ARIZONA,

WITH A SKETCH OF THE GEOLOGY.

By Kirk Bryan.

INTRODUCTION.

The Papago country, as the term is used in this paper, is a region of about 13,000 square miles in southwestern Arizona. As shown on Plate IX (in pocket), it is bounded on the north by Gila River, on the east and west by Santa Cruz and Colorado rivers, respectively, and on the south by the boundary between the United States and Mexico. It is a part of the vast region of northern Sonora and southern Arizona which was known to the early Spanish explorers as the Pimería Alta, but which later, when the Spaniards came to distinguish the Papagos or nomadic agriculturists of this region from the more sedentary Sobaipuris and Pimas of the Santa Cruz and Gila valleys, they called the Papagueria.

The mean annual rainfall of the Papago country ranges from $3\frac{1}{2}$ to 11 inches. The mean annual temperature at Tucson is 67° F. and at Yuma 72° F. The mean temperature at Ajo in 1919 was 70° F. The small rainfall and high temperature with consequent excessive evaporation combine to make the region one of the driest parts of the United States, and much of it is, in fact, true desert. The climatic conditions have produced a striking assemblage of plants, in which large cacti, numerous trees, and woody shrubs are characteristic.[1] This arboreal desert has a deceiving verdure that is in great contrast to the scarcity of watering places.

The broad expanse of desert has a characteristic topography. Small and large groups of mountains are separated by broad basins or valleys. The mountains are so arranged that the valleys have a general north-south trend, but the parallelism of the ranges is by no means so complete as indicated on earlier maps. From the southeast corner, near Nogales, where the valleys have elevations that range from 3,200 to 3,500 feet, there is a general slope to the north

---

[1] MacDougal, D. T., Botanical features of North American deserts: Carnegie Inst. Washington Pub. 99, 1908. Spaulding, V. M., Distribution and movements of desert plants: Carnegie Inst. Washington Pub. 113, 1909. Mearns, E. A., Mammals of the Mexican boundary of the United States, part 1: U. S. Nat. Mus. Bull. 56, 1907.

and west, each valley to the west being lower than that to the east, until on the western edge of the region the Yuma Desert, bordering Colorado River, has a general elevation of 250 feet.

The field work of which the present paper is a by-product was begun September 4 and continued to December 23, 1917. The primary object of the investigation was to prepare a guide to the desert watering places of the Papago country.

To the intelligent and observing traveler the surface features of a desert region like the Papago country are a constant source of enjoyment and interest. The bold slopes of the mountains, brown and desolate if composed of schist or gneiss, variegated, cliffed, and pinnacled if composed of thick and massive lava flows, present a remarkable contrast to the generally smooth and verdant slopes of mountains in more humid lands. The broad plains which support orchardlike forests of strange trees and cultivated fields yet contain no watering places impress him with the majesty and mystery of the desert. These land forms, interesting in themselves and excreasing complete control over the lives of the Papagos and the movements of travelers, are the result in large part of processes peculiar to a desert region. The object of this paper is to describe these land forms and to discuss their probable mode of origin. The rock framework is briefly described in a sketch of the geology of the region in order to form a setting for a discussion of processes that have been active in the production of the desert landscape.

The investigation was conducted under the direction of O. E. Meinzer, geologist in charge of the division of ground water, and to him the author is indebted for a free hand in developing the problems discovered in the course of the work. C. P. Ross, who conducted a similar investigation north of Gila River, has cooperated harmoniously along an extensive boundary line. C. G. Puffer, field assistant, contributed largely to the success of the work by his ability to make desert travel easy if not luxurious. The writer is indebted for generous criticism to Dr. H. H. Robinson and Prof. Chester R. Longwell. M. R. Campbell and other members of the physiographic committee of the United States Geological Survey have been most helpful in the critical reading of the text.

[*Editor's Note:* A section on geology and stratigraphy and structure has been omitted. Enough of this material is included elsewhere in the paper to permit understanding of the pediment ideas expressed.]

---

[2] Bryan, Kirk, Routes to desert watering places in the Papago country, Ariz.: U. S. Geol. Survey Water-Supply Paper 490–D, 1922; The Papago country, a geographic, geologic, and hydrologic reconnaissance: U. S. Geol. Survey Water-Supply Paper — (in preparation; contains detailed descriptions of the mountains and valleys).

FAULTS IN THE GROWLER MOUNTAINS.

The portion of the Growler Mountains north of Growler Pass is a rather simple monoclinal fault-block mountain trending northwestward and about 20 miles long. In the picturesque western escarpment are exposed between 1,200 and 1,500 feet of Tertiary lava, tuff, and conglomerate resting on the crystalline complex, which crops out in places at the foot of the mountain front. The eastern slope is gentle, almost without canyons, and conforms to the dip of the lava sheet.

In Growler Pass there are at least five faults that trend almost due north, but the intervening blocks are rotated in different directions. In the easternmost block the beds are horizontal. In the others the dips are easterly in two and westerly in two. The complex structure of Growler Pass continues in the southern part of the Growler Mountains, which consist of large lava plateaus, separated from one another by narrow canyons. The beds in each plateau dip at a different angle or in a different direction from those in neighboring plateaus, so that there must be a complicated system of faults.

GENERAL CHARACTER OF THE MOUNTAINS.

The mountains of the Papago country consist either of more or less isolated elevated regions separated by broad valleys underlain by alluvium, or of groups or chains of mountains separated by rather narrow alluvium-filled valleys or canyons. To the traveler crossing the country on the Southern Pacific Railroad the ranges appear to be rather monotonous in their characteristics, showing a recurrent sameness which implies a common geologic history. They seem to consist wholly of small detached sierras having very similar topographic forms, composed of granite and other coarse-grained crystalline rocks, and to be largely buried in alluvium. This appearance, however, is the result of the accidental distribution of certain types of ranges along the route of travel. A further exploration of the Papago country shows that many of the mountains are capped with lava beds much younger than the granite and crystalline rocks already mentioned and have the typical fault-block form common to the ranges of Nevada. Other ranges consist of exceedingly complex plateaus, peaks, and pinnacles, generally carved from thick lava beds. The ranges are thus by no means uniform in their rock composition and topography and they may be separated into groups according to their composition and structure. There are 70 mountain ranges and groups of hills in the Papago country,[37] and 11 more that lie east of Santa Cruz River are represented on the geologic map (Pl. IX). Of these 81 ranges 17 can not be described, because they have not yet been sufficiently explored. Of the remaining 64 ranges 2 are old volcanoes; 12 are rather simple fault-block or horst mountains, composed mostly of lava beds; 11 are similar faulted mountains of more complex structure; 21, composed mostly of pre-Tertiary rocks, have large or small areas of Tertiary rocks so disposed that they show that the mountains have been elevated since the Tertiary volcanic period, and many of them were already in the old-age stage of erosion when the Tertiary deposits were laid down and have been revealed by removal of the cap, as well as rejuvenated by uplift; 18 ranges are not known to have outcrops of Tertiary rocks, but at least two of them, the Tinajas Altas Mountains and Sierra Estrella, have been reuplifted in Pleistocene time.

PHYSIOGRAPHY.

INFLUENCE OF ARIDITY.

Erosion and sedimentation, the active geologic processes of the Papago country, are affected in their degree, methods, and results by the intense aridity of the region. Temperature and insolation vary but little, but with respect to average annual precipitation the Papago country may be divided into three parts. West of the Growler Mountains the annual precipitation ranges from $3\frac{1}{2}$ to 5 inches; between the Growler and Baboquivari mountains it ranges from 5 to 10 inches; east of the Baboquivari Mountains it is more than 10 inches, although not much more except in the Tumacacori Mountains. Though precipitation increases with altitude, the smaller mountains, averaging 2 to 4 miles in width and 1,000 to 1,500 feet in height, appear to have little effect on storms, and their vegetation does not indicate any large increase in effective rainfall over

---

[37] Bryan, Kirk, Geology and physiography of the Papago country, Ariz. (abstract) : Washington Acad. Sci. Jour., vol. 10, pp. 52–53, 1920. Different figures are given because a smaller area was included in the " Papago country."

that of the adjacent plains. This is particularly true in the area west of the Growler Mountains. The mountainous district including the Sand Tank and Sauceda mountains probably receives a somewhat greater precipitation than the adjacent plains, as indicated by gramma grass and a slightly more luxuriant vegetation at the higher altitudes. In the same way greater precipitation on the Baboquivari Mountains is shown by scattered live oaks and other small trees not characteristic of the plain. Increase of rainfall due to increase of altitude is best shown in the Tumacacori Mountains, where there are orchard-like forests of live oak and a thick cover of perennial grasses. Even here the ease with which soil is formed on tuffaceous rocks and the consequent slow run-off may have as great an influence in producing the relatively heavy vegetative cover as increased precipitation.

[*Editor's Note:* A section on mountain sculpture is omitted as not being directly related to the development of pediments at their feet.]

MOUNTAIN PEDIMENTS.

CHARACTER.

In general, the mountains of the Papago country rise from plains which are similar in form to the alluvial plains that commonly front mountains of an arid region, but large parts of the plains

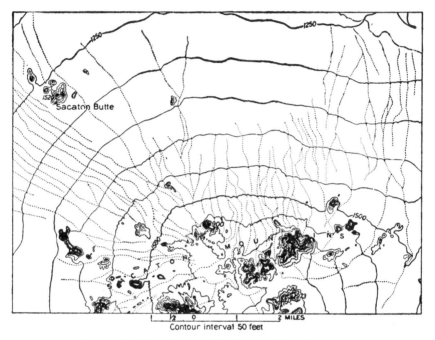

Contour interval 50 feet

FIGURE 13.—Map of part of the Sacaton Mountains, Ariz.

are without alluvial cover and are composed of solid rock. These plains are called "mountain pediments," a term suggested by Mc-Gee's usage,[44] although he applied it to only one of the many similar plains cut on rock in the Papago country that he described. It corresponds to the terms "subaerial platforms" and "sub-alluvial benches" used by Lawson,[45] but certain differences in concept, as well as the awkwardness of his terms, make it advisable to substitute a term which does not imply any particular origin of the topographic form. The normal mountain pediment has smooth slopes, broken only by scattered hills that rise abruptly from the

[44] McGee, W J, Sheet-flood erosion: Geol. Soc. America Bull., vol. 8, pp. 92, 110, 1897.
[45] Lawson, A. C., The epigene profiles of the desert: California Univ. Dept. Geology Bull., vol. 9, p. 34, 1915.

plain and are more or less strung out in lines which are prolongations of the intercanyon ridges of the mountains.   Unfortunately no maps that cover large areas of normal pediment are available. The northern slope of the western part of the Sacaton Mountains is shown in figure 13.   The plains around the hills and small mountains are cut on rock, but the northern part of the area shown has a heavy burden of alluvium.   Schrader[46] states that the "subaerial or nearly subaerial eroded bedrock floor * * * seems to continue to within 3 miles of Casa Blanca," or to the vicinity of Sacaton

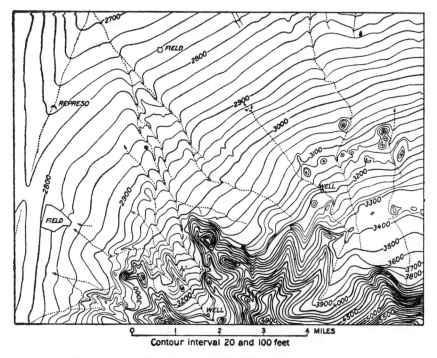

Contour interval 20 and 100 feet

FIGURE 14.—Map of the northern border of the Baboquivari Mountains, Ariz. Redrawn from map by the United States Indian Service.

Butte.   (See fig. 13.)   Similarly figure 14 shows a pediment dissected by small canyons.   This dissection is not so great, however, as to obscure the general form of the pediment.

The angle of slope of pediments ranges from about 50 feet to 200 feet to the mile.   It is noticeable, however, that in any one mountain range the slope is steeper opposite the smaller canyons and very much flatter opposite the large canyons.   The parts opposite the intercanyon portions of the mountain front are steeper than the parts opposite canyons, and they commonly slope not only outward from the mountains but toward the stream channels, which emerge from the canyon mouths.

---

[46] Schrader, F. C., unpublished manuscript.

Exceptions to this general condition occur, for at the mouths of certain canyons the pediments have steep slopes, especially in the direction of the axes of the streams. This exception seems to be due to especially resistant boulders which have been brought down by the stream and dropped at the canyon mouth, causing the stream to spread and lose its carrying power. The boulders, until they weather into fragments small enough to be moved, protect and preserve a slope steeper than is normal to the pediment.

ORIGIN.

The mountain pediment is a plain developed at the foot of the mountains under the processes of erosion normal to the desert. The causes that produce the pediment are obviously different from those which produce the mountain slope. The key to these causes is found in the character of the loose material which covers the respective slopes. The mountain slopes are covered with loose boulders ranging in diameter from 6 inches to 5 or 6 feet. The pediment, on the other hand, is covered with fine fragments one-sixteenth of an inch to 6 inches in diameter.

Davis [47] recognizes that in the closing stages of an erosion cycle in desert regions graded rock floors will intervene between the alluvial slopes and the mountain slopes of an intermontane basin, just as he had previously recognized and shown these plains in a diagram in his discussion of the basin ranges.[48] Under conditions of extreme aridity the development of plains cut on rock does not necessarily indicate old age, but such plains may be produced earlier in the cycle and are dependent for their position on the upper limit of alluvial deposition in the intermontane valleys. Rock plains or pediments develop at or above the edges of the valley fill and are limited in size only by the available area and the factor of time.

As explained on page 42, the angle of slope of the mountain is controlled by the resistance of the rock to the dislodgment of joint blocks and the rate at which these blocks disintegrate. The angle of slope of the pediment, however, is due to corrasion by the streams, and this corrasion is controlled by the ability of water to transport débris, for the pediment is a slope of transportation intervening between the mountain slopes and the alluvial plain in the middle of the valley. Fine rock débris is moved down the mountain slope by rain wash and carried away from the foot of the slope by rivulets and streams that form through the concentration of the rain wash. The supply of this

---

[47] Davis, W. M., The geographical cycle in an arid climate: Jour. Geology, vol. 13, pp. 381–407, 1905; reprinted in Geographical essays, pp. 296–322, Ginn & Co., 1909.

[48] Davis, W. M., Mountain ranges of the Great Basin: Harvard Coll. Mus. Comp. Zoology Bull., vol. 42, pp. 129–177, 1903; reprinted in Geographical essays, pp. 725–772, Ginn & Co., 1909.

fine material on the mountain slope is, however, so small that it is readily removed, and the angle of the slope is determined by the size of the boulders. In the same way the fine material temporarily accumulating at the foot of the mountain slope by the breaking up of boulders, which have rolled down the slope, is usually carried away by the streams of the pediment as rapidly as it is supplied. The sharpness of angle between the mountain slope and the pediment is one of the most remarkable results of this division of labor between rain wash on the mountain slopes and streams on the pediment. The result, however, merely confirms the dictum of Noë and De Margerie [49] : " The slope of an element of the surface is the more gentle as the force of the current active on that element is the greater."

In many localities the transition between mountain slope and pediment takes place in a belt which ranges in width from 100 to 200 feet. This change in gradient seems to be due to the abrupt concentration of rain wash from the mountain slope into streams on the pediment, though doubtless near the mouths of canyons lateral planation by streams is also an effective process, as suggested in a later paragraph.

In Lawson's excellent description of the forms developed by desert erosion he assumes that the débris from erosion of the mountains is poured into an inclosed basin, and consequently that the middle of the valley constantly rises through filling.[50]   He states in substance that the alluvial slopes rise step by step with deposition in the middle of the basin, and their upper edges advance toward and finally engulf the mountains. On the same assumption Paige [51] first stated the results in the following words:

The rising edge of the gravel sheet acts as an effective control below which erosion can not take place. The result is unavoidable if the time factor and the factor of area are sufficiently large. A process tending toward leveling with respect to the gravel sheet will proceed. But the gravel sheet has been gradually rising; therefore the leveled surface is a sloping plain thinly veneered with gravel.

In other words, erosion can take place only above the level of the alluvium and has its maximum value at the upper edge of the alluvial slope. On this hypothesis a plain is cut in the rock whose position and form are governed by the successive positions of the upper edge of the alluvial slope. With equal or diminishing increases in elevation of the middle of the valley in equal times the plain would have a curvature convex upward. With accelerated increase in elevation of the middle of the valley, which, however, must be slow enough to permit erosion of a plain and not so rapid

[49] Noë, G. de la, and Margerie, E. de, Les formes du terrain, p. 22, Service Géog. Armée, Paris, 1887.

[50] Lawson, A. C., op. cit., pp. 30–31.

[51] Paige, Sidney, Rock-cut surfaces in the desert ranges : Jour. Geology, vol. 20, p. 449. 1912.

as to bury the original surface, a plain of concave curvature would result. The convex type of plain which Lawson called a suballuvial bench is shown graphically in figure 15.

When, however, streams are able to maintain through drainage and there are no inclosed basins, the level of the middle of a desert valley may remain essentially stationary. A slope is established on the alluvial plain sufficiently steep for the transportation of the available débris by the available water. The edge of the alluvium then becomes relatively stationary, and as the mountain slope recedes a slope cut on rock is formed, which also is a slope of transportation. As the débris decreases in size with movement away from the mountain it can be moved on flatter slopes, and hence a pediment is concave upward.

The formation of pediments in the Papago country has been very generally interrupted by a new cycle of erosion, and the processes may now be observed only in localities of small extent. These

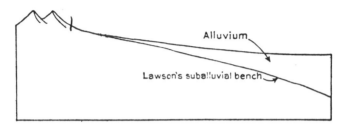

FIGURE 15.—Diagram showing the structural relations resulting from the slow but continuous filling of a structural basin.

processes will be briefly stated. At the base of a mountain front débris is swept outward by ephemeral streams which head in the mountain canyons and by small rills which originate through the concentration of rain wash at the base of the mountain slope. These streams are not permanent or even intermittent; they operate only during or immediately after rain.

From the larger canyons streams move pebbles as much as 6 inches in diameter, but this is the approximate limit in size, except on the east side of the Baboquivari Mountains and around the Tumacacori Mountains. These mountains are the highest and largest in the area and consequently have more rainfall and larger streams, in which boulders from 2 to 3 feet in diameter are not uncommon. In the stream channels a very rigid selection of material takes place; the finer particles move with every flood; the coarser material can be moved only by the larger floods. As the stream emerges from the mountains and its water spreads, the larger particles tend to be dropped and only the smaller ones are carried forward. Because there is an interval of time between floods, often a long interval, the larger pieces are reduced in size by weathering and may then be moved by smaller floods.

At the base of a mountain slope the fine débris washed down by rains is moved forward by little rills toward the larger streams. As the supply of débris is small these rills are not fully loaded and are effective erosive agents, tending to reduce the height of interstream areas, but the grades on which they can work are steeper than those of the larger streams. Thus the pediment has a steeper gradient near the mountains between the canyon mouths and these portions also slope to the streams on either side.

The pediment is greatly increased in size by lateral migration of the streams at and below the mouths of the canyons. The irregularities of the pediment are removed, the higher places being eroded and the lower filled with débris and protected. The lower parts of the canyons are also widened by undermining the slopes of the intervening spurs.[52] When the spurs become narrow they are cut through by slope recession on both sides, and hills are left standing as outliers on the pediment. These solitary hills are worn away with extreme slowness. Their erosion depends entirely on the gradual disaggregation of the rock which composes them and on the movement of the débris over the pediment during rains. The hills retain the same steep slopes as the original mountain but grow gradually smaller until the last remnants are masses of boulders or single rocks projecting above the general level.

The development of the pediment is therefore due to erosion, which may be summed up under three heads—(1) lateral planation by the streams issuing from the canyons, (2) rill cutting at the foot of mountain slopes, (3) weathering of outliers and unreduced remnants, with transportation of the débris by rills. The processes can not operate below a level determined by the grade necessary to transport débris away from the mountains. All depressions below this level will be filled up just as those above it are eroded. Thus the pediment is a slope of transportation and is usually covered with a veneer, from 18 inches to 5 feet thick, of débris in transit.

As transportation of relatively fine material by water is the essential factor in the formation of the pediment, in contradistinction to the erosion of mountain slopes, which is largely controlled by the movement of large boulders, it is obvious that the pediment grows most rapidly along the major streams. In every indentation in the mountain front and in places where streams emerge from the canyons onto the plains the rate of formation of the pediment is rapid, and consequently extensions of the pediment into the mountains are com-

---

[52] Paige, Sidney, Rock-cut surfaces of the desert ranges: Jour. Geology, vol. 20, p. 450, 1912.

mon.   These extensions consist of branching valleys, many of which
are 2 miles or more in width and reach far into the interior of the mountain mass.   The erosion of the mountains at the headwaters of many
streams is much faster than in the lower portions of the same streams,
for there is obviously more water pouring down the mountain slopes
of the larger mountain masses, and feeble streams are incapable of
much planation.   Consequently the headwater slopes may recede
more rapidly than the side walls of valleys.   The extension of the
pediment may thus divide the original mountain into groups of detached hills separated by relatively broad surfaces cut on rock, as
in the Sacaton Mountains (fig. 13).

CONCEALED PEDIMENTS.

If the débris eroded from a desert mountain range and poured
into the adjacent valleys is carried away by some through-flowing

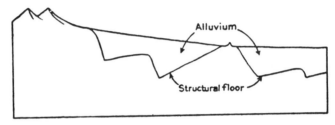

FIGURE 16.—Diagram showing the structural relations resulting from rapid and continuous filling of a structural basin.

stream and the base-level of erosion remains constant for an indefinitely long period, the pediment becomes simply the floor over which
transportation is carried on, and as it is in a balanced condition between degradation and aggradation it can not be further lowered.
On the other hand, if the adjacent valleys have no outlets or insufficient outlets sediment accumulates in them and the local base-level
of the streams flowing from the mountains is raised.   Deposition begins in the lower part of the stream courses, and the sediment gradually increases in thickness and extends toward the mountains.   The
edge and finally a large part of the mountain pediment may thus become concealed beneath a covering of alluvium.[53]   A mountain pediment buried in alluvium may be called a concealed pediment.

If alluviation in the valleys proceeds slowly the formation of
pediments will continue at and above the upper edge of the alluvium,
but as this edge advances it will bury a surface convex upward, as
postulated by Lawson.   (See fig. 15.)   On the other hand, accumulation of alluvium may be so rapid that the formation of a mountain
pediment is not possible until a condition of stability is reached, and

[53] Paige, Sidney, op. cit., p. 449.

hence the alluvium covers and conceals a floor largely of structural origin, as shown in figure 16. A pediment may then form above the level of the alluvium, as illustrated in the figure.

A more complicated sequence of events is possible by which there is a slow but continuous filling of a structural basin with the formation of a suballuvial bench, a pause in alluviation during which an extensive pediment is formed, and then rapid filling that buries and conceals the pediment. This sequence is shown in figure 17 and is illustrated in part by an example described in one of the following paragraphs.

Identification of a concealed pediment is difficult, for the surface has the form of a normal alluvial slope and there is no surface indication of the buried rock floor. Large outliers such as characterize a pediment may not be wholly buried, but they can not easily be distinguished from projections above the general level of a structural floor. Positive identification of a concealed pediment must

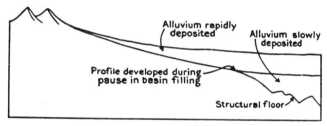

FIGURE 17.—Diagram showing the structural relations resulting from the discontinuous filling of a structural basin.

rest on excavations and wells which show alluvium resting on rock and also the form and character of this rock floor. The thickness of the alluvium must be greater than the ordinary depth of scour by streams, for pediments while being formed may have a shifting mantle of alluvium from 18 inches to 5 feet thick.

The presence of a concealed pediment may be inferred on a general physiographic argument, as illustrated by the following example. Westward-flowing streams in the Sand Tank Mountains, after passing across a large mountain pediment in which they are intrenched in little canyons below the general surface, flow out on grade into the Gila Bend plains (fig. 18). This alluvial slope is dominated by débris moving from the Sauceda Mountains northward to Gila River. The lava plateaus of the Sand Tank Mountains, which rise above the pediment on the east, are bounded on the west by talus slopes from 150 to 500 feet in height, whose lower portions are buried in alluvium. These slopes are the result of a considerable recession of the plateaus, for a continuance of the dip of the lava beds indicates that they once extended farther west, as shown in the cross section, figure 4 (p. 34). The large pediment east of the plateaus (Pl. XIII, A) required for its erosion a period of time more than sufficient to provide for recession of these slopes and the formation of a corresponding pediment on the west side of the plateaus. Since then the pediment within the mountains has been dissected, but the same streams choked by a flood of alluvium derived from the Sauceda Mountains have buried the pediment on the western border.

COALESCING PEDIMENTS.

When a mountain range has been subjected to erosion for a long period of time, mountain pediments formed along individual streams

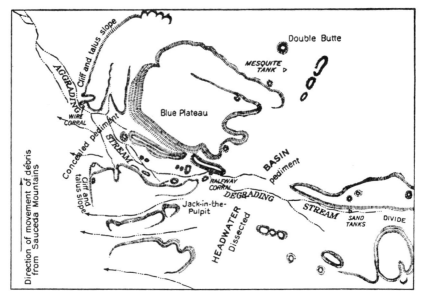

FIGURE 18.—Sketch map of the northwestern part of the Sand Tank Mountains, Ariz., showing location of dissected and concealed pediments.

will tend to unite, and eventually the range will be surrounded by a continuous pediment formed by the coalescing of many individuals. Coalescing pediments begin to be formed in the old-age stage of erosion of desert mountains, and when the projecting hills have been removed the grades will be almost stable.

DISSECTED PEDIMENTS.

If the base-level of erosion of streams flowing from the mountains is lowered, the streams carry their load of débris with ease on the increased gradient and begin to dissect the mountain pediment. Narrow canyons are cut by the streams; the canyons begin to form near the center of the basin within or on the margin of the alluvial plains and increase in length by headward erosion. As the little canyons increase in length and complexity the pediment is carved into a maze of small hills with intervening sharply cut valleys, and the original surface is preserved only in the tops of hills or interstream areas.

Dissection of the pediment may be brought about by uplift of the mountain mass, as in the Burro Mountains, N. Mex.,[54] or the trunk stream may intrench itself and thus lower the base-level. At the north end of the Gila Mountains dissection of the narrow pediment is due to the lowering of Gila River in late Pleistocene time. Similar lowering of Santa Cruz River near its junction with Gila River has caused the dissection of alluvial fans and a very narrow pediment at the north end of the Estrella Mountains.

However, dissection of pediments due solely to lowering of a trunk stream is the exception in the Papago country; yet the pediments are commonly dissected, as shown in Plates X, *A*; XII, *B*; and XIII, *A*. Near or within the mountain borders each stream passes over falls or through rapids into a steep-walled canyon. The walls are rock, in many places capped with a few feet of gravel, and average about 10 feet high (Pl. XII, *B*). Near the mountain they may be from 20 to 150 feet high, but downstream they become lower, and near the outer border of the pediment they gradually fade out in broad plains of alluviation, as shown in figure 19.

This type of dissection of the pediment has taken place generally and irrespective of the conditions of erosion and sedimentation in the adjacent valleys. In some valleys, such as the upper part of Altar Valley, where the axial stream runs in a trench, the canyons of the pediment connect with similar canyons in the alluvial plains and with the axial trench. Dissection of the pediment in this locality may not, however, be ascribed wholly to lowering of local base-level, but the dissection of the pediment and the dissection of the valley by its axial stream are doubtless due to the same cause. In the northern part of Altar Valley there is no middle trench, yet the pediment on the northern border of the Coyote Mountains and also in the Tucson Mountains is dissected. Thus the trench of the axial stream fades out downstream, as do also the little canyons of the pediment. (See p. 73.)

The lower margin of a mountain pediment is generally mantled with the alluvium, which can be seen resting on the rock, both being dissected by the streams. This alluvium is therefore older than the undissected alluvial plains characteristic of the northern part of the Altar Valley and the interior valleys of the Papago country. What is the relation of this older alluvium to the extension of the pediment under the valley and to the original rock floor? The relations which are thought to be general in the Papago country are shown in figure

---

[54] Paige, Sidney, op. cit., pp. 444 et seq.

A. PASS WEST OF TULE TANK, ARIZ.

Niches in a granite mountain slope of the cliffy type due to massive jointing.

B. MOUNTAIN SLOPE AND DISSECTED PEDIMENT.

View north across Sand Tanks, which are in the small canyon in the foreground.

A. HEADWATER BASIN AND DISSECTED PEDIMENT IN THE SAND TANK
MOUNTAINS, ARIZ.

Tilted lavas of Jack-in-the-Pulpit, on the extreme right, rest on the crystalline rocks in the
foreground.

B. EAST SLOPE OF MARICOPA MOUNTAINS, ARIZ.

Undissected mountain pediment, with an outlier. In the outlier lava and gravel rest on an old
erosion surface (pediment) from which the present pediment has been developed by a relatively
small amount of erosion.

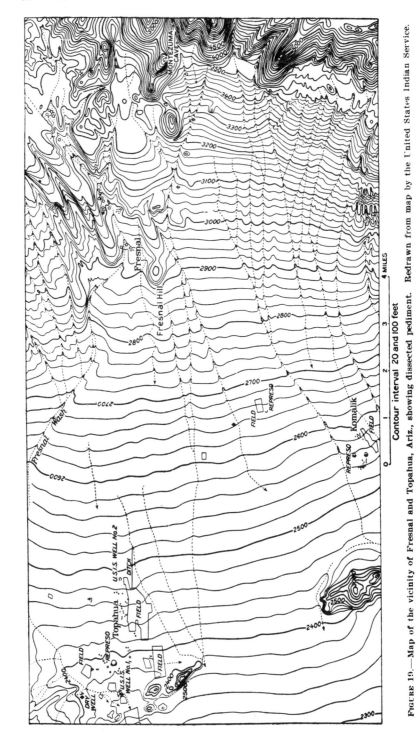

FIGURE 19.—Map of the vicinity of Fresnal and Topahua, Ariz., showing dissected pediment.   Redrawn from map by the United States Indian Service.

20. It is believed that during the deposition of the older alluvium there was a gradual rise by aggradation of the middle portions of the valleys, accompanied by the erosion of a convex suballuvial bench. The alluvium then remained essentially stationary, occupying a slope of transportation of the available débris toward the local base-level. The erosion of the pediment took place during this time. The dissection of the pediment by small canyons took place after the completion of the pediment, and in that process the edges of the older alluvium were eroded, but the great mass in the middle portions of the valleys was either slightly buried or perhaps simply reworked by streams. The strict correlation of this older alluvium with that found in the valleys of Gila and Santa Cruz rivers has not been made but is discussed on page 30.

The two cross sections in figure 21 present the best data available regarding the rock floor of the valleys. Section A–B crosses the area represented in figure 14 to the Santa Rosa well, which is 920

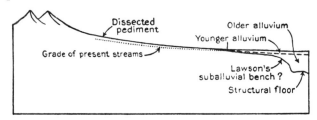

FIGURE 20. Diagram showing the structural relations assumed for a typical valley in the Papago country, Ariz.

feet deep and does not reach the rock floor. To provide the requisite depth of alluvium at the well the rock floor under the alluvium from the edge of the pediment must be convex upward, and an error of as much as 2 miles in the location of the edge of the pediment will not vitiate this conclusion. Section C–D runs from the crest of the mountains westward across the area shown in figure 19 through the United States Indian Service wells. Well No. 2 is 602 feet deep and does not reach rock, but well No. 1 reaches rock at 100 feet. An even greater convexity of the rock floor must be assumed here, and the depression in which well No. 2 is sunk can not drain through the pass in the Artesa Mountains like the present streams.

The rock floors in both sections A–B and C–D have been drawn smoothly convex as if they were suballuvial benches, but it seems probable that their form is, in part at least, structural.

CAUSE OF DISSECTED PEDIMENTS IN THE PAPAGO COUNTRY.

Dissected pediments are widely scattered, lie at various elevations, and border valleys of unlike characteristics. The little canyons

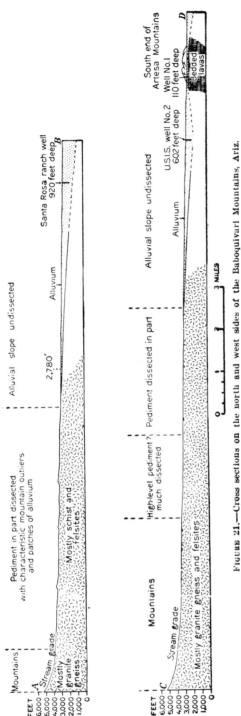

FIGURE 21.—Cross sections on the north and west sides of the Baboquivari Mountains, Ariz.

which dissect them must be due to some cause that affects the characteristics of the streams within the pediments. For a general condition a general cause must be sought, and the following analysis leads to a partial conclusion.

After the establishment of a pediment stream grades are adjusted to the supply of rock débris and the amount of water available in average floods. As the mountain front recedes there is a tendency for the slope of the pediment to become less steep as streams work farther into the mountains at the upper end; at the same time the mantle of alluvium at the lower end tends to increase in extent and still further cover the lower end of the pediment. In the interstream areas weathering and later planation tend to reduce the surface to the level of the stream beds. Conditions, however, are relatively stable.

Canyon cutting by the streams can take place only by an increase in the amount of water in proportion to the sediment carried, either by an increase in the supply

226

of water or by a decrease in the supply of sediment. Because the streams are wholly ephemeral, the quantity of water in them is dependent directly on the amount and rate of rainfall. On the other hand, the supply of sediment depends upon the rapidity of mechanical and chemical erosion of the rocks on the mountain slopes. The principal factor in weathering is expansion and contraction through the influence of changes in temperature. It seems improbable that any general change in temperature could be sufficiently great to make an essential difference in the daily or seasonal changes in temperature in the desert. If these changes in temperature crossed the frost line, it would, of course, make a great deal of difference whether it was colder in any period of years than it had been in the preceding period of the same length. Although freezing temperatures are common throughout the winter, it is doubtful if frost action is an effective factor in the weathering of rocks in the desert ranges. In most of the mountains the rocks do not contain sufficient moisture to make freezing temperatures of any moment in their disintegration. None of the smaller peaks appear to be affected by frost action, but in a few of the higher summits there is some evidence that frost action may aid erosion.

It seems probable, then, that the dissection of the mountain pediments at the present time is due to increased rainfall, or to a greater concentration of rainfall with consequent increased stream action. It is also probable, however, that in this region of extreme aridity minor fluctuations in climate are not recorded with great sensitiveness. The conclusion that the present state of dissection of the mountain pediment is due to increased or more concentrated rainfall does not indicate that the present is a wet period. It implies only that the last change in climate which was effective on the mountain pediments was from a somewhat dryer to a somewhat wetter period.

[*Editor's Note:* The section on valleys has been omitted.]

14

Reprinted from *J. Geol.*, **38**(1–2), 1, 12–14, 24–27, 136–141, 145–156 (1930)

ROCK FLOORS IN ARID AND IN HUMID CLIMATES. I

W. M. DAVIS

ABSTRACT

This paper, after an introduction regarding the fundamental elements of the cycle of erosion, institutes a comparison between the barren mountain sides and the desert rock floors (pediments) of arid regions and the forested mountain slopes and the fertile peneplains of humid regions, with the object of discovering how far these contrasted landscapes are comparable with each other and of defining the various factors on which their contrasts depend. For humid regions the familiar principles of weathering and streams are followed; for arid regions the principles of degradation set forth by Lawson and the examples described by Bryan are accepted as guides. Sequences of forms are sketched for both regions, beginning with those which are largely consequent upon the deformation of a peneplain of massive granite, and ending with those which are wholly consequent upon the far advanced degradation of the deformed mass. It is concluded that a close analogy exists between the two sequences, and that their difference of appearance is due to their difference of climate. The plant cover of humid regions permits the accumulation of abundant soil even on fairly steep slopes; and of still deeper soil on lowlands of degradation; yet under the soil cover of mountain slopes boulders of decomposition are found grading into solid rock, which are closely homologous with the visible boulders resting on the bare rock of arid mountains. Similarly, under the heavy soil cover of a granite peneplain is a buried foundation of decomposed rock which is truly homologous with the gravel-veneered rock floors or pediments of arid degradation.

[*Editor's Note:* Certain introductory matters, discussed elsewhere and later in this paper, are omitted in the interests of brevity.]

HUMID HOMOLOGUES OF ARID ROCK FLOORS

Although the gradual climatic transition that is often found from a region of humid erosion to one of arid erosion strongly suggests that, as noted above, the processes and forms of such regions merge into one another, and that the cycle of arid erosion may therefore be treated as a climatic modification of the cycle of humid erosion, this view has been questioned by an experienced German writer.[1] It would consequently seem that, in spite of the nearly fifty years of age now reached by the simplest scheme of the cycle of erosion, and of the nearly twenty-five years of age reached by the application of the scheme to arid conditions,[2] this application needs fuller explanation.[3] The object of the present essay is therefore twofold. First, to set forth in some detail the facts and arguments which have, during my recent residence in the Southwest and in spite of the adverse opinion just cited, confirmed an earlier

[1] His conclusion is, in effect, that the attempt to explain arid erosion of rock floors as a climatically controlled modification of the cycle of humid erosion must be regarded as a failure. This conclusion is perhaps based on his misunderstanding of the meaning given by American geologists to the term "erosion." We take it to include subaerial degradation in general as well as river corrasion, while German writers ordinarily use separate terms, *Erosion* and *Abtragung*, for these closely associated destructive processes; yet even German writers tell of *Erosionsgebirge*, in the production of which *Abtragung* has done enormously more work than *Erosion*. The German usage of the two words is illustrated in a passage quoted from W. Penck's *Morphologische Analyse* in footnote 3, p. 3.

[2] W. M. Davis, "The Geographical Cycle in an Arid Climate," *Jour. Geol.*, Vol. XIII (1905), pp. 381–451. This essay treats chiefly the gradual integration of initially independent interior drainage areas and their reduction to extensive rock plains as dust is exported from the lowest central basin floor. The form of the surrounding mountains is not particularly considered.

[3] A number of European geographers have failed in various respects to understand the spirit of the cycle scheme. They have criticized it as too largely deductive, without seeing either that its basis is essentially inductive or that the deductive form often given to it is merely a matter of presentation. They have misconceived it as rigid, thus failing to recognize its elastic applicability to all sorts of natural conditions. One of them has objected that it could not hold good for the Alps because of the several movements that those mountains have suffered, thus implying that the scheme always demands a complete cycle and overlooking the plain possibility that interruptions by movements at any stage of its progress are among the most manifest of its many complications. Another refuses to call the valley of a single river mature where it is openly excavated in weak rocks and young where it is narrowly incised in resistant rocks, and protests that a single valley must be of the same age all along its length; thus failing to recognize that young and mature as here used refer to stages of development and not to age measured in years. A third insists that the scheme takes no account of erosion during the uplift of a land mass, although explicit account of such erosion has repeatedly been given. A fourth condemns the explanation of adjusted drainage, inherent in the scheme, because it fails to apply in a case where it really has no application. Finally, the scheme of the cycle has been pronounced dangerous, because it may be misused by students who are not fully trained in its proper use. All these objections seem to me to arise from a too literal following of various brief expositions of the scheme, without grasping the spirit that is behind them, and that has been clearly indicated in fuller expositions.

belief that the cycle of arid erosion may be, with all its peculiar consequences, reasonably treated as a variant of the cycle of humid erosion, even to the point of relating the typical boulder-clad mountain slopes and their débris-strewn piedmont pediments or rock floors in arid regions to the corresponding and truly homologous features in humid regions; and, second, to make clear, by following the teachings of Lawson and Bryan, that although arid erosion resembles humid erosion in many respects, the two kinds of erosion nevertheless differ so much in process and in product that they cannot be clearly understood if they are briefly brought together as examples of normal erosion, as has been done by several of the German writers above cited. The unlikenesses of the two erosions deserve explicit treatment. In order to justify these statements, the processes of arid erosion and the forms that they produce will first be reviewed; the processes of humid erosion and the resulting forms will next be reviewed; and a comparison will then be instituted between the two classes of processes and of forms in order to make clear the striking homologies that exist between them. These reviews and the comparison to which they lead will involve much familiar matter, which it seems advisable to set forth in some detail in order to gain a clear entrance into the problem under consideration.

[*Editor's Note:* Some review of the origin of pediments has been omitted.]

[*Editor's Note:* The material omitted previous to this section has been discussed earlier in other papers.]

VALLEYS DEVELOPED BY HUMID EROSION

Let attention now be directed to the forms developed on a single upheaved fault-block by stream erosion in a humid climate, not that this problem is novel, but that it must be explicitly explained in order to understand its bearings on the question before us. For convenience of illustration let the block be broadly uparched, as near the foreground of Figure 3; part of the block being shown on a larger scale in Figure 4. While elevation is still in progress, each consequent stream excavates a valley draining into a lake or trough; and each valley-cutting, consequent stream may, at an early stage of its work, be considered in three parts or courses, which blend into each other. First a headwater course where, in spite of the considerable altitude to which the stream has there been elevated, its small size prevents it from rapidly eroding a deep valley; hence it will for a time cut only a shallow valley, the sides of which will be weathered well open during the slow deepening. Second, a lower course where, in spite of the good size of the stream, it stands so little above local baselevel that it cannot cut a deep valley even after elevation is completed. Third, an intermediate course where, the stream having fair volume and the land surface gaining fair altitude, a relatively deep and steep-sided valley will be quickly eroded while elevation is in progress, at the bottom of which the stream will flow for a time as an ungraded torrent. However, short streams on the side slopes of uptilted fault blocks, such as here considered, will seldom be large and active enough, even along the intermediate course, to cut a valley with canyon walls so steep that no waste lodges upon them; the valley sides will be more or less graded and waste-covered as the fault-blocks are uptilted and the valleys are deepened. Deep,

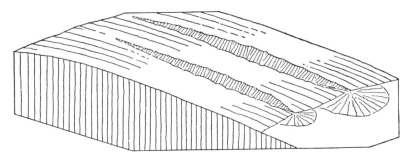

Fɪɢ. 4.—Young consequent valleys in the back slope of an upwarped fault block, on a larger scale than that of Fig. 3.

bare-walled, canyon-like valleys are expectable only along the intermediate course of large rivers.

After block-raising and valley-deepening cease, the side walls along the early graded, lower course of a stream will soon be weathered back to moderate declivity, if they are not so weathered while block-raising is in progress. In the intermediate course, where the valley will be deeper and steeper-sided when elevation ceases and the stream reaches grade, the valley may be somewhat canyon-like, as caricatured on a much-enlarged scale in the back block of Figure 5. At this stage the irregular disposition of rock joints will cause the production of many precipitous and bare rock faces, separated by less precipitous ledges on which, as coarse detritus accumulates, root-hold will be afforded for such plants as can grow in coarse, shallow, and dry soil. Wherever bare walls are thus developed in massive rocks, fragments may break off and roll down from the rock faces in sufficient quantity to accumulate for a time as a veneer or talus at the base of the retreating slope, as shown in the second block of Figure 5. In valleys that are eroded through plateaus of horizontal, cliff-making strata, the retreat of a capping cliff will be long accompanied by a retreating, subcliff talus; but in massive rocks such as are here postulated, the talus marks only a transitory phase of valley-widening. It vanishes with the extinction of out-cropping ledges on the valley-side slopes, as shown in the third block of Figure 5.

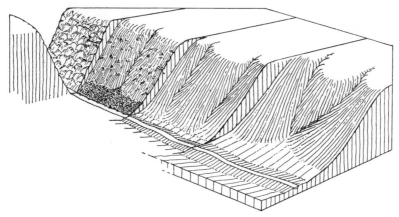

FIG. 5.—Four early stages of valley-widening, on a much larger scale than that of Fig. 4.

During the delay in the recession of the basal slope by reason of its being talus-covered, the more active recession of the bare-rock surfaces in the higher slope will decrease the average declivity of the valley sides; and then, with increasing development of a soil-cover and with the better establishment of plant growth upon it, the declivity of the valley side will be progressively lessened until all the slope above the retreating talus comes to be cloaked with finer and finer detritus and covered with an increasingly dense plant growth, as in the third and fourth blocks of Figure 5. Thereafter, the removal of the soil cover will be more and more leisurely, and the soil will therefore become much refined in texture at the surface and at the same time increased in depth.

As rill-wash and soil-creep slowly carry the finer surface soil down to the stream, in spite of the protection given to it by plant growth, underground weathering will penetrate deeper and deeper along joints and into joint-split blocks, thus dividing the subsoil granitic mass into irregular, boulder-like masses and developing coarse-textured soil in the spaces between them. The boulder-like masses will be increasingly subdivided, rounded, and disintegrated by continued weathering until, as their soil cover is slowly removed, they are themselves converted into soil and new subsoil blocks are weathered out. The subsoil blocks and the coarse interstitial débris stand still, while the overlying finer soil slowly washes and creeps downslope past them; and by the time the removal of the outer soil to the valley bottom has been accomplished, the sluggish under-blocks are themselves disintegrated and their débris is in its turn slowly removed, while new and deeper-lying subsoil blocks are weathered out.

During these early stages of valley growth the cross-profile of an intervalley ridge will still be nearly flat-topped, if the uplifted block be part of a former peneplain as is here assumed to be the case; but as the valleys widen, the cross-profile will become broadly convex, yet retaining for a time a somewhat "square-shouldered" form on either side, where the more pronounced descent into the valley begins. This is all familiar enough, and it is restated here simply in order to secure a clear entrance into what follows.

ROCK FLOORS IN ARID AND IN HUMID CLIMATES. II

W. M. DAVIS

VALLEY FLOORS OF HUMID EROSION

The continued removal of the finer soil from the valley-side slopes causes them to recede from the banks of the graded stream to which they previously descended; and as they do so narrow strips of valley floor, shown by spaced lines in the third block of Figure 5, will be developed at their base, back of whatever flood plain, drawn with close-set lines, is simultaneously formed by the stream.[1] As these strips widen, the concavity of the profile across the valley bottom, initiated by the growth of the talus, is given broader expression, as in the fourth block.

The lateral valley-floor strips are, like the flood plain, underlaid with degraded rock, which may be called the valley-floor basement. It has a somewhat ragged and vague surface, except that where cut by lateral shifts of the stream it may be more even and better defined. The lateral strips will be everywhere covered by detritus, partly derived from local subsoil weathering, partly washed down from the valley sides; and this detrital cover will, in time, constitute a large part of the valley floor. The surface of the cover thus prepared will have the same longitudinal or downvalley slope as that of the stream and its flood plain. If the stream is of good size, its gradient will be so weak that an upstream view of the flood plain between the lateral valley-floor strips will seem nearly level. Each lateral strip of the valley floor will have, besides its downstream slope, a somewhat stronger transverse slope from the valley side toward the stream or its flood plain, so that fine soil washed down from the valley side can be transported across the strip by the agencies, rill-wash chiefly, there available for that duty.

---

[1] Earlier accounts of valley-floor strips of degradation have not been looked up, and I am unable to say who first called attention to them explicitly. They have been alluded to in at least two of my own writings, *Practical Exercises in Physical Geography* (1908), text, p. 41; Atlas, Pl. 8, Fig. 10; and "La vallée de l'Armançon," *Ann. de Géogr.*, XXI (1912), 312–22.

The lateral swinging of the graded stream broadens the flood plain and its rock basement, as in the front block of Figure 5, and thus the valley-floor strips may be encroached upon; but in spite of this they gain width by the retreat of the valley-side slopes. As far as the stream thus swings, it will cover the rock basement with the gravels, sands, and silts of the flood plain, which, besides having a downvalley slope, also slopes transversely away from the stream bank. In time the continued withdrawal of the valley-side slopes will carry their base farther away than the lateral swinging of the stream can reach: thereafter the valley-floor strips will be developed

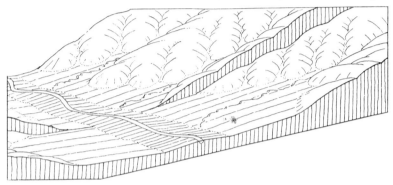

FIG. 6.—Two advanced stages of valley-floor widening with associated branch-floor development.

to greater and greater width, independent of the flood plain. In the meanwhile many brooks, short side branches of the main stream, will excavate ravines in the valley sides, as in the back block of Figure 6. The concavity of the main-valley cross-profile is then increasingly shown on the line of these brooks; it later appears part way up the valley-side spurs also; and by that time the spur crests will have become so gracefully rounded, so free from all hard-line edges, that it is difficult to make a drawing of them in which they do not look like wrinkled bolsters and pillows.

After the side-brook ravines are developed to the stage of having their own flood plains, side branches of the main valley-floor strips, each with its underlying rock basement, will be opened along their margins heading farther and farther up the ravines, as in the front block of Figure 6. The branch floors will have slightly steeper down-brook gradients than the downvalley gradient of the main-valley strips or of the main flood plain; and somewhat steeper still will be the transverse gradient of the parts of the main valley-floor strips which lie between an interravine spur-end and the main stream or its flood plain, because there the gradient is determined only by wet-weather rill-wash instead of by a branch brook: but this delicate modulation of valley-floor slopes is not shown in the accompanying diagrams.

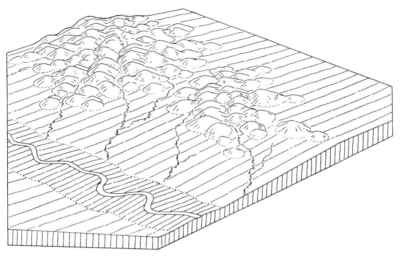

Fig. 7.—A late stage of valley-floor widening with the accompanying dissection of an intervalley ridge and the isolation of its severed parts.

Neighboring streams, subparallel to the one here figured, will develop similar ravines and valley-floor branches; the ridge crests between the ravines will be encroached upon and increasingly dissected as the ravines and their valley floors widen. Later on the ridges may be gnawed through at various points by the brookhead valley-floor branches, and successive parts of an originally continuous ridge may thus be isolated, as in Figure 7.

AGGRADATION OF HUMID VALLEY FLOORS

By this time, the load to be washed from the subdued hills of the dissected intervalley ridges across the valley-floor strips to the main streams has decreased to a less amount than when the hills were higher and steeper; therefore, the valley floors will be worn down to gentler and gentler transverse gradients, just as they and the flood-plain belt between them are also worn down to gentler longitudinal gradients.

It may be here pointed out that several conditions associated with the development of flood plains in humid regions will cause them to increase in the three dimensions of length, breadth, and thickness, so that they come to cover an unexpectedly large part of their degraded valley floors. First, while a river is actively cutting down its valley in an uplifted highland, its channel in the valley bottom tends to remain of small width; not that lateral erosion is then inoperative, but that downward erosion is more operative. Then, as the river approaches grade, downward erosion is so much retarded that channel widening sets in actively, especially if the current is so well charged with coarse detritus that it develops a braided flow; and channel widening thus begun may continue after grade is reached. But with continued widening there must be a loss of depth, of velocity, and of carrying power; hence, in consequence of such channel widening, the fall to which the river reduced its course when grade was first reached may later prove insufficient for its work of transportation. Deposition must thereupon take place in order slightly to increase the fall, the velocity, and the carrying power; that is, some of the load must be laid down, thus steepening the river course a little and accelerating its flow so that the rest of the load may be carried. Aggradation of this kind must be zero at the river mouth and of increasing measure upstream; it may there-fore be called divergent aggradation. The resulting flood plain will tend to spread laterally across the incipient valley-floor strips toward the base of the valley sides, and also headward upon the first-encountered residual riffles of the not yet graded torrential course.

Second, in consequence of this incipient aggradation, the river current, which had all been flowing in its rock-bed channel while valley deepening was in progress, will now lose some of its water to seepage in the deposited detritus; and with this loss of volume there will be still further loss of carrying power, and deposition will have to continue, again in divergent fashion. This might be a cumulative process, were it not that the rapid gain of carrying power due to increase of velocity will soon make up for the slower loss of carrying power due to loss of volume.[1]

In the meantime a third cause of aggradation is arising. With advance of highland dissection by branch brooks, the area of wasting valley-side slopes increases and the discharge of waste from them delivers a greater and greater load of detritus for the main stream to carry. The main stream must therefore once more build up its course by divergent aggradation in order to gain velocity for its increased work. Fourth, with these various processes all conspiring to lengthen and widen the flood plain, the development of meanders is greatly favored; and the more they are developed the longer is the course of the river through the flood plain; and with greater length there must be a loss of fall, of velocity, and of carrying power, unless recourse is had for the fourth time to divergent aggradation. A fifth cause for aggradation is found in delta growth; but this cause produces only parallel aggradation, and is therefore less effective (except close to the river mouth) in producing an increase in depth of flood plain than are the four causes of divergent aggradation.

It is not intended to assert that all these causes of aggradation operate, as here outlined, during the pre-mature stages of all rivers in all humid regions; but it seems probable that, if they operate at all, they will conspire to extend flood-plain deposits beyond the limits that would otherwise be reached. If so, the flood plain will increasingly overlap the side-strips of normally degraded valley floors: but, even then, an upstream view along the flood plain of a fair-sized or large river will seem almost level. Furthermore, after reaching a maximum of aggradation in the maturity of a normal cycle, when the discharge of waste from the valley sides is believed to reach its greatest measure, a graded river will have a decreasing

---

[1] It may be noted in passing that, while the power of running water to carry detritus theoretically increases with the sixth power of the velocity, the carrying power of an actual stream will not increase faster than the fifth power of its velocity, because its cross-section will diminish as much as its velocity increases. That is, if a given stream runs twice as fast as before, it will be only half as large, and it will therefore have only half as many threads of current available for the work of transportation. On the other hand, inasmuch as a stream in flood increases in size as well as in velocity, its carrying power will be increased by more than the sixth power of its velocity.

load, because of decrease of valley-side height and slant; the river will then resume its degradational work and will slowly wear away the flood-plain deposits; yet in so doing it will not terrace them because their surface will be worn down *pari passu* with the slow wearing down of the entire valley floor. The river will thus in its later maturity sweep away all the earlier-formed flood plain and return to the long-postponed task of slowly wearing down the rock basement of its first graded course, and will gradually reduce it to the fainter and fainter gradients of old age. In the meantime the previously formed and temporarily buried lateral strips of valley floor will be laid bare and will gain increased breadth by the recession of valley-side spur ends.

DISSECTION AND DEGRADATION OF A MOUNTAIN MASS

In consequence of retrogressive erosion by the headwaters of various streams and branch brooks which gain steeper gradients and increased activity of erosion as the torrential intermediate course of a river is deepened, an uparched highland will be more and more dissected through advancing and passing maturity, until its central part comes to be resolved into several more or less completely separated mountains, while its interstream spurs will be, as already noted, divided and subdivided, especially in their terminal parts, by branching ravines into isolated groups of mounts or hills. All the chief valley floors and their rock basements, reduced to gentle or faint downstream gradients, will be growing wider and wider at the expense of the spurs, ridges, and mountains; and the soil-cloaked, tree-covered slopes of the mountains, mounts, and hills will be worn down to gentler and gentler declivities, because the downwash of soil from their convex upper slopes will be faster than its removal from the base of the slopes, where it will coalesce with the soil cover of the widening valley floor in a smooth concave curve.

Soil-creep, although acting with increasing slowness as relief weakens, appears to be of growing importance at this advanced stage of the cycle in comparison with stream corrasion and rill-wash; for it is largely by soil-creep that the sharper crests of early-mature mounts, ridges, and spurs are rounded off and subdued to low mounds of gently convex cross-profile in late maturity and early old age.

[*Editor's Note:* The material omitted here is not pertinent to the discussion of homologies in the broad sense.]

COMPARISON OF HUMID AND ARID REGIONS

The erosional processes and the forms of humid and of arid regions may now be compared. Certain of the forms, such as branching valley systems which subdivide mountain masses, are manifestly common to both. Certain others are usually thought to be unlike. Conspicuous among these are, on the one hand, the soil-cloaked and forested slopes of maturely dissected mountains in a humid region together with the fertile valley floors below them, and on the other hand the rocky and boulder-clad slopes of maturely dissected mountains in arid regions together with the barren pediments below them. These contrasted features have, indeed, been regarded by some physiographers as so completely unlike as to indicate that the erosional processes of humid regions must be correspondingly unlike those of arid regions.

The comparison here to be instituted will show that, while various contrasted features are truly enough unlike in certain respects, their unlikeness is rather a matter of degree than of kind. The apparently unlike features are really homologous; their resemblances, once recognized, are much more striking than their differences. The same may be said of the erosional and degradational processes by which the apparently unlike forms are sculptured.

CONTRASTS OF HUMID AND ARID PROCESSES

In both humid and arid regions the processes of weathering are partly physical, partly chemical; more physical than chemical in arid regions, more chemical than physical in humid regions. But let it be noted that, wherever bare-rock surfaces are exposed in humid regions, there surface weathering is largely physical; and, conversely, that wherever detrital deposits containing ground water lie on bed rock in an arid region the subsurface weathering may be largely chemical. In other words, in both regions weathering will vary according to its physiographic opportunity; and this opportunity will vary from place to place at any one stage, and at the same place in successive stages of a cycle of erosion.

In both humid and arid climates, rill-wash and soil-creep are operative, but soil-creep appears to be favored and rill-wash hindered on the heavily soil-cloaked and forested slopes of maturely dissected mountains in humid regions, while rill-wash is favored and soil-creep is weakened in the thin, stony soil on the rocky and barren slopes of maturely dissected mountains in arid regions. In both, the water courses divide and subdivide repeatedly, but their subdivision is carried much farther in arid regions where every rivulet has its little valley than in humid regions where plant growth greatly impedes the formation of rivulets. This contrast is especially striking in the later stages of humid and arid cycles of erosion: for then, as noted in the foregoing, the headwater stream lines are more or less obliterated by the smoothing actions of rill-wash and soil-creep on low, plant-covered forms in the old age of a humid cycle; but they are still abundant and manifest on the worn-down, barren rock floors of a well-advanced arid cycle. In both, the streams vary from time to time in volume of discharge, but the variations are vastly greater in arid than in humid regions; for in humid regions streams seldom fail to flow, except at their very heads during droughts; while in arid regions streams seldom flow anywhere except locally, at one or another part of their courses, for a short time during and after rains. A housekeeper from a humid region once exclaimed, on seeing clouds of dust blown from the dry channel of a river in an arid region, "I never before saw a river whose bed was so well aired!" and it could have been only in a humid region that a poet would write of a brook, "I go on forever."

In both humid and arid regions, streams in time reduce their courses to grade; but, other factors being equal, the graded streams of an arid region, heavily charged with relatively coarse detritus, have a steeper fall than streams of the same size in a humid region, where by reason of the presence of a plant cover the detritus given to the streams to carry is moderate in quantity and fine in texture. Here, again, the arid-humid contrast is most striking in the later stages of the cycle of erosion; for while the aging streams of a humid climate will, because of the small quantity and fine texture of their detrital load, have reduced the peneplain that they drain to a nearly level surface of almost imperceptible slope along the stream lines, the aging streams of an arid climate may still have, by reason of their relatively abundant and coarse load, very perceptible slopes on their barren rock floor.

RESEMBLANCES OF ARID AND HUMID MOUNTAIN SLOPES

As the various processes of arid and of humid erosion are thus seen to differ in degree and manner of development rather than in nature, and as their differences in degree and manner are wholly due to differences in their climates, so the forms produced by the two erosions may be shown to differ in the degree to which certain elements of form are developed rather than in the essential nature of the elements themselves.

For example, the maturely graded slope of a soil-cloaked and forested, granitic mountain in a humid region is really about as well boulder-clad as a maturely graded and barren granitic mountain slope is in an arid region. The difference between the two is that the boulders on the humid mountain slope are hidden under a soil cloak and the soil cloak is hidden under a forest cover, while the boulders on an arid mountain slope are, in the almost complete absence of finer soil and of plants, clearly visible. In both cases, after the graded slope is developed, the weathered boulders rest upon more or less decomposed bed rock. In humid regions, and especially in warm humid regions, the underlying bed rock may be, if of granitic nature, decomposed in mass to depths of 30, 50, or more feet below the sub-surface boulders; in arid regions rock decomposition penetrates to a

243

much less depth. In both cases, the boulders on graded slopes do not roll down to form a heavy talus at the mountain base, but are for the most part destroyed by disintegration while they are resting on or slowly descending the slope.

On graded humid slopes, by the time that the finer surface soil is removed by wash and creep, as if to reveal the underlying boulders of subsoil weathering, the boulders themselves are reduced to soil and new boulders are formed by deeper weathering of the bed rock behind them. As to the boulders on graded arid mountain sides, Bryan's account of the Papago country in Arizona tells us:

> Under conditions normal to the region few boulders reach the bottom of the [mountain] slope. . . . . Great heaps of boulders do not occur at the mountain base. Fine débris forms a loose film over the surface between the boulders. . . . . The bedrock on which the boulders lie is . . . . as thoroughly disintegrated as the interior of the boulders. . . . . Every rain sets in motion down the slope trains of this fine débris. As the slope steepens locally by the removal of the fine débris, the boulders roll down to find new lodgment lower on the slope or at the base. In this movement many boulders already disintegrated within the outer crust are shattered to fragments. . . . . As the bedrock of the mountain slope disintegrates, and the rain washes away the fine fragments, protuberances of the bedrock are left that consist of the most compact rock between the most widely spaced joints. The protuberances are cut loose from the bedrock by the same processes that formed them, and a new crop of boulders comes into existence [pp. 85–86].

Clearly the contrast between the visibility of boulder-clad rock slopes in arid mountains and their invisibility in humid mountains is wholly due to the difference of their climates; and as clearly the visibly boulder-clad rock slopes of arid mountains are closely homologous with the invisibly boulder-clad rock slopes of humid mountains.

DIFFERENCES BETWEEN HUMID AND ARID MOUNTAIN SLOPES

A striking difference between humid and arid (granitic) mountains is found in the decreasing angle of slope of the first as the erosion cycle advances, in contrast to the constant angle of slope in the second; but this, like the invisibility and visibility of boulders on the slopes, is again wholly a consequence of a difference of processes due to a difference of climate. In a humid climate the steep and partly bare walls of a young valley, with scanty and coarse detritus lodged on its ledges, are succeeded by the less steep side-slopes of a mature valley, cloaked with abundant soil and overgrown with forest, the gentler declivity of the slopes being appropriate to the weaker processes of soil removal then in operation. As maturity is slowly followed by old age, the soil cloak is weathered to greater depth and to greater fineness at the surface, and the valley-side slopes are reduced by wash and creep to gentler declivity still.

In an arid climate the young stage of valley erosion, producing steep and bare rock walls if the streams are vigorous enough to erode such valleys, is soon followed by the mature stage of boulder-clad slopes if the rock weathers to boulders; but thereafter the declivity of the slopes remains unchanged, because in consequence of aridity they do not become soil-cloaked and plant-covered but remain persistently barren and boulder-clad until the widening of the valleys consumes them. As Bryan puts it, "By these slow but continuous processes [of weathering and washing on graded granitic slopes] the mountain front recedes, maintaining its angle of slope according to the spacing of joints and the granular structure of the granite" (p. 86). As the mountain mass is increasingly dissected and subdivided into island-like hills, "each of these hills retains the slopes characteristic of the rock and gradually decreases in size. . . . . Solitary . . . . hills retain the same steep slopes as the original mountains, but grow gradually smaller until the last remnants are masses of boulders or single rocks projecting above the general level" of a pediment (p. 96). Evidently, constancy of slope in this case and decrease of slope in the other are wholly determined by differences of climate.

Associated with the contrast between the persistence of a uniform slope in wasting arid mountains and the systematic decrease of slope in wasting humid mountains is the contrast between the abrupt, almost angular change from the slope to the piedmont pediment in (granitic) arid mountains and the curved change from slope to valley floor in humid mountains. But even in a humid mountain, the change from slope to valley floor is abrupt in an early stage of their erosion, when the valley sides are steep and the valley floors are narrow, because there is then a rather abrupt change from the coarse detritus of the slope to the finer detritus of the floor; it is only as time passes and the slope is decreased and its detritus is refined that the slope blends into the floor by a more or less curved descent.

RESIDUAL MOUNDS ON PENEPLAINS AND ISLAND-MOUNTS
ON PEDIMENTS

Although the residual mounds on humid peneplains and the island-mounts on arid pediments are strictly homologous, it is a mistake to treat them as essentially identical, as some German writers seem inclined to do. The first are pale forms of weakening convexity and lessening slopes which merge, without any sharp line of demarcation, into the surrounding lower surface, and which gradually fade away as time passes. The second are of vigorous form and are, especially in granitic areas, sharply separated from the surrounding rock floor; they retain their vigor, in spite of losing size, even to the moment of their extinction. It is therefore inappropriate, to say the least, to confuse the fading mounds on humid peneplains with vigorous island-mounts on arid pediments; for, according to Bornhardt's original definition of Inselberge, they are steep and rocky mounts which rise boldly from the surrounding rock floors like islands from the sea.

To be sure, if a humid peneplain is worn down beneath a land area on which volcanic action had previously taken place, its faintly undulating surface would be here and there sharply interrupted by dikes and stocks; but these special structures have little in common with the pale residual mounds on peneplains of uniform rock structure. It is, indeed, entirely unnecessary to assume that the fading mounds of humid peneplains or the bold island-mounts of arid pediments consist of more resistant rock than that which is smoothly worn down in the surrounding lower surface, as von Stapf seems inclined to do. And there is no more mystery in the survival of bold island-mounts over a widening rock floor than in the surviving of pale residual mounds over a humid peneplain; for inasmuch as the production of arid rock floors is, like that of humid peneplains, a slowly progressive and not an instantaneous process, the survival of residuals here and there at a late stage of the process before the ultimate stage is reached, is in both cases inevitable. The process begins with the incision of narrow valleys in irregularly branching arrangement; it goes on with the widening of the valley floors, in consequence of which the intervalley highlands are progressively and irregularly consumed. It is out of the question that all parts of such irregularly bounded highlands should be simultaneously destroyed; the survival of residuals here and there is an inherent element in their reduction to rock floors.

The two kinds of forms, island-mounts and residual mounds, stand at the opposite ends of an arid-humid series, which must include many intermediate forms; and although the successive members of such a series must grade into each other, the members at the opposite ends of the series are as unlike as early youth and advanced old age are at the opposite ends of the series of life-changes; as unlike as torrid and frigid climates are at the extremes of a series of intermediate climates.

HOMOLOGY OF HUMID VALLEY-FLOOR BASEMENTS AND ARID MOUNTAIN PEDIMENTS

Visibly sloping, piedmont pediments of bare rock, thinly veneered with cover-patches of rock chips and sand, are the most peculiar features of erosion in an arid climate; and they are commonly regarded as altogether different from anything that is to be found in a humid climate. Yet the faintly sloping bed-rock basements of valley floors in a humid climate, invisible because covered with slow-creeping soil and flood-plain deposits, are truly counterparts of the more strongly sloping, bare-rock pediments in an arid climate; just as truly as the invisibly boulder-clad slopes of humid mountains are counterparts of the visibly boulder-clad slopes of arid mountains. Both valley-floor basements and piedmont pediments grow most rapidly along the larger streams of their districts. Such is clearly the case in humid regions, and the same is true, according to Bryan, in arid regions:

The pediment grows most rapidly along the major streams. In every indentation in the mountain front and in places where streams emerge from the canyons onto the plains the rate of formation of the pediment is rapid, and consequently extensions of the pediment into the mountains are common. These extensions consist of branching valleys, many of which are two miles or more in width and reach far into the interior of the mountain mass [p. 97].

Both humid valley-floor basements and arid pediments are developed by the lateral extension of valley bottoms at the expense of the inclosing slopes; the extension of both is aided by the swinging of their streams, as is well known to be the case in humid regions and as Bryan certifies for his arid region. He states explicitly that mountain-slope recession and the associated widening of the pied-mont pediment are "aided by the erosive action of streams in lateral migration" (p. 89). "The pediment is greatly increased in extent by lateral migration of the streams at and below the mouths of canyons. The irregularities of the pediment are removed, the higher places being eroded and the lower filled with débris and protected" (p. 96).

Apart from the concealment of one and the visibility of the other, the chief difference between these two homologous features is the manifest ascending slope of the arid pediment in its extension upward from its alluvial embankment, and the almost imperceptible ascending slope of a humid valley floor from its terminal delta. In consequence of this, the residual mountains that rise from the center of a well-developed pediment seem to have a high-perched base that is reached by a long and smooth, rock-floor ascent from the sur-rounding aggraded basin-plain, while the residual mountains at the head of maturely widened, humid valley floors do not appear to have a high-perched setting.

SUBDIVISION OF HUMID AND OF ARID MOUNTAINS

Earlier paragraphs have described the manner in which humid erosion accomplishes the resolution of a single, uparched fault block into a maturely dissected mountain group, with isolated and sub-dued outliers between the widened floors of the outgoing valleys. A very similar subdivision of arid mountains takes place, but by reason of the persistent abruptness of their slopes, the isolated out-liers and monticules stand forth more conspicuously than do the subdued moundlike outliers of humid (granitic) mountains. Yet the essential similarity of the two cases becomes manifest when it is recognized that the subdivision of an originally single mass into its maturely separated parts is wholly the result of the deepening and widening of stream-eroded valleys, as they branch this way and that, extending their length by headward erosion in the most ordinary manner, and thus cutting the peripheral parts from the central mass.

Bryan gives various details of those progressive changes:

> After the stage of maturity is reached . . . . slope recession and the lateral cutting of the streams are the dominant processes. . . . . The residual masses [between opposing streams] are reduced to isolated island-like hills. . . . . Each

of these hills retains the slopes characteristic of the rock and gradually decreases in size through slope recession assisted by the erosive action of streams in lateral migration over the plain. . . . . The residual hills are strung out in lines between the original canyons [p. 89]. . . . . When the [intervalley] spurs become narrow they are cut through by slope recession on both sides, and hills are left standing as outliers on the pediment. These solitary hills are worn away with extreme slowness. . . . . The hills retain the same steep slopes as the original mountains [p. 96]. . . . . The normal pediment has a smoothly sloping surface . . . . broken only by scattered hills which rise abruptly from its surface . . . . many of which are prolongations of the intercanyon ridges of the mountains [p. 94]. . . . . The extension of the pediment may thus divide the original mountains into groups of detached hills separated by relatively broad surfaces cut on rock [p. 97].

THE SLOPES OF HUMID VALLEY-FLOORS AND OF ARID PEDIMENTS

In no respect are humid valley-floors and arid pediments more homologous than, as already noted, in the relation of their gradients to the transportation agencies working upon them. The gradients of humid valley-floors are of low value because their streams are supplied with water in abundance and with rock waste in moderation. The gradients of arid pediments are of high value because, although their water supply is occasionally abundant, the rock waste to be carried is great in quantity and coarse in texture. The variation of gradients on different parts of a humid valley floor have already been detailed; they are gentlest along the course of the flood plain of the largest streams, a little steeper along the flood plains of the minor streams, and steepest on the transverse slopes of the valley-floor strips from spur ends to flood-plain border. Precisely similar variations are found on arid pediments.

As Bryan states, the transportation agencies "cannot operate below a level determined by the grade necessary to transport detritus away from the mountains" (pp. 96–97). The slope of a pediment away from a mountain base is such that "débris is swept outward . . . . by small rills which originate through the concentration of rain wash at the base of the mountain slope." The rills "operate only during or immediately after a rain" (p. 96).

Transportation of relatively fine material by water is the essential factor in the formation of the pediment [p. 97]. . . . . In any one mountain range the slope of the pediment is steeper opposite the smaller canyons and very much flatter opposite the larger canyons. The parts opposite the inter-canyon portions of the mountain front are steeper than the parts opposite the canyons [p. 95]. . . . .

The grade of the rills and of the smaller streams is steeper than that of the larger streams. . . . . As a result of this relation of stream grades, the pediment, cut and molded by these streams, has a lower slope opposite the larger canyons than opposite the smaller canyons. Also, the parts of the pediment opposite the inter-canyon spur ends have a steeper slope away from the mountains

than the parts that lie along the stream courses (p. 96).

WEATHERING OF HUMID BASEMENTS AND OF ARID PEDIMENTS

The bed-rock basements of valley floors in humid regions and the piedmont pediments in arid regions are more or less completely covered with detritus that is intermittently on its way downstream. The detritus in humid flood plains is so deep and continuous that the underlying basement is seldom thought of. The same is true for the basement under the lateral strips of the valley floor, beyond the reach of the flood plain: they are always soil-covered, and the cover, although in slow progress toward the stream, is so persistently present that the basement is never naturally exposed. Moreover, all across a humid valley floor the water table lies near the surface and the surface is abundantly plant covered; there is nothing in sight to suggest that the concealed valley-floor basement is the homologue of the visible arid pediment; yet such it truly is.

In contrast with the soil-covered basement of humid valley floors, the rock pediments in arid regions are usually discontinuously covered with a thin veneer of detritus: yet the intervening and some-times smaller areas of completely bare rock may more strongly im-press the observer, especially if he comes from a humid region, by their strangeness, and he may thus be led to overlook the close resemblance of the visible arid pediment of the unfamiliar desert to the hidden rock basement of the humid valley floors with which he ought to be familiar.

Humid valley-floor basements as well as arid pediments are ordinarily weathered more or less below the detritus that covers them. This is manifest enough in arid pediments, where the exposed rock surface is very generally split, chipped, and crumbling. It is much less manifest in humid, valley-floor basements, because the bed rock there is so universally buried. But when the rock is reached in excavations, it is found to be more or less decomposed, especially under the valley-floor side-strips. Beneath a flood plain, firmer bedrock may be found where the stream has lately scoured the basement at time of flood.

FLOOD SCOURING OF BASEMENTS

Flood-scouring of basement bed rock beneath a flood plain may take place even when no rock in place is to be seen in the river bed at low water. Observations made on the Colorado River at Yuma under the United States Bureau of Reclamation show that at low water in September, 1921, the river surface stood at 113 feet, gage-height, with a breadth, partly occupied by a 200-foot sand bar, of 480 feet, and a maximum depth of from 6 to 8 feet in a mid-channel about 100 feet wide; the total discharge was then 1,200 cubic feet a second. These figures appeared to be typical of previous low-water seasons. During a flood in the previous June, the water surface rose to 131 feet, gage-height, with a breadth of 570 feet, a general depth of 60 feet, and a maximum depth of 66 feet in a channel about 300 feet wide; the total discharge then being 186,000 cubic feet a second, or 155 times the low-water discharge. The significant item here in connection with the scour of buried valley-floor basements is that, while the flood surface rose 18 feet, the flood bed deepened 41 feet, or nearly two and a half times the rise.

It may be pointed out that the deepening of a river channel during a flood involves the lifting of a large volume of detritus into the flood current, and that while thus lifted, the detritus is shifted downstream over a considerable distance before it is laid down again as the flood subsides. The flood-water lift and shift of detritus thus accomplished down the slope of a valley-floor basement in a humid region corresponds closely with the sheet-flood lift and shift of débris down the slope of a rock pediment in an arid region.

SIMILARITY OF HUMID FLOOD PLAINS AND ARID ALLUVIAL EMBANKMENTS

The aggradation of a flood plain, with its resulting extension farther and farther upstream on the rock basement of a degrading valley floor in the earlier stages of a cycle of humid erosion, as described in the foregoing, shows a certain similarity to, although hardly a homology with, the upbuilding of an alluvial embankment, with its headward extension over the suballuvial rock bench or pediment which widens as it is worn down, as explained by Lawson. The similarity is chiefly between the parallel aggradation of the flood plain, due to forward delta growth by reason of the early downwash of abundant detritus by active young rivers, on the one hand, and the parallel aggradation of the alluvial embankment, due to upward filling of its distal intermont basin by reason of the early inwash of abundant detritus by active young stream-floods, on the other hand.

Again, the past-mature degradation of an aggraded flood plain, caused, first, by reason of the lessened load coming from the subdued slopes of its headwater hills, and, second, by reason of the retrograding abrasion of its delta front, may be compared with the past-mature degradation of an arid alluvial embankment, caused, first, by reason of the lessened load coming from the lowered island-mounts at its head, and, second, by reason of the lowering of its intermont basin by wind exportation of dust; for such exportation becomes effective when an arid region is worn down and filled up to low relief in the later stages of its history, and when the detritus supplied to the basin is therefore smaller in quantity and finer in texture than in the earlier stages.

Finally, a true homology may be drawn between the slow degradation of a humid and soil-covered peneplain of hardly perceptible declivity after all its mountains and mounts are laid low and the slow degradation of an arid and thinly veneered pediment of very visible declivity after its retrogressive extension has consumed all its surmounting peaks and pinnacles. But these somewhat transcendental considerations need not be pursued farther for the present.

[*Editor's Note:* Material has been omitted at this point.]

Editor's Comments
on Papers 15 and 16

15 SHARP
Geomorphology of the Ruby-East Humboldt Range, Nevada

16 SCHUMM
Erosion on Miniature Pediments in Badlands National Monument, South Dakota

EVALUATION OF MODES OF ORIGIN

In Paper 15 Sharp makes a useful comparison between pediments working toward open drainage on one side of a mountain range and those working toward interior drainage on the opposite side. He indicates that the stream regime on the west mountain flank favors the operation of lateral planation, but not to the exclusion of wash between stream areas. The generally narrower pediments cut on the east flank granites are less smooth and have a thin sediment cover. Here the emphasis is thought to be on slope wash rather than stream wear.

Sharp's review of the pediment problem in general and his evaluation of the origin of the local pediments are most instructive. Space limitations restrict us to this limited extract.

In Paper 16 Schumm describes and analyzes the results of a simple field experiment, designed to show the nature and amount of scarp retreat and pediment regarding over a period of several years. In reading this short, but excellent, paper on pediment measurement, it should be remembered that, although the author revisited the scene at intervals to record the successive stake exposures, he apparently did not see the water in action. Thus, his conclusion that sheetwash is the probable agent would be an inference, and not a direct observation. In view of the fan-shaped alluvial veneer referred to by the author, which is well shown on Plate 1, Figures 2 and 4, one wonders whether this might mask a rock fan and that at least part of the erosion might be the result of stream work.

The other contribution bearing on the cycle of erosion is the series of profiles surveyed across segments of the pedimented area. These profiles show that (1) these pediments have concave profiles, (2) in less advanced stages, residuals have convex profiles, and (3) in later stages (at pediment passes) small convex summits occur at the head of the pediment.

15

Reprinted from *Bull. Geol. Soc. America*, **51**, 338–339, 361–369, 370–371
(Mar. 1940)

GEOMORPHOLOGY OF THE RUBY-EAST HUMBOLDT RANGE, NEVADA

BY ROBERT P. SHARP

ABSTRACT

A middle or late Pliocene open-valley stage in the Ruby-East Humboldt Range is recognized and described. The anomalous position of the drainage divide in places east of the range crest is explained in terms of the geomorphic evolution of the range. Pediments and terraces on the range flanks and in the adjoining basins are described, and those on the west side formed under a régime of through-flowing drainage are compared with those on the east side formed on the borders of hydrographically closed basins.

Seven surfaces, the two highest being pediments and the others partial pediments and terraces, are recognized on the west flank of the range. The dissection of these surfaces is shown to be related to successive rejuvenations of the Humboldt River drainage. The exposure and dissection of surfaces on the east flank of the range are attributed chiefly to relative uplift of the range by faulting. These pediments and terraces have been formed between the middle Pleistocene and Recent.

The origin of pediments under contrasting conditions is considered, and the conclusion is reached that different geologic, climatic, and topographic conditions impose a difference in the efficacy of processes of pedimentation. In this region lateral planation is dominant in areas of permanent streams and soft rocks; and rill wash, rain wash, and weathering are dominant in areas of hard rocks, ephemeral streams, and low mountain masses.

Lateral erosion by streams accounts for approximately 40 per cent of the retreat of the mountain front, and approximately 60 per cent is due to weathering and various types of wash.

INTRODUCTION

INTRODUCTORY STATEMENT

This paper is devoted chiefly to consideration of pediments and terraces flanking the Ruby-East Humboldt Range and in the adjoining basins. The geomorphic forms on the west side of the range, developed under a régime of exterior drainage, are compared with those on the east side, developed in an area of interior drainage. The geomorphological evolution of a progressively uplifted and tilted fault block, such as the Ruby-East Humboldt Range, would be of great interest, but, unfortunately, extended dissection of the mountain block makes this area unsuited for such studies. The glaciation of the range and some of the forms produced by block faulting have already been described (Sharp, 1938, 1939b) and will not be discussed further.

A total of 5 months during the summers of 1936 and 1937 were devoted to field work in the Ruby-East Humboldt region with particular emphasis on its Cenozoic history.

Sincere appreciation and acknowledgment are extended to Professors Kirk Bryan and Marland P. Billings, of Harvard University, for their interest, aid, and suggestions in the field and laboratory. Professor H. A. Meyerhoff, of Smith College, thoroughly criticised the manuscript and offered several excellent suggestions. Edward A. Schmitz aided in the preparation of illustrations. Profit has been derived from the interest and suggestions of Dr. Leland Horberg and Jean Todd Sharp.

[*Editor's Note:* Material has been omitted at this point.]

ORIGIN OF PEDIMENTS

The origin of pediments remains one of the foremost problems of geomorphology. Gilbert (1877, p. 126-132) is credited with the initial recognition and description of one type of that group of erosion surfaces which we now call pediments. Some geomorphologists believe that they are carved dominantly by lateral planation of streams. Gilbert (1877, p. 126-132) suggested this origin, and it is supported by Johnson (1931, 1932a, 1932b), Blackwelder (1931), and Field (1935). Paige (1912) was inclined toward this view, and Mackin (1937) is one of its latest and most enthusiastic adherents.

Bryan (1923, 1925, 1935-1936, 1936) maintains that in many areas lateral planation is dominant, particularly in the early stages of the cycle, but that it is considerably less effective as the cycle progresses and may be subordinate to weathering, rain wash, and rill wash in the later stages. Lawson (1915) would seemingly emphasize weathering and rill wash as more important than lateral planation. Davis (1930, 1933, 1936, 1938) favors the process of "backwearing" (weathering, rain wash, and rill wash) aided by "downwearing" (downcutting by streams) of the mountain block as the more significant factors in pedimentation, the relative efficacy of "backwearing" and "downwearing" being dependent upon the initial form and stage of the cycle. Gilluly (1937, p. 345-347; 1938, p. 123) states that rill wash and weathering have been more effective than lateral planation in forming the pediments around the Little Ajo Mountains of southwestern Arizona, an area in the mature stage of the cycle. Rich (1935) favors wasting (weathering and transportation) and sheet washing as the most essential factors in pedimentation.

McGee (1897) long ago suggested that pediments were the result of sheetfloods. This view has not found ardent supporters in recent years, although some workers, particularly Davis (1930, 1933, 1938) and perhaps Rich (1935), tend to think of sheetfloods or related phenomena as having a part in shaping and grading pediments after they have been formed by other means. Stheeman (1932, p. 1-17) suggests that the mountain or valley slopes waste away chiefly by weathering, and the detritus so derived is moved across the "arena" (pediment) by "flood sheets" and creep.

A review of the studies of pediments in widely separated regions brings the conviction that different processes of pedimentation are dominant under different climatic, topographic, and geologic conditions. The relations in the Ruby-East Humboldt region are such that pediments developed under somewhat different conditions can be compared, though the degree of contrast is not so great as could be desired. These differences may be outlined briefly.

The west flank of the range is included in an area of exterior drainage; the east flank drains to a series of hydrographically closed basins which contain aggrading playas. For this reason the west flank surfaces have been graded to a stable or progressively lowered base level, and the east flank surfaces to a slowly rising base level. Bryan (1936, p. 769-772; Bryan and McCann, 1936, p. 148-152) has given the most complete analysis of the effects of various types of base level on pedimentation, and reference should be made to those papers for details.

Most of the storms in this region come from the west, and, since the crest of the range is high enough to cause most of the precipitation to fall on the west slope, the area east of the crest lies in a rain shadow. In addition, the west slope streams have much larger drainage areas. For these reasons the ratio of permanent to ephemeral streams is greater on the west than on the east side, and the west slope streams are larger and more constant in flow.

The term pediment as now used (Johnson, 1931; Bryan, 1936, 1938; Mackin, 1937) includes surfaces like those west of the range, which are cut mostly on the soft deposits of the basin block and only in small part on the mountain block. The east flank pediments, with one exception, are cut entirely on the hard rocks of the mountain block. This difference is fundamental and accounts for some of the differences noted between the pediments on opposite sides of the range. It is due in large part to the structural relations and to the exterior drainage of the west flank as contrasted to the interior drainage of the east flank.

Remnants of pediments on the west flank extend as far as 5 miles from the foot of the mountain slope, and the undissected surfaces were probably even more extensive. The east flank pediments are much narrower, at least as exposed, and seldom extend more than 1½ miles from the mountains. The west flank pediments have gentler gradients and are more nearly smooth than those of the east side. The pediments of the west side are mantled by a comparatively uniform cover of stream gravel;

those of the east side have only local patches of stream gravel and are either bare rock or mantled by alluvium, in large part slope wash or detritus developed *in situ*.

It seems reasonable to conclude that these differences are attributable to differences in the efficacy of the several processes of pedimentation as determined by different geologic, climatic, and topographic conditions. The facts that terraces cut largely by lateral planation along adjacent streams coalesce to form partial pediments; that the largest and smoothest areas of pediment are adjacent to the largest streams; that the surfaces indent the mountain front along the streams and are restricted in areas of hard rocks suggest that the west flank pediments have been cut dominantly by lateral planation of streams. A cover of stream gravel on a surface is not proof that the surface was cut by lateral planation; it may merely show that a stream has migrated over the surface which actually may have been cut by other processes. Of the several features cited, the formation of partial pediments by coalescing terraces, obviously cut by lateral planation as shown by meander scars and related features, is considered the most convincing argument favoring the lateral planation theory. The detailed evidence favoring pedimentation by lateral planation of streams has been given by Mackin (1937, p. 826-828, 834, 877-878).

At the base of the mountain slope the west flank pediments lap over onto the hard rocks of the mountain block. In these places the cover of stream gravel is restricted to patches along the streams which cross the pediment, the gradients of the surfaces are steeper, the surfaces are less smooth, and the influence of differential rock hardness on weathering is apparent. The lateral planation of streams appears to have been less effective owing to increased rock hardness, and the work of weathering, rain wash, and rill wash on the interstream areas is proportionately more effective. These relations demonstrate what might be expected— that lateral planation is more effective on soft than on hard rocks. The areas between streams not attacked by lateral planation must be reduced wholly by local processes. Furthermore, if local processes are able to reduce hard rocks, their ability to reduce soft rocks cannot be dismissed. It seems likely that the broader the surface the greater the time and opportunity for processes other than lateral planation to act. Therefore, it seems logical to conclude that, even though lateral planation is dominant along large permanent streams where they flow on soft rocks, local processes such as weathering and rill wash are also effective.

The relatively narrower pediments on the east flank of the range may be grouped into three classes: (1) Pediments exposed and dissected by rejuvenation of streams either by faulting or by some other less common cause; (2) pediments which are covered by a veneer of alluvium and whose presence is inferred from isolated outcrops and exposure by local dissection; (3) bare-rock pediments which are slightly dissected and which show no evidence of ever having been covered by gravel or alluvium.

The first class includes the well-exposed pediment extending from Smithers Creek to Battle Creek along the east base of the central Ruby Mountains (Fig. 10) and the pediment at the north end of the East Humboldt Mountains along Willow and Clover creeks. These pediments are dissected to depths of 50 to 200 feet. The dissection in the first case is related to faulting and in the second to rejuvenation of streams flowing into the Humboldt River. The Ruby Mountains pediment is cut entirely on the hard rocks of the mountain block, and the East Humboldt pediment almost entirely on soft basin deposits. The Ruby Mountains pediment has an uneven surface upon which low residuals of resistant rock rise above the general level, and in many places it is a bare-rock surface. Elsewhere, it is covered by a thin veneer of slope wash or residual material, except along the major streams where it is mantled by stream gravel. The East Humboldt pediment is smoother and mantled by stream gravel.

Pediments formed chiefly by weathering and related processes compared to those formed chiefly by lateral planation are less smooth, show notable relations to the resistance of the underlying rocks, show only minor indentations of the mountain front along stream courses, and are largely bare or mantled by slope wash or residual detritus.

A comparison of the various features of the two surfaces just described suggests that lateral planation has been more effective on the surface cut on soft rocks despite the fact that the streams crossing the hard-rock pediment are greater in size and number. Thus, lithology appears to outweigh stream size in determining the relative efficacy of the processes of pedimentation in the particular examples considered.

Most of the piedmont slope east of the range as indicated by isolated outcrops and in a few places by local stream dissection is an alluvium-mantled surface bevelling the hard rocks of the mountain block. Following Lawson's analysis (1915), this surface is a pediment, graded to a rising base level, which has been left at the foot of the mountain front as it retreated more or less parallel to itself under the attack of weathering and rain and rill wash. The writer would join Bryan and others in adding lateral planation by streams to the above processes. This surface is undoubtedly modified by the scour and fill action of the streams which flow across it as they shift their courses. Furthermore, it is extended back into the mountain block along stream courses by the lateral erosion of these streams. However, the major work of the streams after they emerge from the mountain block appears to be largely that of transportation and only in minor part that of erosion.

As pointed out by Lawson (1915, p. 37) and Davis (1933, p. 223-224), the surfaces of the second group—alluvial veneered pediments—barring climatic or diastrophic change, will be mantled by alluvium until late in the erosion cycle when the amount of detritus supplied by the range is too small to cover completely the surface as it is formed. At that time the mountainward part of the surface will be left bare. According to Davis (1933, p. 219-222; 1936, p. 707), sheetflood robbing may uncover parts of the previously buried surface late in the cycle. However, the Ruby-East Humboldt Range is still so youthful that there is not the slightest evidence of sheetflood action along its flanks. In addition, the geologic and climatic conditions are not particularly favorable for sheet-floods. Pediments of this type can be stripped of their alluvial cover by rejuvenation owing to faulting or some other cause, as exemplified by the pediment of that nature already described.

The third class of east flank pediments is represented by a single example—the bare rock, slightly dissected pediment which encircles the south tip of the East Humboldt Mountains. In detail this surface appears relatively uneven and rough with an average relief of 20 to 40 feet. It is crossed by a number of shallow, narrow-floored, detritus-filled stream channels which form a dendritic pattern and which are separated by low rounded divides of bare rock. This pediment is concave in a section parallel to the mountain front, and this concavity is probably a reflection of the coalescent drainage (Gilluly, 1937, p. 335).

A series of processes somewhat like those outlined by Bryan (1935-1936) and Davis (1933, p. 215; 1936, p. 710; 1938, p. 1397) may be considered as having a part in forming this pediment. The streams are ephemeral and run only in times of flood. At present they are largely agents of transportation engaged in carrying detritus brought to them from the mountains and from the interstream areas by rill and rain wash. These streams are able to scour their beds to only a minor degree, owing to the great amount of detritus which they carry. Undoubtedly, at times of especially heavy floods these streams do scour their beds. However, the interstream divides, rounded and deeply disintegrated, are probably being worn down at a rate comparable to the scouring of the stream channels. This situation will prevail until the detritus supplied from the mountains and interstream divides is insufficient to prevent the streams from scouring their beds more deeply. A delicate balance must exist between the processes so that at all times the pediment has a relief fitted to the climatic conditions and rock types involved. The lack of meanders, undercut banks, and oversteepened slopes, and the narrow floors of the stream channels would seem to indicate that lateral stream corrasion is only a minor process, though the floods of water which pass down the channels must gnaw at their banks to some degree, like any moving mass of water seeking to obtain a wider channel. The concavity of the pediment transverse to the stream channels cannot be used as an argument against lateral planation, however, because lateral planation might take place along the individual channels which are part of a dendritic system and at the same time produce a concave pediment. In brief, the pediment at the south tip of the East Humboldt Mountains is thought to have been developed by weathering and transportation by rain wash, rill wash, and streams aided by some lateral and downward erosion of the streams.

We may ask with Gilluly (1937, p. 343) if pediments of this type may not have been "born dissected". The field evidence in this particular case is not convincing, for rejuvenation of the streams cannot be dismissed with certainty; but the conception that at least some pediments have always had a low relief is worthy of consideration.

This surface has been developed on a low subdued part of the range where the streams are ephemeral and the detritus supplied by the mountains does not form an alluvial cover for the whole surface. The pediment is probably formed not only by the retreat of the mountain front but also by "downwearing" of the mountain block under the attack of weathering and corrasion by the transporting agents. It differs from the pediments heretofore described in this article in its form and mode of origin and resembles those described by Davis (1938) on broadly upwarped or slowly uplifted masses which have low initial relief. Thus, in an area of relatively hard rocks, ephemeral streams, and low relief, the processes of weathering and transportation by rill wash, rain wash, and streams, with only minor lateral erosion appear to dominate. It is possible that as a mountain mass is reduced by erosion its pediments arrive at a stage where they resemble the surface just described. These pediments could be dissected, so to speak, without a diastrophic or climatic change, as already noted by Gilluly (1937, p. 343) and Rich (1935, p. 1023). In a sense a climatic change does occur as the mountain mass is reduced, for, in the Great Basin, streams once permanent may become ephemeral if the range is greatly lowered by erosion. From the foregoing discussion it appears that the form and nature of pediments and the processes producing them depend in part upon the stage of the erosion cycle. The east face of the range is considerably higher and steeper than the west slope. This may explain why the east flank pediments, although relatively narrow, have extended somewhat farther back onto the mountain block than the west flank surfaces.

In the field the work of rills and small streamlets at the base of the mountain front in the interstream areas is impressive. The larger streams issuing from the mountain block do cut laterally to some degree, but the arc over which they swing appears limited on the average to 60

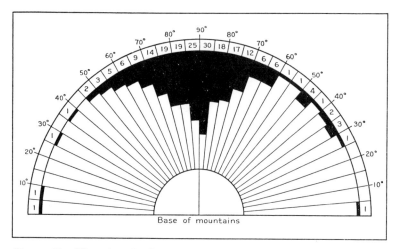

FIGURE 12.—*Plot of angles between stream courses and base of mountains*
Number of streams in each 5-degree interval is recorded just inside the outermost circle.

or 70 degrees. In Figure 12 the angle that stream courses make with the base of the mountain block has been plotted for over 200 major stream courses of the Ruby-East Humboldt Range. The base of the diagram represents the base of the mountain block. Over 90 per cent of the streams make angles of 50 degrees or more with the mountain front. There seems no reason why the density distribution of streams would have been much different in the past. Assuming that the plot (Fig. 12) represents the life history of the average stream, an assumption supported by field observation, lateral planation would be dominant in approximately 40 per cent of the area of Figure 12, and in the remaining 60 per cent other processes would be dominant, although in neither case would any of the processes be wholly excluded. A mountain mass can be cut back by adjacent streams which swing over arcs of 60 to 70 degrees, but the base of the mountains would have a jagged "sawtooth" trace which is not observed in the field. The points of interstream ridges and spurs are cut back dominantly by the attack of weathering and various types of wash including rain and rill wash, so that the range front maintains an essentially linear trace with minor indentations along the streams. Figure 12 shows that only a few of the spurs are trimmed back by lateral stream erosion parallel to the mountain front. Furthermore, the small gullies on the interstream areas are shallow and narrow bottomed and show no evidence of lateral erosion. No features in these interstream areas resemble the miniature rock fans described by Johnson (1932a, p. 399; 1932c, p. 546), probably because the geologic and climatic conditions in this region are not comparable to those of the areas which he describes.

The data for Figure 12 were taken from both flanks of the range and represent predominantly permanent streams in areas of hard rock. The analysis shows that the retreat of the faces of this particular range is due approximately 40 per cent to lateral erosion by major streams and approximately 60 per cent to weathering and the various types of wash.

The effects of past climatic conditions different from those of the present cannot be evaluated in this region. The late Cenozoic in the Great Basin has been a time of generally increasing aridity. (*See* references in Sharp, 1939a.) However, in at least part of the Pleistocene more water was flowing from the range, and the general climatic conditions may have been more humid. As shown later, the pediments of this region date from the latter half of the Pleistocene; hence, climatic conditions at the time they were developed may have been more humid, and lateral planation may have been more effective, but the relic features of the surfaces give no indication of this.

The conclusions reached as to the formation of pediments under the various geological, topographic, and climatic conditions of the Ruby-East Humboldt region are as follows:

(1) Pediments are formed by lateral planation, weathering, rill wash, and rain wash. The relative efficacy of these various processes is different under different geologic, topographic, and climatic conditions.

(2) Lateral planation is most effective along large permanent streams and in areas of soft rocks.

(3) Weathering, rill wash, and rain wash are most effective in areas of ephemeral streams, hard rocks, and a low mountain mass.

(4) All variations from pediments cut entirely by lateral planation to those formed entirely by the other processes are theoretically possible, although in this area the end members of the series were not observed and perhaps do not actually exist in nature.

The above outline contains nothing startlingly new or different from suggestions advanced heretofore, but an attempt has been made to evaluate the various processes of pedimentation. As emphasized particularly by Bryan (1936, p. 775; Bryan and McCann, 1936, p. 149), the efficacy of the processes must change as the cycle progresses, for newly uplifted mountains of considerable height are likely to have permanent streams powerful enough to erode laterally once they reach grade. The same mountains in a later stage of the cycle when reduced to lower height may have ephemeral streams and conditions relatively more favorable for the other processes of pedimentation.

[Editor's Note: Material has been omitted at this point.]

WORKS TO WHICH REFERENCE IS MADE

Blackwelder, Eliot (1931) *Desert plains,* Jour. Geol., vol. 39, p. 133-140.

Bryan, Kirk (1923) *Erosion and sedimentation in the Papago country, Arizona,* U. S. Geol. Survey, Bull. 730, p. 19-90.

―――― (1925) *The Papago country, Arizona,* U. S. Geol. Survey, W. S. Paper 499, p. 1-436.

―――― (1935-1936) *Processes of formation of pediments at Granite Gap, New Mexico,* Zeitschr. für Geomorphol., Band 9, Heft 4, p. 125-135.

―――― (1936) *The formation of pediments,* 16th Intern. Geol. Cong., Rept. (1933), vol. 2, p. 765-775.

――――, and McCann, F. T. (1936) *Successive pediments and terraces of the upper Rio Puerco in New Mexico,* Jour. Geol., vol. 44, no. 2, p. 145-172.

――――, ―――― (1938) *The Ceja del Rio Puerco; a border feature of the Basin Range province in New Mexico, part II, Geomorphology,* Jour. Geol., vol. 46, p. 1-16.

Davis, W. M. (1930) *Rock floors in arid and humid climates,* Jour. Geol., vol. 38, p. 1-27, 136-158.

―――― (1933) *Granitic domes of the Mojave Desert, California,* San Diego Soc. Nat. Hist., Tr., vol. 7, no. 3, p. 211-258.

―――― (1936) *Geomorphology of mountainous deserts,* 16th Intern. Geol. Cong., Rept. (1933), vol. 2, p. 703-714.

―――― (1938) *Sheetfloods and streamfloods,* Geol. Soc. Am., Bull., vol. 49, no. 9, p. 1337-1416.

Fassig, O. L. (1932) *Climatic summary of the United States,* U. S. Dept. Agric., Weather Bureau, Sec. 19, p. 1-33.

Field, Ross (1935) *Stream carved slopes and plains in desert mountains*, Am. Jour. Sci., 5th ser., vol. 29, p. 313-322.

Gilbert, G. K. (1877) *Report on the geology of the Henry Mountains*, U. S. Geog. Geol. Survey Rocky Mt. Region, p. 1-160.

Gilluly, James (1937) *Physiography of the Ajo region, Arizona*, Geol. Soc. Am., Bull., vol. 48, p. 323-348.

——— (1938) *Physiography of the Ajo region, Arizona* [Abstract], Geol. Soc. Am., Pr. 1937, p. 122.

Johnson, Douglas (1931) *Planes of lateral corrasion*, Science, n. s., vol. 73, p. 174-177.

——— (1932a) *Rock fans of arid regions*, Am. Jour. Sci., 5th ser., vol. 23, p. 389-416.

——— (1932b) *Rock planes of arid regions*, Geog. Rev., vol. 22, p. 656-665.

——— (1932c) *Miniature rock fans and pediments*, Science, n. s., vol. 76, p. 546.

Lawson, A. C. (1915) *The epigene profiles of the desert*, Univ. Calif. Publ., Bull., Dept. Geol., vol. 9, no. 3, p. 23-48.

Mackin, J. H. (1937) *Erosional history of the Big Horn Basin, Wyoming*, Geol. Soc. Am., Bull., vol. 48, p. 813-894.

McGee, W J (1897) *Sheetflood erosion*, Geol. Soc. Am., Bull., vol. 8, p. 87-112.

Paige, Sidney (1912) *Rock-cut surfaces in the desert ranges*, Jour. Geol., vol. 20, p. 442-450.

Price, W. A. (1933) *Reynosa problem of south Texas, and origin of caliche*, Am. Assoc. Petrol. Geol., Bull., vol. 17, no. 5, p. 488-522.

Rich, J. L. (1935) *Origin and evolution of rock fans and pediments*, Geol. Soc. Am., Bull., vol. 46, p. 999-1024.

Sharp, R. P. (1938) *Pleistocene glaciation in the Ruby-East Humboldt Range, northeastern Nevada*, Jour. Geomorphol., vol. 1, no. 4, p. 296-323.

——— (1939a) *The Miocene Humboldt formation in northeastern Nevada*. Jour. Geol., vol. 47, p. 133-160.

——— (1939b) *Basin-Range structure of the Ruby-East Humboldt Range, northeastern Nevada*, Geol. Soc. Am., Bull., vol. 50, p. 881-920.

Stheeman, H. A. (1932) *The geology of southwestern Uganda*, M. Nijhoff, The Hague, p. 1-144.

Woolnough, W. G. (1928) *Origin of white clays and bauxite and chemical criteria of peneplanation*, Econ. Geol., vol. 23, p. 887-894.

UNIVERSITY OF ILLINOIS, URBANA, ILLINOIS.
MANUSCRIPT RECEIVED BY THE SECRETARY OF THE SOCIETY, MARCH 27, 1939.

16

Reprinted from *Bull. Geol. Soc. America*, **73**, 719–724 (June 1962)

Erosion on Miniature Pediments in
Badlands National Monument, South Dakota

S. A. Schumm

Abstract: Measurements of erosion on miniature pediments in Badlands National Monument, South Dakota, show that during almost eight years, the pediments were lowered by sheetwash. The adjacent hillslopes have retreated leaving a belt of newly formed pediment from 6 to 12 cm wide at their bases.

The hillslopes are rough and relatively permeable; the miniature pediments are smooth and less permeable. Calculations based on the Manning equation suggest that the velocity of overland flow on the pediments may be of the same magnitude as that on the hillslopes. The decrease in roughness from hillslope to pediment compensates for the decrease in slope angle. The pediments are swept free of debris and are regraded by a more effective utilization of runoff energy.

CONTENTS

INTRODUCTION

The literature concerning pediments is voluminous. No attempt is made here to review this literature; however, the reader can consult Tator (1952), who summarized observations and conclusions from 43 of the most significant papers on this topic.

Miniature pediments have been recognized and described in several areas (Bryan, 1936; Bradley, 1940; Higgins, 1953; Schumm, 1956a; Smith, 1958). These authors describe the miniature pediments as having formed at the base of a retreating hillslope, after which the pediment is modified by sheet–and rill–wash. Johnson (1932) stressed the importance of lateral planation by streams in the formation of rock fans and miniature pediments, but neither previous studies nor present observations support this hypothesis.

As part of a general study of the development of drainage systems and slopes in badlands (Schumm, 1956b), segments of 3/8–inch reinforcing rods, 46 cm long, were driven into slopes and miniature pediments in Badlands National Monument, South Dakota, during July 1953. The stakes were driven flush with the ground surface, and their progressive exposure was measured in October 1954 and August 1955. The measurements revealed that the pediments were being eroded and that the badland hillslopes were retreating, thereby enlarging the miniature pediment (Schumm, 1956a). Since 1955 the writer has remeasured the continued exposure of the stakes twice—in June 1959 and May 1961. The additional measurements confirm the author's previous conclusions on slope retreat (1956a), but they allow an expansion of the comments on miniature pediment formation.

ACKNOWLEDGMENTS

The writer wishes to acknowledge the assistance of G. C. Lusby and H. F. Matthai in the selection of *n* values and the suggestions for improvement of the manuscript made by G. E. Harbeck, Jr., and R. W. Lichty, all of the U. S. Geological Survey.

DESCRIPTION

The badlands of Badlands National Monument are formed by erosion of the flat-lying clays and silts of the White River Group, Brule and Chadron Formations, of Oligocene age. The pediments studied were cut on the Chadron Formation.

The miniature pediments appear concave (Pl. 1, fig. 1) and are graded to shallow drainage channels. Except where obviously dissected by rills, the pediment surfaces are very smooth (Pl. 1, fig. 2). This smoothness is probably enhanced by a thin layer of alluvium which covers all but the uppermost portion of the pediment. Where measured, this deposit increases in thickness away from the head of the pediment from about 3 to 15 mm. At distances varying from 3 to 16 meters from the hillslope-pediment junction, vegetation in the form of grasses and forbs appears on the pediment surface. The presence of the alluvium and vegetation suggests that the pediments either are being aggraded or are stable slopes of transportation.

The miniature pediments range in slope from a minimum of 0.5 to a maximum of 34 per cent (Figs. 1, 2). The average of the minimum slopes is 2.5 per cent, and the average of the maximum slopes at the junction of hillslope and pediment is 13 per cent. A short distance below the junction the average slope decreases to 8.5 per cent.

On the hillslopes the Chadron Formation weathers to form a highly permeable surface of relatively loose aggregates (Pl. 1, figs. 2, 4). This material creeps downslope causing the slope angles of the Chadron hillslopes to decline as they retreat (Schumm, 1956a). During the spring much of the runoff from these slopes flows beneath the layer of aggregates to reappear on the pediment at the base of the slope (Pl. 1, fig. 2; Schumm 1956b, Pl. 6, fig. 2). However, the proportion of overland flow increases during the summer as the slopes are compacted by rainbeat (Pl. 1, fig. 3) and become less permeable.

Only a few rills, which deliver small streams of water to the pediment surface, are present on the badland slopes; the streams of water are dispersed and move over the pediment as sheetwash (Pl. 1, figs. 2, 4). Where well defined rill channels exist on the miniature pediments, they are obviously being dissected because of a change of baselevel.

Total rainfall during the period of study,

seven years and ten months, was 2.67 meters as measured at Badlands National Monument headquarters (J. W. Jay, written communication). Rainfall amounts are not related to erosion rates for the four periods of measurement, probably because the study are was about 18.5 km from the weather station.

EROSION

The erosion and deposition which occurred on miniature pediments during the period of study was measured by the exposure or burial of 15 stakes. Only two of the stakes were buried; the remaining 13 stakes were exposed by erosion from 10 to 75 mm.

On Figure 1 the total amount of erosion during the period is given above the position of each stake on the longitudinal profiles of the pediments. Erosion is greatest on the badland slopes; on the pediment, erosion decreases away from the junction of pediment and badland slope. In each case the badland slope has retreated 6–12 cm, leaving behind an extension of the pediment which is steeper than the older parts of the pediment. Initially the uppermost stakes on the pediments were emplaced at the junction of badland slope and pediment; now they are located on the pediment surface proper (Pl. 1, figs. 3, 4). The data show that the weathering and eroding back of the hillslope leave at its foot a strip of newly formed pediment, which is relatively steep, having an inclination of between 15–34 per cent.

Perhaps of most interest is the erosion at the junction of hillslope and pediment. Except for profile 4 (Fig. 1), the erosion of the pediment causes a slight lengthening of the hillslope at its base. The sudden transition from steep hillslope to gentle pediment might be expected to cause deposition rather than erosion at this point. The sediment, moved downward by creep and by overland flow, does not accumulate at the slope base as might be expected, for the runoff keeps the base of the slope clear of sediment.

One factor that determines the erosive power of overland flow at the slope base is the abrupt transition from the rough hillslope, down which runoff moves as subdivided and surge flow (Horton, 1945, p. 312–313; Schumm, 1956b, p. 628–629), checked and deflected by irregularities of the slope, to the relatively smooth pediment surface (Pl. 1), which offers few obstructions to flow. It is not difficult to understand why sediment does not accumulate

Figure 1. Pediment pass between two badland residuals. Note sharp break between hillslope and pediment and concavity and smoothness of pediment. Dissection of miniature pediment is in progress at the upper left, owing to degradation of stream to which pediment is graded.

Figure 2. Flow traces on miniature pediment. Discharge from small rills, much of which flows beneath surface, spreads as sheetwash across pediment. Note change in spacing of cracks where pediment is covered by veneer of alluvium.

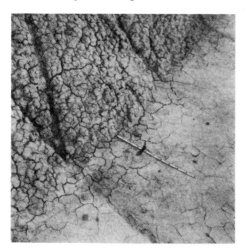

Figure 3. Position of stake (Fig. 1, profile 3) in June 1959. Stake was placed initially at junction of hillslope and pediment. Comparable position of stake in this photograph would be at upper end of scale.

Figure 4. Position of stake (Fig. 1, profile 4) in May 1961. Contrast condition of slope and pediment with Figure 3 of Plate 1. Frost action during winter months loosens surface of hillslope, and some hillslope materials roll downslope onto pediment. In Figure 3 of Plate 1, the hillslope surface has been compacted and partially sealed by raindrop impact, and sheetwash has removed slope debris from pediment.

MINIATURE PEDIMENTS IN BADLANDS NATIONAL MONUMENT, SOUTH DAKOTA
SCHUMM, PLATE 1

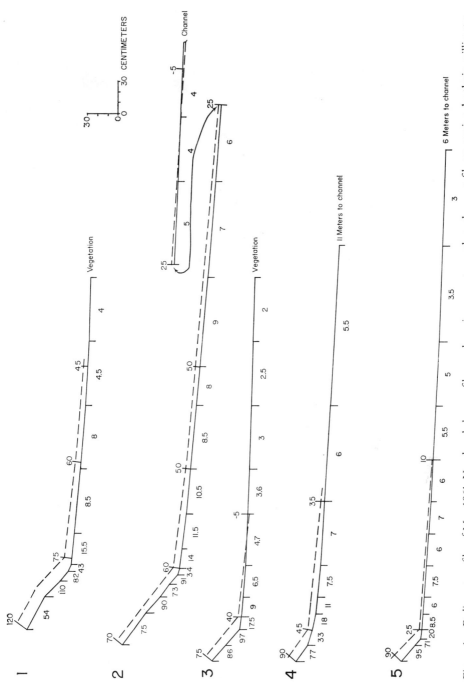

Figure 1. Pediment profiles of May 1961. Numbers below profiles are slopes in per cent; numbers above profile are erosion depths in millimeters at position of stakes. Dashed lines above profiles indicate inferred position of profiles in July 1953.

on the relatively flat miniature pediment and why the pediment is regraded, when a comparison of velocities of flow is made for the hillslope and pediment. Manning's equation for velocity of open-channel flow can be adapted to the overland flow situation by letting the hydraulic radius equal mean depth of flow (Linsley, Kohler, Paulhus, 1949, p. 272). The Manning equation is

$$V = \frac{1.49}{n} D^{\frac{2}{3}} S^{\frac{1}{2}}, \qquad (1)$$

where V is the mean velocity in feet per second, n is a roughness factor, D is the mean depth, and S is slope. Assuming that the depth of a sheet of water moving over the hillslope is the same as that of a sheet moving over the pediment, we can for purposes of comparison drop D from the equation. A relative velocity then would equal:

$$Vr = \frac{1.49}{n} S^{\frac{1}{2}}. \qquad (2)$$

The assumption that the average hillslope has an inclination of 0.8 and that the average miniature pediment slope is 0.1 is reasonable (Fig. 1). The selection of values for n is much more difficult. The materials forming the slope and the pediment have essentially the same grain size, but surface irregularities are probably the controlling factor. Chow (1959, p. 109) has compiled data on the magnitude of n in open channels for type of material, degree of irregularity, effect of obstructions, and other factors. Utilizing Chow's tables, a value of 0.02 was selected as a representative n value for the pediment (earth, smooth with negligible obstructions); 0.06 was selected as a representative n value for the hillslope (earth, moderate irregularity and appreciable obstructions). These values can be used in equation (2) to obtain the relative magnitudes of the velocity of overland flow on the pediment and hillslope. The relative velocity on the pediment is about 24 and on the hillslope about 23. Therefore, water movement may occur on the smooth pediments at velocities of the same magnitude as those on the steep but rough hillslope. The decrease in roughness apparently compensates for the decrease in slope angle.

It was assumed for simplification in the above analysis that mean depths of flow were the same on hillslope and pediment. In fact, the mean depth of flow should be greater on the pediment because of the downslope increase in flow. This greater depth of runoff on the pediment will increase the velocity of flow over the pediment surface relative to that on the hillslope (equation 1).

The veneer of alluvium over the pediment is probably a deposit left by the last traces of runoff. The alluvium on the upper part of the pediment is probably removed during runoff only to be replaced by renewed deposition at the close of the runoff episode. In any event, the suggestion that overland flow may move over the steep hillslope and gentle pediment at about the same velocity explains the removal of sediment and the erosion of miniature pediments by the more efficient utilization of runoff energy on the pediment.

EROSION CYCLE

The progressive headward extension and coalescence of miniature pediments is illustrated in Figure 2 by a series of profiles surveyed across a small badland residual and a pediment pass (Pl. 1, fig. 1). The pediments are concave during their headward extension. The only major convexity on the lower part of a pediment is on the right side of profile 3, which occurs at a belt of dead vegetation apparently hindering erosion of the pediment above that point. When the badland residual disappears, the broadly concave pediments coalesce; in each case, a convexity has developed at the junction. This might have been expected, for any sharp junction of the miniature pediment surfaces would probably be reduced by rain-drop impact. The upper convexity, however, is only a small part of the total profile, which is in accordance with Sharp's (1957) observations at the classic locality of Cima Dome.

CONCLUSIONS

Miniature pediments are formed at the base of retreating badland slopes, as there is no accumulation of slope debris there. Sediment from the hillslope is swept away from the slope base, and the pediments are degraded, because the relatively smooth pediments afford little hindrance to water movement. Computations based on an extension of the Manning equation to overland flow suggest that the velocity of runoff on the smooth pediments may be of the same magnitude as that on the relatively rough hillslopes. Therefore, as the pediments are formed by retreat of the hillslope, they are lowered by sheetwash erosion. The 15–20–per cent slope at the junction of hillslope and pedi-

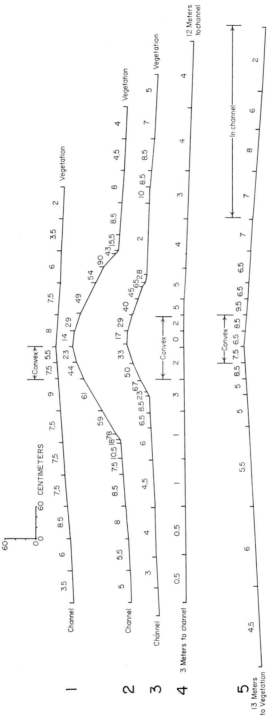

Figure 2.   Profiles illustrating headward extension and coalescence of miniature pediments. Profiles 2, 3 surveyed across a small residual; profiles 1, 4 surveyed across coalesced pediments at both ends of this residual; profile 5 surveyed across the pediment pass (Pl. 1, fig. 1). Numbers above profiles give slope in per cent.

ment is reduced to as little as 2 per cent at a distance from the junction.

It is unlikely that the formation of these miniature pediments is analogous to pediment formation in general; nevertheless, these limited observations and those of King (1953, p. 733–734) may stimulate additional work designed to describe the hydraulic characteristics of these landforms.

REFERENCES CITED

Bradley, W. H., 1940, Pediments and pedestals in miniature: Jour. Geomorphology, v. 3, p. 244–255

Bryan, Kirk, 1936, Processes of formation of pediments at Granite Gap, New Mexico: Zeitschr. Geomorphologie, v. 9, p. 125–135

Chow, Ven Te, 1959, Open channel hydraulics: New York, McGraw-Hill, 680 p.

Higgins, C. G., 1953, Miniature pediments near Calistoga, California: Jour. Geol., v. 61, p. 461–465

Horton, R. E., 1945, Erosional development of streams and their drainage basins; Hydrophysical approach to quantitative morphology: Geol. Soc. America Bull., v. 56, p. 275–370

Johnson, D. W., 1932, Miniature rock fans and pediments (Abstract): Science, v. 76, p. 546

King, L. C., 1953, Canons of landscape evolution: Geol. Soc. America Bull., v. 64, p. 721–752

Linsley, R. K., Kohler, M. A., and Paulhus, J. L., 1949, Applied hydrology: New York, McGraw-Hill, 689 p.

Schumm, S. A., 1956a, The role of creep and rainwash on the retreat of badland slopes: Am. Jour. Sci., v. 254, p. 693–706

—— 1956b, Evolution of drainage systems and slopes in badlands at Perth Amboy, New Jersey: Geol. Soc. America Bull., v. 67, p. 597–646

Sharp, R. P., 1957, Geomorphology of Cima Dome, Mojave Desert, California: Geol. Soc. America Bull., v. 68, p. 273–290

Smith, K. G., 1958, Erosional processes and land forms in Badlands National Monument, South Dakota: Geol. Soc. America Bull., v. 69, p. 975–1008

Tator, B. A., 1952, Pediment characteristics and terminology: Assoc. Am. Geographers Annals, v. 42, p. 295–317

MANUSCRIPT RECEIVED BY THE SECRETARY OF THE SOCIETY, OCTOBER 1, 1961
PUBLICATION AUTHORIZED BY THE DIRECTOR, U. S. GEOLOGICAL SURVEY

Editor's Comments
on Papers 17 Through 20

Paper 17, by Howard, introduces the idea that many planation surfaces may have been produced under arid conditions and that they show pediments rather than "peneplains."* In present-day terminology it would be better usage to compare peneplains with pediplains rather than with pediments wherever the latter are of regional extent.

The great exponent of pediplains on a worldwide basis is Lester C. King, from the University of Natal, South Africa. King sees the pediplain as the widespread universal planation surface of the plainlands of the earth, at the same time denying even the existence of the peneplain.

Parts of King's "Canons of landscape evolution" are reprinted in Paper 18. Here King totally rejects the Davis concept of a peneplain, and indicates why he considers pediment and pediplain to be the fundamental planation surface elements of the

*Students of Douglas Johnson may prefer "peneplane," rather than "peneplain." Johnson saw the logic of considering an almost level surface made by erosion as being "planed" as distinguished from a filled flat surface or "plain." The original Davis spelling has been almost universally adopted mainly on the basis of priority. It is a rather wry commentary that many workers retain the Davis spelling but would like to throw out the idea itself.
In defense of Johnson's suggested spelling, we note that Mescherikov (1968), a Russian worker, felt the need of the word "sediplain" to denote a sedimented surface, as distinct from a "pediplain" or "peneplain."

earth. Since not all King's ideas expressed here are universally accepted, it will be well to remember that the 50 canons cited are *"opinions derived* from the analysis here presented," as stated in the abstract (emphasis added). A canon is a rule or law issued by an authority. To this editor it would have been of greater value if these items had been presented as possible stipulations or areas of general agreement. In any case, King must be applauded for at least inferentially pointing out in his "canons" that there must be general agreement on certain aspects of a problem before firm conclusions can be established. The reader may wish to consider which of the canons are agreeable to him. Substantial dissent could cast doubt on the author's general conclusions.

King makes frequent reference to Davis's 1930 paper (Paper 14) and agrees with the homologues indicated there between rock plains in humid and arid regions. In fact, King sees Davis as on the right track at last, but getting lost in his 1932 paper, "Piedmont Benchlands and Primärrumpfe (not reproduced). The reason cited by King is apparently not related to the homologues themselves but to interpretations concerning Walther Penck's *Primärrumpf,* which according to Davis may express itself in a variety of ways. The shift of emphasis by King from intracyclic homologues to what amounts to intercyclic matters, such as the shape of a surface following uplift, is rather puzzling.

King, in rejecting the idea of a "normal" cycle of erosion together with its peneplain, implies a certain parochialism on the part of Davis. It is curious, however, that King sets up what amounts to a "normal" four-component hillslope, which may "degenerate" (canon 9) into a rounded two-component slope. Again, we may ask, what is normal? The four-element slope appears to depend on a structural cornice rock. If true, this would contradict that part of "canon 1" which indicates that structure is the least important member of the geomorphic trilogy. Perhaps the terms "normal" and "degenerate" should be excised from usage.

In Paper 19 Holmes makes a strong effort to reconcile the views of Davis and King (and incidentally Penck) by discussing two types of slopes: a "wear" or gravity slope, which acts as a source of sediment, and a "wash" slope, which serves as a transporting slope for the sediment produced on the wear slope. These terms had already been used by H. A. Meyerhoff (1940).

In the later stages of the cycle, the wash slopes would seem to dominate the landscape. In a humid region, with a more complete drainage net, many wash slopes should give the undulatory character associated with peneplains. The fact that Davis recognized a peneplain in the uplands of eastern Montana in 1883 (Davis, Paper 1), a semiarid region, as consisting of long wash slopes with some lava-capped gravity slopes indicates that the word "peneplain" finds broad application in both semiarid and humid regions.

The reader will be interested in Holmes's suggestion that "peneplain" be retained for all low-relief surfaces of planation, except in those special cases where the surface is demonstrated to consist of coalescing pediments, or a pediplain. This is the opposite view of L. C. King, who believes that Davis saw pediplains but misidentified them as peneplains.

Paper 20, by Bigarella and De Andrade, should be read with the author's closing statement in mind: "Our approach to the problem is new, our conclusions are preliminary, and much more needs to be done . . . ."

The novelty of the Brazilian situation revolves around three points made in the paper: (1) pediments and pediplains do in fact exist in what are now Brazilian forest areas, (2) stratigraphic correlation of scattered continental sediments may be possible through the relation of sediment to pediment, and (3) the morphogenesis requires climatic change in post-Cretaceous time.

Parenthetically, we note that not too long ago the areas described here as pediments adjacent to and sloping toward the ocean might have been described as marine terraces (Paper 4).

Elsewhere in South America, French Guiana, for example, planated areas on granite have developed laterites modified to resemble karst landscapes in limestone. In 1957, Boris Choubert's "Essay on the Morphology of Guiana" indicates that, in places where the laterite cap is permeable, water sinks to the impermeable residual clay and flows laterally, picking up clay particles as it moves toward nearby valleys. In this way caves several meters high may be developed in the clay. Collapse of cavern roofs produces surface furrows or more equidimensional depressions that somewhat resemble the dolines of a karst area. Stunted vegetation on the unbroken laterite crust is taken as an indication of the presence of bauxite.

REFERENCES

Choubert, B., 1957. Essai de la morphologie de la Guyanne. Memoir explaining the detailed geologic map of France, published by the French Government.

Davis, W. M., 1932. Piedmont benchlands and Primärrumpfe. Bull. Geol. Soc. America, 43, p. 399–440.

Mescherikov, Y. E., 1968. Plains. *Encyclopedia of Geomorphology* (R. H. Fairbridge, ed.), Van Nostrand Reinhold Co., New York, p. 850–855.

Meyerhoff, H. A., 1940. Migration of erosion surfaces. Assoc. Amer. Geogr., Ann., 30, p. 247–254.

17

Reprinted from *J. Geomorph.*, **4**(2), 138–141 (1941)

Rocky Mountain Peneplanes or Pediments

ARTHUR DAVID HOWARD

New York University

Two conspicuous upland surfaces characterize many of the ranges of the Rocky Mountains. They are referred to herein as the summit and subsummit surfaces, rather than as the Flattop and Rocky Mountain peneplanes, names which were applied to them in the Front Range of Colorado.[1] The summit surface is separated from the more extensive eastward-sloping subsummit surface by a ragged east-facing scarp locally more than two thousand feet high. Monadnocks rise above both surfaces. Van Tuyl and Lovering[2] have recently described additional upland erosion levels in the Colorado Front Range and interpret all the high surfaces as peneplanes. Rich[3] believes, however, that at least some of the additional levels may be due either to recent stripping of a sedimentary cover or to differential erosion, and hence may have no cyclical significance. The presence of additional upland levels does not affect the suggestions herein made.

On the assumption that the surfaces are true peneplanes, great uplifts of the Rocky Mountains have been postulated. Keith,[4] as a result of his Appalachian studies, suggested the possibility that peneplanes may form simultaneously at different levels without uplift. Davis[5] applied Keith's suggestions to the Blue Ridge and proposed that the high, westward-sloping upland surface was formed at the same time as the lower Piedmont surface. He attributed the greater height of the former to the longer distance to baselevel of the western streams. It seems strange that this interesting possibility has not been considered in connection with the geomorphology of the Rocky Mountains. Johnson[6] has suggested that the lower surface in the Rockies may be a pediment,

1. Lee, W. T. Peneplains of the Front Range and Rocky Mountain National Park, Colorado. Bull. U. S. Geol. Surv., 730, 1-17, 1923.

2. Van Tuyl, F. M., and Lovering, T. S., Physiographic Development of the Front Range. Bull. Geol. Soc. Amer., 46, 1291-1350, 1935.

3. Rich, J. L., Comment on the paper by Van Tuyl and Lovering. Bull. Geol. Soc. Amer., 46, 2046-2054, 1935.

4. Keith, A., Some Stages of Appalachian Erosion, Bull Geol. Soc. Amer., 7, 519-525, 1896.

5. Davis, W. M., The Stream Contest along the Blue Ridge. Bull. Geog. Soc. Phila., 3, 213-244, 1903.

6. Johnson, D. W., Planes of Lateral Corrasion, Science, 73, 177, 1931; and Rock Fans of Arid Regions, Amer. Jour. Sci., 23, 410, 1932.

formed at its present high altitude and hence requiring no great sub-sequent uplift. To these possibilities the writer wishes to add another, namely, that both of the high-level surfaces in the Rocky Mountains are pediments, and that these pediments were formed simultaneously in a manner similar to that in which comparable surfaces are now being developed in some of the ranges of the southwest. Conceivably this explanation may apply to the Blue Ridge as well, providing the climate there was arid when the surfaces were formed. Contemporaneous pedi-ments are now being formed at different levels in the Sacaton Moun-tains of Arizona. The relations are described below and may well serve to illustrate this particular type of geomorphic development.

The eastern half of the Sacaton Mountains is a linear unit trending approximately north. It is encircled by a remarkably well-developed pediment. The wet-weather streams which drain the eastern flank have a much shorter horizontal distance to baselevel than those draining the western flank. The baselevels have about the same elevation. The streams are degrading only in the mountains; they are at grade on the encircling pediment. The head of the pediment thus forms an approximate bound-ary between the ungraded and graded portions of the streams. Surveyed profiles show little difference in the slope of the eastern and western pediments, but the head of the pediment on the west side of the range is at a much higher elevation than the head of the pediment on the eastern side. The difference in elevation amounts to nearly two hundred feet. The eastern streams have locally extended embayments of the lower surface into the area of the western pediment. At such places a steep slope usually separates the contemporaneous surfaces. At these places, too, the drainage divide has been pushed appreciably west of the crest-line of the range, and it is quite clear that in time the divide will lie entirely west of the range axis. A similar discordance in contempo-raneous levels might result where the distances are similar, but where the elevation of the baselevels differ.

The situation in the Front Range, in spite of the difference in scale, if sufficiently comparable to justify the presentation of the "contempo-raneous pediments" hypothesis. The relatively steep slope of the sub-summit surface and its abrupt junction with the uplands are pediment characteristics, although they admittedly do not prove that the surface is a pediment. The junction of the summit surface and its surmounting monadnocks is not everywhere so abrupt, yet is not such as to exclude this possibility. It is interesting to note in this connection that Van Tuyl and Lovering[7] report minor hills rising sharply above the level of the summit surface at many localities. They also report[8] that all of the erosion surfaces below a certain level "appear to be at a distinctly higher elevation on the west side of the range than the equivalent surfaces on the east side." Hence, until it is demonstrated that one or both of the upland surfaces are not pediments, the possibility must be entertained that they had such an origin and that, in spite of their different elevations, they were formed contemporaneously. If such a relation could be demonstrated, no uplift need be postulated to explain the vertical interval separating them.

Admittedly the high level surfaces of the Rockies are not everywhere as level as pediments are expected to be. The inequalities, however, may represent original pediment relief, including inselbergs, or may be the result of subsequent deformation or incomplete regrading. The phrase "pediment relief" may require clarification. Van Tuyl and Lovering describe the summit surface (Flattop peneplane) in the Front Range as undulatory. Even at Flattop Mountain, the type locality, the surface is described as rising from 11,900 feet at its eastern edge to 12,300 feet at the divide and then descending several hundred feet to the west. Pediments, however, may also undulate to some extent. The undulations may represent rock fans or minor pediments formed during the degradation of individual hills or groups of hills isolated from the main mountain mass. Such protuberances are less common away from the range divide, for the more distant features were probably isolated first, and hence have been undergoing degradation for a longer period of time. Thus the remark[9] that the summit surface "shows less relief with increasing distance from the Divide itself" might apply to a pediment as well as to a peneplane. The fact that the upper surface at its type locality also displays an eastward slope does not preclude the pediment possibility, because recession of the scarp at the head of the subsummit pediment might occasionally bring it against the eastern slope of one of the minor fans of the higher pediment. The additional high-level surfaces mentioned by Van Tuyl and Lovering may be due to stripping of a sedimentary cover or to differential erosion as sug-

7. Van Tuyl and Lovering, Op. cit., 1301.
8. Op. cit., 1314-1315.
9. Van Tuyl and Lovering, Op. cit., 1300-1301.

gested by Rich, or to incomplete regrading subsequent to the formation of the major surfaces. The regrading need have no diastrophic significance; it might have resulted from climatic changes, from changes in the load volume relationship in streams due to piracy or other causes, from intermittent lowering of temporary baselevels, or from other causes.

18

Reprinted from *Bull. Geol. Soc. America,* **64**(7), 721, 742–751 (1953)

CANONS OF LANDSCAPE EVOLUTION

By Lester C. King

Abstract

The manner in which epigene landscapes evolve is examined and discussed. Slope flattening as a general process of landscape evolution is rejected, and with it Davis' concept of the peneplain. Landscape evolution by scarp retreat and pedimentation is accepted, and several of its consequences are examined.

The opinions derived are expressed at the end of the paper as a series of canons of landscape development

[*Editor's Note:* The earlier parts of this paper contain reviews and comments on the work of Davis and others. It also deals rather extensively with hillslope development, a subject dealt with in the Benchmark volume, *Slope Morphology*.

The latter part, here reproduced, starts with King's view of the "Epigene Cycle of Erosion" (p. 742) and continues to the end. Only the first two plates are included.]

Epigene Cycle of Erosion

The secret of landscape evolution lies, evidently, in the mode of development of hillslopes. Now hillslopes may be initiated either as valley sides consequent upon stream incision, or as tectonic features the result of faulting, monoclinal warping, or even gentle tilting in a landscape. This is an important distinction upon which the evolution of the landscape as a whole afterwards depends.

In the first case, after a region has been uniformly uplifted, nickpoints run rapidly up the rivers and tributaries so that stream incision and the production of youthful valleys appear powerfully in the landscape. The interfluves are at first scarcely affected, and, if the initial surface was a plain, there comes into existence a landscape of the dissected plateau type, such as the Appalachian plateaus around Pittsburgh, Pennsylvania. Where the streams were already organized in a previous cycle upon structural controls, an incised trellised pattern develops like that of the folded Appalachians, or the joint-controlled pattern of the Matopo granite area, Southern Rhodesia.

The second case is illustrated by Southern Africa where continental uplifts terminated near the coasts in great monoclines facing toward the sea. These tectonic scarps are directed athwart the drainage lines (*e.g.*, Natal, Fig. 3). Once generated, the scarps that are large and therefore retreat rapidly make a series of risers which separate different cyclic land surfaces ascending step-wise successively away from the coast. The scarps retreat virtually as fast as nick-points advance up the rivers, so that the distribution of successive erosion cycles bears no relation to the drainage pattern whatsoever (Fig. 3). Under erosion, the scarps may not always remain clean and wall-like. Some degenerate into zones of dissected country separating an upper and earlier planation from a lower and later surface;

but many of the cyclic scarps of Africa impress precisely because they have not so degenerated but have remained clear and distinct throughout their history. The oldest date even from the Mesozoic, and these are no exception—*e.g.*, the Drakensberg (Pl. 1, fig. 1).

The destruction of earlier cyclic landscapes may be less rapid under the retreat of continental erosion scarps than when the country is first divided up under stream incisions and the proportion of hillslopes available for active retreat is thereby greatly multiplied. In this factor probably lies the explanation for the great ages of many African landscapes as compared for instance with the cyclic landscapes of North America. Smooth erosional landscapes such as the South African highveld or the Rhodesian plateau belong to the Mesozoic cycles, yet standing above them and separated by steep scarps are remnants of even older, conceivably Paleozoic, planations, some of which have apparently been exposed continuously to the weather since that time—*e.g.*, southern Congo.

Nothing like this appears to exist in North America, where few traces of very ancient cycles remain unless their preservation has been aided by burial for a time. A Mesozoic surface is described from Minnesota, and there are one or two other probable remnants of Cretaceous landscapes, but the fundamental surface from which the eastern half of the United States was later carved is the mid-Tertiary land surface, found emerging from beneath the Ogalalla covering formation in the Llano Estacado of the West, and appearing as the summit uplands of the Appalachian Plateaus, Schooley "peneplain", and the New England upland in the east.

We may now follow certain points in the evolution of landscape after the youthful stage. When the streams are already well and widely incised, parallel retreat of hill and valley sides reduces the areas of initial surface upon interfluves. If the relief is low the natural curve of the pediment may soon take it up to meet the waxing slope, so that concavoconvex valley sides appear and slope development becomes moribund. If, on the other hand, the relief is high, shrinkage of the interstream plateaus continues until opposed pediments meet and a biconcave transverse section to the interfluve results. A noteworthy fact emerges here. The upper surface of the interfluve remains in this case virtually without alteration until the hillslope elements of the new cycle encroach upon it. I have seen on outliers of upland surfaces quite unmistakable summit areas of as little as 2 acres surviving virtually without alteration though all the country for miles around had been consumed by scarp retreat in a newer cycle of erosion. The hill called Showe in the Shamwa district of Southern Rhodesia is a superb example of this. Where jointed granites form the terrain, and hillslopes stand steeply, bornhardts are a typical landform under this process (Pl. 3, fig. 3).

Where primary dissection of the country is not by stream incision, but by the retreat of great cyclic scarps (Fig. 3), earlier cyclic land surfaces may survive long after the initiation of the new cycle. They remain practically unaltered until destroyed utterly by the encroachment of the fresh cyclic scarp (Pl. 1, fig. 1). Here is one of the most important canons of landscape development.

Whereas the cyclic erosion of the older Davisian conception, with its emphasis upon slope flattening and universal downwearing under the action of weathering, early obliterated all trace of the initial surface and reduced everything to a peneplain, the newer concept involving cyclic scarp retreat permits the initial surface, itself a record of an earlier cycle of erosion, to remain much longer in the landscape. Inherent in the doctrine is the persistence of the older surface without significant alteration for a long period into the currency of the new cycle. Two cyclic surfaces are for long co-existent, the older above the retreating scarp and the younger below.

The far-reaching consequences of this have already been demonstrated in studies of African and other continental landscapes (King 1949a; 1951b). It forms indeed the primary canon for deciphering the past histories of landscapes, and with its aid the first, tentative correlations of dominant cyclic landscapes have been made from continent to continent. On the down-wearing hypothesis, leading to Davisian peneplains, the very possibility of this achievement would be denied.

After scarps have retreated, both the major continental cyclic scarps transverse to the drainage and the minor scarps or valley sides, a wide development of pediments becomes apparent in the landscape. From either side of every stream large and small twin pediments extend laterally and make by coalescence pediment plains upon which stand unconsumed, steep-sided residuals (Pl. 4, fig. 3). In Southern Rhodesia (excluding the sand country of the west) pediments constitute probably 65 per cent of the landscape; the rest is residual hill masses. As the pediments provide the areas for cultivation their social and economic importance is also great.

As they widen they are subject to regrading, especially the steeper part toward the foot of the overlooking scarp. From an initial slope of perhaps 10°, they are reduced commonly to declivities of half a degree or so. This regrading is accomplished chiefly under sheet wash, and even ancient and very flat pediments often show beneath a thin veneer of transported material a smoothly cut, unweathered rock surface. Fair remarked of the general smoothness of pediments in the semiarid Karroo that "sheet-wash and rills are capable of regrading them with little difficulty in response to minor changes of base-level". But such changes are not necessary to regrading as Rich (1935) has also noted: "Drainage diversions and the dissection of abandoned fans and pediments are normal and to be expected. They do not require the intervention of diastrophic or climatic changes".

Conflict of pediments from neighboring streams, in which the stronger (usually the larger) extends its area at the expense of the weaker, is also a very real process causing regrading. The "development of master pediments", as we may call it, has most important consequences when we come to the dating of land surfaces.

In the later stages of the cycle, when the hills

are reduced to small, rocky koppies and the pediments stretch perhaps for miles (Pl. 4, fig. 3), a *multi-concave* landscape is characteristic. In Southern Africa, this advanced evolutionary stage is widely developed with respect to stable, or gently falling base levels. Extensive alluvia or even flood plains are seldom present, and the soils may be quite different on opposite sides of even small streams. We may describe as typical the landscape on the east and south sides of Manda (Concession), Southern Rhodesia, where the pediments sweep right down to the swampy area of the streams themselves. The Marodzi valley west from Concession is larger and possesses a strip of flood plain 200 to 400 yards wide. But the bordering pediments are each over a mile in width before they abut against the rocky slopes of koppies. The general slopes of these koppie sides are nearly 30° inclination, but the lower 300 to 500 feet may approach a slope of 45°. The inner, middle, and outer zones of the pediment have slopes of 1°40′, 1°10′, and 30′ respectively, and these estimates are more or less standard for the district. Where the pediments pass as rock fans into the mouths of narrow side valleys they become steeper, sometimes 5° or 6°. The pediments themselves do not seem to exceed slopes of 3°, even when they are narrow and not very concave. In the Glendale-Bindura district of the same colony, the Mazoe Valley, though miles in width, nowhere possesses extensive flood plains but always a narrow river channel with wide, complex bordering pediments. Small residual hills rise from between the various pediment surfaces related to sundry tributaries. And so the descriptions of maturely pedimented landscapes in Africa could be repeated for innumerable instances.

At senility a continental erosion surface is not simple. It consists of a complex assemblage of closely related surfaces (chiefly pediments) referable to a multitude of drainage lines and basins, large and small, and united only in reference to a single, widespread, long-stable base level.

Even in this stage there may be further, accidental complications such as the multiplication of cyclic surfaces by structure (split nickpoints), local land movement, and so forth; but these factors have been discussed separately (King, 1947; 1951b) and shall not detain us here.

Observational data suggest that even at the senile stage there is little weathering of the firm rock beneath the pediment veneer, but this conclusion may not be valid. In the African theater, Pleistocene climatic changes were expressed in an alternation of arid and relatively more humid phases. As has been remarked by Cooke, the superficial materials overlying rock-cut pediments, when revealed in the sides of dongas, in many places contain Middle Stone Age implements. Following a previous phase of pedimentation there was thus

---

PLATE 1.—EROSIONAL SURFACES AND SCARPS IN SOUTH AFRICA

FIGURE 1.—NATAL DRAKENSBERG SOUTH OF CHAMPAGNE CASTLE

View north. Relief over 4000 feet. This scarp began at the Natal coast as a structural feature, the Natal monocline, in the late Jurassic. It has since retreated westward under erosion the full width of Natal, 150 miles. Its steepness is maintained by innumerable gully heads. The ancient, Mesozoic landscape above has remained virtually without alteration throughout the late Mesozoic and Tertiary. Anon. photo.

FIGURE 2.—TYPICAL LANDSCAPE OF A SEMIARID REGION, THE KARROO, SOUTH AFRICA

There is no flood plain, and there are wide pediments backed by steep hillslopes on which are minor structural effects caused by horizontal, resistant sandstones and dolerites. T. J. D. Fair photo.

PLATE 2.—EROSIONAL FEATURES OF GRANITE TERRAINS, SOUTHERN AFRICA

FIGURE 1.—SMALL GRANITE HILLOCK

Showing waxing slope, free face, and incipient pediment. In the absence of detritus the sharp angle between free face and pediment is clearly shown. Hluhluwe Game Reserve, Zululand.

FIGURE 2.—RETREAT OF HILLSLOPES AND GROWTH OF PEDIMENTS

In a region of jointed granite (Matopc Hills, Southern Rhodesia). Airphoto from 15,000 feet. Aircraft Operating Co. photo.

FIGURE 1

FIGURE 2

EROSIONAL SURFACES AND SCARPS IN SOUTH AFRICA

FIGURE 1

FIGURE 2

EROSIONAL FEATURES OF GRANITE TERRAINS, SOUTHERN AFRICA

a widespread phase of deposition while the Middle Stone Age folk occupied the scene. Still later came the phase of incision when the dongas were trenched through the depositional blanket. All these changes, covering an enormous area, appear to have been climatically controlled. They obscure the ultimate stage of pedimentation over much of Africa.

DATING OF LAND SURFACES

Land surfaces are customarily dated by the discovery of superficial deposits whose age can be defined. But with the regrading and conflict of pediments such deposits, usually thin anyway, may be wholly or partially removed. The net alteration in the landscape produced by one pediment cutting shallowly across another may be small indeed and quite irregular, but from the point of view of dating it becomes highly important.

Let us assume that a pedimented landscape of low relief except for residual koppies and small plateaus acquired a thin cover, partly detrital and partly residual, of Cretaceous age. The original landscape must therefore have been either Cretaceous or pre-Cretaceous. The region is not invaded from outside by incisions or other features of later continental erosion cycles, but continues to develop solely by the processes of pedimentation under which certain struggles for mastery go on, more-

favored pediments cutting shallowly across their neighbors and so removing the cover and perhaps a few inches or feet of bedrock. Over the area so replaned, Eocene deposits may accumulate. So from time to time during the Tertiary, portions of the original area are remodelled in this way, having a shaving taken off them as it were and acquiring new and younger superficial deposits.

At no time does the landscape as a whole depart materially from the original planed condition, yet its parts are manifestly of several different "actual" ages,[8] as defined by the various superficial deposits. Unquestionably, the planed landscape is at one point Cretaceous, at another Eocene, at another Pliocene, and at yet another Pleistocene, so that the subjacent bedrock surface is a compound Cretaceous-Eocene-Pliocene-Pleistocene surface. This seems to be essentially the history of the Gobi, and other ancient landscapes e.g., parts of the Kalahari.

But dating land surfaces in this manner is unsatisfactory except for local purposes, and to derive the fundamental or "comparative" age of such a compound landscape we can argue as follows. The landscape, viewed as a whole, was planed first in Mesozoic (Cretaceous)

---

[8] The age of the surface at any actual spot where datable deposits actually lie upon it.

287

time. From the planed aspect then achieved, it has never since materially departed. What we see now is essentially a Cretaceous surface that has since undergone minor modification only. Its fundamental or "Comparative" age, by which it should be compared with other planed surfaces, is Cretaceous; and so we regard it.

The number and extent of planed Cretaceous land surfaces surviving thus as remnants in landscapes of the present day, especially those of the southern (Gondwana) lands, is remarkable. The survival of these surfaces affords a direct negative to the concept of significant vertical downwearing of landscapes—*i.e.*, to the so-called "Normal Cycle of Erosion".

The rate of retreat of major continental cyclic scarps (*e.g.*, Fig. 3) affords another aspect of the dating of ancient land surfaces. Thus the Natal Drakensberg originated along a coastal flexure which formed the margin of South-East Africa during the Jurassic period (King, 1940; 1944). Wall-like in form, it has retreated over 150 miles westward to its present position. The wall is still 4000 feet or more high with black, forbidding precipices in its upper part along the borders of Basutoland where the crest rises above 10,000 feet. It is quite unmistakable, and shows little sign of flattening in its 120 million years of existence. The rate at which it has migrated averages about a foot in 200 years.

Knowing the date and place of initiation of the scarps and their present position, similar calculations can be made for several other of the world's major cyclic scarps—*e.g.*, the frontier scarp of Moçambique, the scarp of western Mexico, and so forth. Many of these furnish a rate of retreat between 1 foot in 150 years and 1 foot in 300 years. This shows that over long periods, *ceteris paribus*, the rate of scarp retreat is sensibly constant. In other words the erosive processes responsible for such retreat act at much the same rate the world over.

Only when scarps are small (low relief) and the amount of free face and debris slope is much reduced does the rate of retreat slow up. As we have seen, the rate of parallel retreat upon concavo-convex slopes from which these elements are absent appears to be almost negligible.

QUANTITATIVE METHODS OF LANDSCAPE STUDY

However landscapes originate, it is important that their distribution should be known, and for this there is no better method than routine mapping. Erosion-cycle maps, or morphological maps exist of many small areas and even of certain countries—France, Belgium, European Russia, Southern Rhodesia, etc.—but many more such maps are necessary. The data they provide are valuable to engineers, soil conservationists, and others as well as geomorphologists. There is no substitute for such maps, even though useful data may be gleaned from the ordinary topographic sheets.

Methods involving frequency of elevations, spot heights on grids, and so forth have been used (Maze, 1944) to bring out the cyclic facets present in multicyclic landscapes, and several methods involving superimposed and generalized profiles have been employed for similar purposes (*e.g.*, Wooldridge and Morgan, 1937); all have yielded helpful results more or less directly.

The application of statistical analysis is still more recent, and the pioneer has been A. N. Strahler (1950a; 1950b). Statistical analysis is essentially the method of the bulk sample, and is admirable for the study of complex phenomena and processes into which enter a large number of variable factors. As yet few geomorphic topics provide data suited directly to statistical treatment, and methods may have to be adapted to the new field of inquiry, so that too facile results should not be expected. The net result must be, however, a greater precision in our thinking.

At present the method of statistical analysis is in danger of lending an air of truth to erroneous conclusions. Decisions concerning type and admissibility of data need to be made most carefully, or the results will be falsified. No amount of statistical work can improve upon dubious assumptions and we must be ever mindful of Huxley's dictum concerning the use of mathematics (Mathematics is a mill which will grind you meal of any degree of fineness, but the quality of the meal depends upon the quality of the grain used). Unless the original data and assumptions are above suspicion, the conclusions lack certainty and perhaps validity. Fundamental errors of judgment made in the sampling process are not eliminated in the method, but they may be corrected in the course of further field work.

Strahler (1950a), using valley sides in the Verdugo hills and assuming that they were no longer undercut by stream action, concluded that hillslopes, in general, flatten progressively with the passage of time. Criticisms here would be: (1) The material adduced is too small in bulk to found a general principle upon. (2) It places a relationship between slope angle and stream gradient which may cease to exist after a certain stage is reached in the landscape cycle (*i.e.* slopes may become stabilized in gradient). (3) The manner of slope development is governed by forces active upon the slope and is independent, except initially, of the river. (4) Even upon the slope, erosional agencies do not act uniformly. (5) There are four different slope elements possible, all of which are modified independently. (6) Equilibrium between various features and factors is assumed whereas such equilibrium patently does not necessarily exist in nature. (7) The general conclusion that slopes flatten progressively is denied in nature by the close agreement of maximum slopes in many districts with a mean angle, and the retention of freshness and activity of major cyclic scarps that have existed from mid-Tertiary or even Mesozoic times. This last, under the Euclidean method, reduces the proposition to an absurdity.

Nor can one accept Strahler's general proposition of the equilibrium theory ("decline of stream gradient is accompanied by slope reduction") for it fails to recognize the great diversity of factors operative in the carving of a landscape. No simple relation can exist between hillside slope and thalweg. After the stage of extreme youth they are modified quite independently. What finer examples need we quote than some of the plates reproduced herewith?

To follow the argument further would profit us nothing. Our object here is not the rebuttal of the Equilibrium Theory of Erosional Slopes or any other variant of Davis' early views

upon hillslope evolution; it is merely to emphasize that no amount of statistical analysis will compensate for poor geology. Geologists will be grateful for such practical results as will undoubtedly accrue from the use of the new method, but they will not regard it as a substitute for honest field work, nor will they assign to mathematical expressions an importance equal to facts which can be verified and reverified in the field.

Baulig (1950) has made another point: "Engineers will always use more or less empirical formulas for practical purposes. But there remains to be proved that mathematics has ever revealed in geomorphology an actual relationship that had not been discovered without its aid".

What seems necessary now is a multitude of further observations on the nature of land-forms, with special reference to hillslopes and perhaps following the lines indicated by Wood, Fair, and Horton (1945). Quantitative work upon pediments and similar topics is needed, and this will lead to statistical studies. An increased volume of measurement upon topographic forms is necessary and should be undertaken directly in the field rather than from maps in the laboratory.

Also required are extensive observations of the processes operating upon landscape, perhaps stemming from the manner of dispersal of rainfall, through water flow to soil creep and weathering. These need to be correlated with different types of climate and terrain. Studies of stream grade, such as those by Kesseli and others, are of less direct application than processes affecting wide areas of landscape simultaneously. Equally required is a multitude of field studies in which cyclic and noncyclic erosion surfaces are mapped in ever-increasing detail. As the number of morphological maps increases, and landscapes are accurately dated, both "actually" and "comparatively", so the erosional histories of the lands will be deciphered and compared, perhaps upon the lines of world-wide correlations.

CANONS OF LANDSCAPE EVOLUTION

(1) Landscape is a function of process, stage, and structure. The relative importance of these is indicated by their order.

(2) The word *epigene* as applied to landscapes means "at the surface" or "subaerial". It does not include landscapes moulded beneath a solid cover of ice, and certain modifications are understood to be necessary in regions of permafrost.

(3) There is a general homology between all epigene landscapes. The differences between landforms of humid-temperate, semiarid, and arid environments are differences only of degree. Thus, for instance, monadnocks and inselbergs are homologous.

(4) Four elements may occur in a hillside slope. From the top, these are: the waxing slope, the free face, the detrital slope, the waning slope (usually pediment). Each or any element may be suppressed on a given hillslope.

(5) Each of the four elements of hillslope may evolve more or less independently, although each affects the others in some degree.

(6) The most active elements of hillslope evolution are the free face and the debris slope. If these are actively eroded, the hillside will retreat parallel to itself.

(7) In planed landscapes, pediments are the most important features. In stable regions like Southern Africa, pediments may occupy more than half the whole landscape, and locally may exceed nine-tenths of the landscape.

(8) The waxing slope is developed under weathering and soil creep.

(9) When the free face and debris slope are inactive, the waxing slope becomes strongly developed and may extend down to meet the waning slope. Such concavo-convex slopes are degenerate.[9]

(10) Parallel retreat of slopes is aided by (a) high relief, tending to maintain a clear free face and a debris slope; (b) resistant formations, tending to make cliffs (a good free face); (c) horizontal structure in sedimentary rocks; (d) generation originally as a tectonic scarp (*e.g.*, fault or monoclinal scarp).

(11) Erosion of the free face and the debris slope is accomplished chiefly by rill wash forming gully heads.

---

[9] Dr. Fair demurs "Rather does the waning slope extend *upwards* to meet the waxing slope and so give the pediment its predominantly concave form". He is thinking chiefly of African conditions, I of European. Both statements may be true.

(12) Whereas the transport of debris upon the waxing slope, free face, and detrital slope is governed by both gravity and water work, that upon the pediment is solely accomplished by water work.

(13) Davis' old deduction of continuous lowering of hillside gradients, a feature also of Strahler's "Equilibrium Theory", is incorrect, and never existed as a general process of landscape development apart from terrains of rocks so weak that they cannot maintain a free face and detrital slope.

(14) Rock floors in epigene landscapes appear commonly between the base of hillslopes and stream channels. These rock floors, which are found under all three climatic regimens, originate by retreat of the hillslopes behind them and are subsequently modified by the passage of water across them which confers upon them a concave profile. Such rock floors should, in all cases, be called *pediments*.

(15) Stream spacing is closer in humid than in nonhumid regions so that an evenly distributed rainfall is largely discharged by channel flow. Wider spacing of streams and heavier incidence of rainfall favor sheet flow and also allow room for wider pediments.

(16) Pediments are normally veneered with detrital material which is in process of transport across them. But pediments themselves are essentially cut-rock surfaces.

(17) Pediments which have ceased to evolve may show weathering of the bedrock.

(18) A pediment is the ideal landform for the rapid dispersal of surface water, encouraging sheet flow and with a proper hydraulic profile. Pediments are, indeed, moulded under sheet flow.

(19) The pediment is the fundamental landform to which epigene landscapes tend to be reduced the world over.

(20) Gullying may appear upon pediments where laminar flow of water is changed to linear flow.

(21) The break in profile between pediment and hillside may be abrupt if little detritus is supplied from above.

(22) Quantitative study of both slopes and processes provides a sequence of landscape forms different from those propounded by W. M. Davis.

(23) The early studies in the erosion cycle

were conducted in Europe and northeastern North America, both of which regions were previously subjected to a glacial or periglacial climate. These areas and landscapes came to be cited as "normal" for the globe, a misconception that should no longer be tolerated.

(24) The standard or "normal" type of landscape, both now and in the geological past, is the semiarid type with broad pediments and parallel scarp retreat.

(25) Processes of erosion and evolution of landforms can, as a consequence of the above, be best observed in semiarid regions.

(26) Semiarid landscapes are the most efficiently developed. Deviation from the semiarid norm results in less efficient transport of waste, seen on the one hand in the broad, alluviated valley floors and smothered hillslopes of humid regions culminating in moraine and till under glaciation; and the abundant fans and bahadas, or even desert dunes of extremely arid regions.

(27) Water may flow across landscapes either in threads or in sheets. In thin sheets, water may flow in laminae. Such laminar flow is nonturbulent and nonerosive.

(28) In storms of moderate intensity, the manner of water flow appears to best advantage. Rill flow on the steep hillsides is powerfully erosive; in the upper pediment thin laminar flow (nonerosive) may occur, with deeper sheet flood lower down accompanied by turbulence and erosiveness.

(29) Laminar flow past an obstacle shows a depressed water surface, linear flow banks up the water surface against the obstacle.

(30) Only in late Tertiary time have smooth concavo-convex slopes become common. This is a result of retardation of surface wash by a carpet of grass, and consequent enhancement of soil creep.

(31) Before mid-Tertiary time, landscapes generally were of the semiarid (scarp and pediment) type.

(32) Stream work affects the nature of adjacent hillslopes directly only during the early stages of the landscape cycle. After the streams are graded the dominant agencies are the processes acting directly upon the hillslopes, which evolve in an appropriate manner. The streams are, however, affected by the evolution of hillslopes through the nature of the

detritus which these supply for transport by the streams.

(33) New cycles may penetrate inland either by nickpoints and incision of the rivers followed by retreat of the valley sides producing "flanking pediments", or by the retreat of wall-like scarps which are independent of the drainage lines "mountain pediments".

(34) On the whole, a landscape is dissected and reduced in a new cycle more rapidly following widespread river incision than by retreat of cyclic scarps originating tectonically. Conversely, the history of a landscape may be deciphered more readily where cyclic surfaces rise steplike between major cyclic scarps than where the landscape has been gutted by stream dissection.

(35) The ultimate cyclic landform is the *pediplain*, consisting dominantly of broad coalescing pediments. Residuals are steep-sided and have concave slopes. Flood plains may or may not be extensive.

(36) A pediplain is multi-concave upward. When the streams begin to incise themselves due to tilting, uplift, or climatic change, convexity enters the landscape adjacent to the stream channels. The interfluves then become transversely convex, and a landform morphologically different from the pediplain is produced. It is an initial stage of the cycle of erosion; the pediplain represents a senile stage. This expresses Penck's *Endrumpf-Primärrumpf* concept insofar as land form is concerned. We do not necessarily follow his further argument relating these differences to decreasing or increasing rates of land movement.

(37) A peneplain in the Davisian sense, resulting from slope reduction and downwearing, does not exist in nature. It should be redefined as "an imaginary landform."

(38) Davis has quoted low granitic domes of arid regions as though their form resulted specifically from long-continued erosion. Observations in Southern Africa show, however, that these forms are normally functions of structure; that, where the vertical systems of jointing are strong, bornhardts and castle koppies (large and small inselbergs) appear, and where flat or gently dipping joint systems are paramount "ruwares" or flat domes of granite appear in the landscape. That such flat domes do not result from the erosion of bornhardts is, for

this region, certain (Pl. 2, fig. 1; Pl. 3, fig. 3); they follow the broadly convex joints formed apparently in plutonic rocks by "unloading" as superincumbent rock systems were removed under erosion.

(39) Monadnocks are concave in profile as a rule (Pl. 4, fig. 3) (including Mt. Monadnock itself) and have originated by surface wash rather than by downweathering and soil creep. They are not necessarily sited upon outcrops of more resistant rocks.

(40) Inherent in the pediplanation cycle, with scarp retreat, are *two* cyclic land surfaces, the older above a retreating scarp, the younger below.

(41) Many of the major cyclic erosion scarps originate tectonically as fault or monoclinal scarps, especially along outwardly tilted coast lines.

(42) Major cyclic erosion scarps retreat almost as fast as the nickpoints which travel up the rivers transversely to the scarp. Such scarps therefore remain essentially linear and do not have very pronounced re-entrants where they cross the rivers.

(43) Major continental erosion scarps in many lands retreat at a rate of about a foot in 150 years to a foot in 300 years.

(44) A landscape once reduced to a pediplain may remain in that state for an indefinite time with only minor alteration, until some change, tectonic or climatic, is introduced.

(45) Notwithstanding the above, small changes continually take place by regrading and conflict of pediments. These changes, insignificant in the landscape as a whole, perhaps amounting to the removal of only a few inches or feet of material, produce great differences in the superficial deposits of the pediments.

(46) Land surfaces may be dated by the deposits upon them. "Actual" ages are local ages fixed by directly dating the deposits in any given locality. "Comparative" ages refer to the dates at which land surfaces were originally bevelled, and are obtained from the oldest superficial deposits.

(47) Land surfaces may bear deposits of any age from the oldest, used for "comparative" dating, to the present day.

(48) A tentative approach, using "comparative" datings, has been made toward the correlation of major cyclic landscapes from continent to continent.

(49) More use of quantitative methods is necessary in landscape study; especially needed is more morphological mapping.

(50) When more suitable data are available, statistical analysis may become a useful tool in landscape study.

References Cited

Baulig, H. (1950) *William Morris Davis: Master of Method*, Assoc. Am. Geog., vol. 40, p. 188–195.

King, L. C. (1940) *The monoclinal coast of Natal, South Africa*, Jour. Geomorph., vol. 3, p. 144–153.

—— (1944) *Geomorphology of the Natal Drakensberg*, Geol. Soc. S. Africa Tr., vol. 47, p. 255–282.

—— (1947) *Landscape study in Southern Africa*, Geol. Soc. S. Africa Pro., vol. 50, p. xxiii–lii.

—— (1949a) *On the Ages of African landscapes*, Geol. Soc. London Quart. Jour., vol. 104, pp. 439–459.

—— (1951b) *South African scenery*, Edinburgh, 379 p.

eastern area of New South Wales, Royal Soc. N.S.W., Jour., vol. 78, p. 28–41.

Maze, W. H. (1944) *The geomorphology of the central*

Rich, J. L. (1935) *Origin and evolution of rock fans and pediments*, Geol. Soc. Am. Bull., vol. 46, p. 999–1024.

Strahler, A. N. (1950a) *Equilibrium theory of erosional slopes approached frequency distribution analysis*, Am. Jour. Sci., vol. 248, p. 673–696.

—— (1950b) *Davis' concepts of slope development viewed in the light of recent quantitative investigations*, Assoc. Am. Geog., Ann., vol. 40, p. 209–213.

Wood, A. (1942) *The development of hillside slopes*, Geol. Assoc., Pr., vol. 53, p. 128–138.

Wooldridge, S. W., and Morgan, R. S. (1937) *The physical basis of geography*, London, 445 p.

University of Natal, Durban, South Africa.
Manuscript Received by the Secretary of the Society, July 21, 1952.

19

Reprinted from *Amer. J. Sci.*, **253**, 377–390 (July 1955)

GEOMORPHIC DEVELOPMENT IN HUMID AND ARID REGIONS: A SYNTHESIS

CHAUNCEY D. HOLMES

ABSTRACT. Arid-climate landscapes are best explained in terms of Penckian geomorphic concepts and have never been satisfactorily integrated with the Davis system of land-form exposition, although integration seems to be quite possible and certainly desirable. Attention should be directed to those landscape elements common to both arid and humid climates.

The fundamental elements in subaerial landscape evolution are the so-called wash slopes and gravity (or derivation) slopes. These are best displayed in arid regions. In humid regions the grass and forest cover have the effect of breaking up these elements into small units, though their essential functions and relationships remain unchanged. Varying ratios of these small wash- and gravity-slope units give the observed range of hillside slopes characteristic of humid regions, tending toward gradual reduction of overall slope angle. In arid regions these two types of slope remain more nearly unbroken and are therefore more conspicuous. When applied broadly through the climatic range, these two fundamental types are regarded as constituting end members of a continuous series. This point of view provides a unified basis for description and interpretation of all landscapes developed by fluvial erosion.

INTRODUCTION

Geomorphic development in arid regions has always had an anomalous status in the Davis scheme of geomorphic interpretation. In establishing his geographical cycle (as it was called in the early years) Davis drew extensively upon the early works of Powell, Dutton, and Gilbert in western and southwestern United States for most of the basic principles of the cycle concept. Yet in those same arid regions, the contrasts with humid-climate landscapes have so impressed most later workers that a separate scheme of interpretation has seemed necessary. Even Davis himself believed in this necessity, and he is perhaps largely responsible for setting the American pattern of thought and viewpoint on topographic development under arid climatic conditions (Davis, 1905). Maintenance of the separate category in popular geology thus established may have offered better professional opportunity than would have been the case had more effort been directed toward harmonizing the two schemes, inasmuch as the humid-cycle principles then appeared to have been thoroughly exploited. At any rate, American geomorphology has continued to carry the dual scheme of "normal" and "arid" cycles without apparent embarrassment.

No strong opposition to the Davis doctrines developed in America while the great author was living. The case for pedimentation instead of "normal" peneplanation in the Southwest was satisfactorily established through the work of Bryan (1922) and others, and earlier workers among whom McGee (1897) deserves special mention. But all this was accepted within the Davisian arid cycle without serious misgivings. Opposition from abroad in the form of Penckian doctrines has not been taken seriously by most American geologists until recently. Formal statement of the Penck system (Penck, 1924) came while Davis was still active, and his writings seemed to refute satisfactorily the tenets of the new school. Consequently the realization that the principles of pedimentation are in essence those of the Penckian system has come as a surprise to many. Now that an authoritative translation of Penck's chief work is available (by Czech and Boswell), its virtues and defects should become more generally appreciated.

The Association of American Geographers sponsored a symposium on Penckian geomorphology in 1939, at which time some of the currently used English equivalent terms were introduced. However, the belated attention of American geologists was drawn more effectively to these geomorphic problems by the publication of von Engeln's *Geomorphology* (1942). Von Engeln had arranged the symposium, and therefore to his efforts we owe a large measure of the present awareness of the Penckian principles among American geologists. But conservatism is strong, and one may detect a perhaps unconscious partisanship in the common tendency to criticize the obvious errors in some Penckian principles or to dwell on their less essential aspects.

Meanwhile other developments have been in process. A recent well-prepared textbook (Gilluly, Waters, and Woodford, 1951) scarcely mentions the time-honored regional young-mature-old Davisian topography, and the omission seems to have created very little comment. This can be taken as evidence of serious and growing suspicion that a goodly part of traditional Davisian teaching is obsolete. Still more specifically, King (1953) has completely rejected both the principle of downwasting of divides and the peneplain as the penultimate land form, in favor of scarp retreat and pediment extension leading to a compound pediment or pediplain. Even Davis himself (1930, p. 136) seems virtually to have acknowledged the existence of pediments in humid climates (he called them valley-floor basements), with the offhand suggestion that he had taken them for granted since about 1908.

Because the primary aim of the Penck system was directed toward interpreting regional diastrophic history, some of the geomorphic principles therein expounded must be disengaged from the tectonics in order to evaluate their significance. Whether the change from convex to concave slopes, in so far as this change can be demonstrated, reflects chiefly rate of uplift may be questioned on the ground that other factors also partially control the angle of slope. For strictly geomorphic interpretation, the diastrophism involved here is as incidental as is the Davisian assumption of rapid uplift followed by stillstand. Likewise the problems of *Knickpunkte* and piedmont benchlands lie only partly in their alleged modes of origin. Of equal or greater importance is the history of these forms once they have come into existence. This is a part of the great central problem of the evolution of valley-side slopes.

ACKNOWLEDGMENT

The writer is indebted to Professor Richard Foster Flint for helpful criticism and suggestions regarding preparation of the manuscript.

THE PROBLEM AND THE APPROACH

To harmonize the concepts of "humid" and "arid" cycles (and incidentally to combine the best elements of the Davis and Penck interpretations), one must go beyond the obvious field facts that parallel slope retreat prevails in the more arid regions and that gradual reduction in slope angle is equally the rule in the more humid regions. As a first approach, two commonly held alternative propositions may be stated: Either (1) the tendency to parallel slope retreat is inherent in the hydraulic processes and is present under any climatic conditions (Penck); or (2) such tendency arises because of some nonhydraulic factors peculiar to arid climates and absent in humid climates (implied in Davisian theory). A third proposition, namely, that fundamental hydraulic principles depend on the vagaries of climate, is manifestly absurd; yet a hint of this uncritical view is inherent in any insistence that erosion goes on differently in arid and humid regions.

Two related problems seem to hold the possibilities of finding a unifying solution. The first is to recognize in the two climatic domains the fundamental erosion processes and their homologous effects. The second is to account for the differences in terms of controls on the erosion processes.

Most earlier attempts (Davis, 1930; Wood, 1942) to harmonize the phenomena of erosion forms in the two contrasting environments have begun with those of humid climate and proceeded to the arid-climate forms. In reality, the simplest expressions of water-erosion effects are those in arid climates. Therefore on the principle that clarity is best achieved by proceeding from the simpler to the more complex, the following analysis begins with a consideration of arid-type landscapes, followed by those under semi-arid conditions, and then to the humid-climate landscapes. King (1953) has followed this approach in part, but his generalizations have led to revolutionary, rather than evolutionary, conclusions. Nevertheless, all attempts to solve this problem should receive sympathetic consideration.

ADVANCE SUMMARY

This preliminary summary of evidence and deductions is offered in order that the subsequent discussion may be followed more readily.

Slopes are best classified into two fundamental types, in accord with Penck geomorphic interpretation: (1) *wash slopes* (graded surfaces of sediment transport), and (2) *gravity* or *derivation slopes* (surfaces which supply the sediment). In arid regions these two types occur in large-scale units dominating the landscape as pediments of various dimensions, and prominent scarps and mountainsides. Ideally, the wash-slope gradients are controlled only by the local requirements of sediment transportation on them.

In humid regions vegetation impedes both erosion and rate of runoff, resulting in significant contrasts with arid-climate landscapes. (1) Equilibrium wash slopes are under the *dual control* of vegetation and sediment transport over them, and in consequence are steeper than corresponding vegetation-free slopes; whereas valley sides (the chief gravity slopes) are generally less steep. (2) The more complete drainage network creates a correspondingly greater total length of valley sides, and therefore a more varied and intricate areal pattern in which slope retreat may go on. (3) From the beginning of an erosion cycle, rills and sheet runoff slowly lower the upland surface and round off the upper edges of valley walls, creating the characteristic convex portion of the transverse profile. (4) The strong but partial and inconstant control by vegetation gives rise to slopes of various intermediate degrees of steepness. In detail, these slopes consist of minute gravity-slope units alternating with correspondingly small segments of wash slopes. Local and temporary failure of vegetation on wash slopes permits gullying, the gully sides being new gravity slopes, with new and lower-gradient wash slopes eventually developing.

A notable intermingling of small wash- and gravity-slope units occurs in the zone of rounding at the tops of the valley sides. In effect, the upper part of the valley side removed in the rounding is distributed as inconspicuous minor units through the zone of rounding. The degree of steepness of the rounded slope as casually observed at any one place is an expression of the ratio between the two types of small units. Toward the upland divides, wash-slope units become longer and more numerous, and the uplands themselves may or may not be quite devoid of small gravity-slope units. Toward the valley, gravity-slope units prevail increasingly, and the steepest part of the wall may be uninterrupted gravity slope though modified by vegetation. Similarly, at the foot of the steep valley side a few short wash-slope units appear and become increasingly numerous and continuous as the valley floor is approached. This lower, concave portion of the transverse profile is often erroneously ascribed chiefly to alluvial-fan building. While it is true that considerable amounts of alluvium commonly accumulate at these sites, the surface is nevertheless essentially a wash slope.

All gravity (derivation) slopes of whatever unit-size tend to retreat at their characteristic angle of declivity as determined by local conditions. As these conditions change, so may the angle of declivity inasmuch as it develops originally under the control of existing conditions. The tendency to parallel retreat is believed to be everywhere present, but in humid areas it is masked or modified by the effects of more frequent (though less intense) runoff, by vegetation, and by more complete chemical weathering than is characteristic under arid conditions.

The foregoing analysis leads to the conclusion that the fundamental differences between fluvially developed landscapes in humid and arid climates are chiefly those of proportion and arrangement of the basic slope elements; and that, with appropriate adaptation of terminology, the Davisian framework of topographic interpretation may be applied to any landscape thus developed.

The following discussion makes no mention of soil creep and related processes, but such are assumed to go on as is generally understood.

TWO FUNDAMENTAL PROCESSES AND THEIR DOMAINS

It seems scarcely necessary to state that, following weathering of the surface rocks, landscape development by sediment removal involves two distinct and fundamental processes: erosion (or sediment derivation), and transportation. Yet inasmuch as these two processes result in the mystery of parallel slope retreat under some conditions and gradual slope reduction under others, their natures and their particular domains must be carefully noted.

Geomorphically the most significant effect of transportation by running water is the development of a graded slope, whether it be a stream bed or a sediment-mantled rain-washed surface (Mackin, 1948; Fenneman, 1908; Bryan, 1922; Holmes, 1952). So long as an abundance of sediment in various grade sizes within the competence of the current is available, a stable transportation profile will be established leading down to a local baselevel of one kind or another, and changing gradually in response to changes in sediment supply and volume of water from upslope, or to any shifting of the controlling baselevel.

The domain of sediment derivation lies chiefly in the area above the graded transportation surface. In this domain stream beds are being constantly lowered, and slopes leading to the streams are being stripped of whatever sedimentary particles are within the dislodging power of rills and sheet runoff. In both these domains various controls, notably vegetation and rock resistance to weathering and abrasion, introduce endless variation in overall slope declivities.

These two domains have their clearest expression in arid climates where vegetation controls are at a minimum or lacking altogether, and are ideally represented by the pediment and its limiting mountainside or scarp. For purposes of discussion, it seems best to use tentatively the terminology offered by Meyerhoff (1940). The scarp face or mountainside is the *gravity slope* (Penck's *Steilwand* or *Böschung*), which may also be considered as the *derivation slope*. The graded transportation surface is the *wash slope*, equivalent to Penck's *Abflachungshang* or *Fusshang* (Penck, 1953, p. 418-419).[1] The *Haldenhang* or sub-talus rock slope apparently marks the upper limit of steepness of the wash slope because it evidently indicates approximately the angle of rest of the weathered fragments resting upon it.

As a graded transportation surface, the wash slope has much in common with a braided stream channel. Particles at the edge of a rill or minor channel are moved forward into the rill, shallowing it and allowing it to widen correspondingly. Efficiency of the rill channel is thereby decreased, and the widening causes the rill to subdivide. So long as the amount of available sediment exceeds that which the rills can move simultaneously on any and all parts of the slope, no channel cutting or gullying can take place (Lawson, 1932). In Fenneman's words (1908, p. 746), the potential channel-cutting power of the runoff "is *prevented* rather than *withstood.*" The entire surface may be slowly lowered or aggraded, but without losing its graded condition. Davis (1938) has described conditions of incipient channeling of a wash slope during a severe rainstorm, but the abundance of available sediment kept the runoff spread in a manner which well illustrates this principle.

WASH SLOPES UNDER SEMI-ARID CONDITIONS

In its simplest development the wash slope should be completely free from vegetational influence, and controlled only by the abundance of granular sediment. In arid regions, sagebrush and other similar types obstruct runoff to some extent, but vegetational control becomes significant where the

[1] Meyerhoff regarded *Haldenhang* as the wash-slope equivalent.

rainfall is sufficient to sustain a grass cover. As observed on the ground surface, the cover may appear thin and incomplete, but the roots form a continuous mat underneath. The stems, whether standing or fallen, impede and spread the flow of rills, and under these conditions the runoff is presumed to carry its maximum load of sediment. The result is an even-surfaced slope that appears essentially no different from that of typical pediments in the more arid regions. At their upper borders they begin against a scarp or hillside from which comes the bulk of sediment transported across them. An essential characteristic is that of dual control—sediment transportation and vegetational resistance. These controls have operated jointly on these slopes from their beginning, and therefore the gradients represent an equilibrium adjustment at the limit of available transportation energy.

The degree of control inherent in the process of sediment transportation is constant, but that of the grass cover is variable and dependent upon the completeness of that cover. If, when, and where the grass cover is broken, gullying can begin. Then the sides of the gully become small gravity slopes, and the gully widens by slope retreat as well as by stream erosion along its base. But the gully floor, being lower and more moist than the adjacent slopes, acquires a stronger vegetation growth that tends to trap much of the sediment, and such gullies may therefore gradually fill or heal. If they do not heal, the entire slope eventually becomes lowered, with consequently lessened gradient on which the need for vegetational control in maintaining equilibrium is correspondingly reduced. Slopes that illustrate these principles are numerous in the area of Fort Union rocks in southern Montana, in Wyoming, and in many other places. However, in many of these same areas most of the slopes are now being dissected and re-established at lower elevations apparently in response to the present, lowered local baselevel.

Around the northern borders of the Laramie Range, Wyoming, are wash slopes that truncate both the Precambrian granite and the overlying Oligocene-Miocene strata without any topographic expression at the unconformity.[2] Their gradient is slightly higher than that of typical Arizona pediments, but they obviously have developed as equilibrium transportation surfaces. Frye and Smith (1942) have described "pediment-like" surfaces in western Kansas, and more recently Frye (1954) has described similar surfaces in the same region as pediments. All are widening by scarp (valley-side) recession and are in all respects typical wash slopes. Whether the term pediment should be reserved for those wash slopes controlled primarily by the water-spreading power of abundant granular sediment is not discussed here. The alternative term would seem to be valley-floor basements (Davis, 1930, p. 136).

In summary thus far, wash slopes are most conspicuous in arid regions but are equally characteristic of semi-arid to sub-humid regions where impeding vegetation absorbs some of the potential hydraulic energy. The remaining problem is therefore to recognize the corresponding slopes in areas of humid climate. First, however, some features of the gravity slope should be noted.

[2] Blackstone, D. L., personal communication.

THE GRAVITY SLOPE

Although the term gravity slope suggests a surface on which gravity is the chief, or sole, agent in moving loosened fragments to its base, its Penckian equivalent seems to have been applied to landscape elements, at least in the alleged piedmont benchlands, which are far from precipitous. Therefore the term seems applicable to any sediment-yielding surface whose slope is steeper than that of the bordering transport-equilibrium surface with which it merges at its lower edge. Transportation slopes in equilibrium under dual control (vegetation and sediment transport) remain in the category of wash slopes only so long as vegetation continues to be effective. Vegetation failure causes such surfaces to revert to the gravity-type category, though generally the reversion is gradual, spreading with the recession of minor scarps that descend to a new surface of lower gradient. A universal feature of the gravity slope is that the transporting energy of water flowing over it is in excess of that which is taken up in moving the available sediment. Any sediment left stranded on such a surface, following an interval of precipitation runoff, does not form part of a depositional equilibrium surface such as is characteristic of the wash slope.

Distinction between gravity and wash slopes rests therefore on a relative and functional basis rather than on an absolute degree of steepness or gradient; though no wash slopes can be steeper than the sub-talus rock slope (Haldenhang), and only wash slopes may occur at or below the gradient locally controlled by abundance of available sediment, free from vegetational influence. Between these limits, ranging in general from approximately 6°

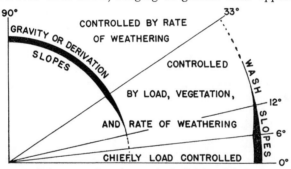

Fig. 1. Diagrammatic interpretation of range in declivities of derivation and wash slopes.

to 30°, both gravity and wash slopes may occur (fig. 1.) In other words, gravity slopes are generally steeper than 6° and wash slopes are never steeper than about 30°. The basis for these distinctions seems clear, necessary, and fundamental; though the possible low angle of the gravity slope makes that term seem inappropriate. *Derivation* slope often seems preferable because it is free from any connotation as to steepness. Gravity slope would then be a special case of the broader category of derivation slopes. At any rate, after the higher gravity slopes have been destroyed, the gradual lowering of wash slopes without loss of graded status involves sediment derivation below the earlier limits of the gravity or derivation surface, but within the limits of equilibrium variation during any one episode of precipitation runoff.

SCARPS AS GRAVITY SLOPES

Except in arid mountain areas where steep slopes extend upward to the ridge summits typical gravity slopes are of moderate height and rise to the resistant edge of a higher surface (wash slope). Most authorities are agreed that such a plains-region scarp cannot retain a sharp or abrupt crest without some kind of resistant capping stratum (Rich, 1938). The degree of resistance required is only relative (and may be inconspicuous), depending on the amount of sheet runoff or rill discharge passing over the brink, and on the rate at which the scarp face beneath it yields to weathering and erosion. The Big Badlands along the White River in South Dakota afford excellent illustrations. Some of the White River strata there are slightly more resistant than others, but the chief brink-forming element is the tough prairie sod. Naturally it cannot withstand the attack of streams flowing from the upland, but it evidently does withstand considerable rill wash. Moreover, it becomes reestablished at successively lower elevations and caps many small tabular surfaces at intermediate levels along the general descent to the valley floor. The steepness of scarps beneath these sod brinks is a measure of their control relative to the rate at which the cliff faces recede beneath them. Summits without sod cover either are sharp ridges or are rounded. Obviously a sod cover is far less resistant than most caprocks that come to notice, but the illustrations afforded by the Badlands introduce some important clues regarding scarp retreat.

HUMID-CLIMATE DIVERSIFICATION

The simplest gravity or derivation slope is the ideal straight or linear scarp, which is probably most nearly realized in arid regions. With increasing runoff and humidity of climate, such a scarp would become notched along its weaker segments, with the result that the length of scarp surface is increased. With more surface thus exposed, a more voluminous supply of sediment is weathered and removed from it; and its removal is conditioned by the increased precipitation and runoff. Some of the notches extend ever farther and become enlarged to the dimensions of valleys. This is especially the case with any streams that rise in the hinterland and flow over the scarp face. In effect, therefore, with increasing humidity of climate, the gravity slope is extended along the valley walls throughout the ramifications of the branching and developing drainage system and is best developed in rock at least moderately resistant to weathering. Slopes steeper than about 30° are entirely of the gravity type; and if all valley sides and the slopes around the valley heads were of this degree of steepness, the domain of gravity slopes in humid areas would be easier to discern. Typically, however, the tops of the valley walls lack a sufficiently resistant caprock. They lose their scarp-like appearance by rounding and merge imperceptibly with the lesser gradients toward the divide crests. The basic reason for such rounding is clear. Vegetational resistance to erosion may be approximately uniform over the entire upland; but outward from the divide summit, each unit distance adds its increment of runoff. Therefore the erosive effectiveness of runoff is correspondingly augmented, with the result that increasing steepness of valley-side slopes, outward from the summit, is inevitable. In large part, this convex zone of pre-dominantly wash slope corresponds to any wash slope lying above (and leading down to) a gravity slope, such as the upland above the White River Badlands.

Loss of scarp-like brink at the top of a young valley wall does not mean that the portion of the scarp or gravity slope thus affected is lost. Rather,it is *distributed* as minor scarps over the lower part of the convex area, each minor scarp being a small-scale gravity-slope unit. At the divide, rainwash creates a graded transportation slope, but within a short distance the rills descend individually over miniature scarps (gravity slopes) controlled by tufts of grass, resistant aggregates, fragments too large to move, or some other comparable temporary and very local baselevel. Then may follow more short distances of graded (wash) slopes, and so on. With increasing distance from the divide, the intervals of wash slope become shorter and fewer and the miniature gravity slopes become correspondingly more numerous until the steepest part of the valley side has been reached. Here erosion is at its maximum, and if any straight element of slope occurs anywhere, it is im-mediately below this line. Lawson (1932) has interpreted such straight segments as indicating removal of sediment as fast as it is loosened by weather-ing. Also Wood (1942) has called it the constant slope; but it is the most conspicuous and uninterrupted part of the gravity or derivation slope.

Below the limit of uninterrupted gravity slope is the area directly under local baselevel control of the stream channel in the valley. The transition begins with the appearance of small graded rill segments, which become dominant as the rills reach the gentler gradient that completes the concavity of the lower part of the valley wall. Ideally this slope is graded to the top of the stream bank, but with some secondary slopes leading to the stream surface; and if the stream were to follow a permanent course along the valley axis, this ideal might be generally attained. However, with meander shifting and continued valley deepening, the actual elevation of effective baselevel control varies from time to time and from place to place along the valley. Where the stream, having impinged against the valley side, shifts away from that site, the margin of the valley floor there must be aggraded. Immediately upslope, gullying has meanwhile been in progress because of the proximity of the main stream as it cut laterally against the base of the valley side. Then as the fans extend across the new valley floor, the upslope gullies may be filled at least in part in response to the requirements of the longer (restored) graded slope. The amplitude of these fluctuations is commonly beyond the limits of immediate adjustment by the slow process of rill- and rain-wash transportation. This tends to obscure the development of the graded bedrock surface (valley-floor basement of Davis) rising to the base of the retreating valley wall. However, the broad uniformity of these slopes reveals a close approximation to the ideal gradient under local conditions. The amplitude of local baselevel fluctuation caused by meander shifting and lateral planation is seen in the thickness of the alluvial-fan deposits. Were it not for these fluctuations, no true fans would develop except in response to possible ac-celerated erosion on the uplands. (In this connection, see King, 1949).

As pointed out by Lawson (1932) and by Wood (1942), the zone of maximum steepness becomes narrowed from both above and below. As this zone is worn back, the transportation surface (wash slope) leading to the valley floor rises against its lower edge while downwasting of the upland reduces its upper limit. In approach to mature dissection, valley widening in this manner and the expanding drainage net carry these wash slopes close to the corresponding network of divides, the limit being established by the local conditions governing the divide profile. The central Appalachian Plateau is still the classic example, and typical badlands are essentially the same except in scale and proportions. Height of the gravity slopes at this stage depends on available relief (Glock, 1932) and on texture of topography. These gravity slopes will remain the zone of most active erosion until the convexity of the lowered divides merges directly with the lower wash slopes. At that stage the gravity slopes as such are eliminated from the landscape, leaving only wash slopes. This should mark the full attainment of Davisian old-age topographic development. Peneplanation represents only the continued reduction of the wash slopes (Abflachungshänge).

SLOPE TYPES AS END MEMBERS

Ideally the terms gravity or derivation slope and wash slope should be defined in terms of the processes which they connote. But in practice it seems feasible to apply either term to those slopes or parts of slopes where the particular process is dominant, with perhaps minor occurrences of the other type. Thus the wash slopes in semi-arid and humid regions have innumerable small vegetation-controlled scarps or non-graded spots. Likewise gravity slope may apply to surfaces on which some graded rills occur. But where the proportion of such minor units approaches half-and-half, neither term should be used without appropriate qualification. As applied to broad slope areas, the terms represent end members of a gradational series, which thereby compels recognition of the minor component units (fig. 1).

In summary thus far, dissection of an upland to the stage of Davisian maturity in a humid climate greatly extends the area of gravity slopes along the valley walls. In further consequence of greater humidity, the stronger vegetational control of erosion results in gravity slopes that are less steep, and in wash slopes that are steeper than their respective equivalents in semi-arid regions. The wash slopes are therefore more vulnerable to gullying where vegetation weakens or fails, causing partial and temporary reversion to gravity slopes. Portions of the gravity and wash slopes become broken up into small and inconspicuous units that are further obscured by the everchanging pattern of vegetation, but that are nevertheless real and that have essential characteristics identical with such surfaces elsewhere. Wash-slope units predominate along the divides and, for the most part, control the lower valley-side slopes. Until well past maturity in the ideal Davisian cycle, the steepest parts of the valley sides are gravity slopes which diminish in prominence as they retreat. They disappear when the divides have been lowered sufficiently to come under the direct baselevel control of the valley floors (fig. 2).

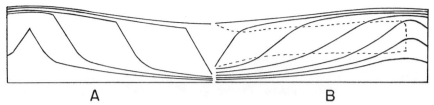

A   B

Fig. 2.   Composite diagram representing valley-side profiles at successive stages in an ideal erosion cycle or sequence. Modified from Davis and from King.

A. Semi-arid grassland environment such as the western High Plains of the United States.

B. Humid environment of the "normal" Davisian cycle. Dashed line indicates zone of predominant gravity or derivation slopes, gradational to predominant wash slopes both above and below.

PENCKIAN-DAVISIAN COMPLEMENTS

The foregoing analysis shows that the fundamental landscape units developed by running water are derivation slopes and transportation slopes, and that these are present under humid as well as arid climatic conditions. The derivation or gravity slopes, wherever they occur, *tend* to maintain their characteristic gradients as they are worn back, the gradient being governed by local conditions of vegetation, if present, and by bedrock response to weathering.

Differences in gross aspects in the two contrasting climatic environments are basically those of unit size and distribution of the two fundamental landscape elements. Both are the less interrupted and therefore the more conspicuous where vegetation is scarce or absent, the effect of vegetation and greater humidity being to decrease the contrast in characteristic slope angle of the two, and to mingle the two types in small units on the slopes of intermediate average steepness. Progressive elimination of the small gravity slopes, with concomitant extension and integration of the wash-slope units, gives the overall effect of decreasing angle of valley-side slopes as downwasting of the divides goes on.

Therefore the essence of Penckian geomorphic principles amplifies and clarifies an aspect of running-water erosion which Davis left obscure. A possible reason for the long-continued obscurity may be the captivating contrast between aspects of arid- and humid-climate topography. Landscapes in humid climates are the more complex, and they have been treated without realizing that they are a composite of small, unlike elements which, in arid regions, occur as the gross or major units of the landscape. To perceive these slope elements in areas of humid climate requires close attention to details of the surface where erosion is in progress. The derisive caricature of a geomorphologist perched on a breezy hilltop and scanning the far horizon for another peneplain has been well earned; and yet perhaps it was inevitable in the development of a field so rich and varied.

PENEPLAIN AND PEDIPLAIN

With the foregoing analysis as a basis for interpretation, the penultimate land form has fundamentally the same topographic elements under any climatic conditions, the differences being mostly in the proportions and distribution of those elements. In arid regions the wash slopes are well consolidated and the few remaining gravity slopes are therefore relatively conspicuous; whereas on a humid peneplain the wash slopes are still partially under vegetation control and show relatively more range in gradient, with possibly a few minor units of gravity slope still in evidence along the low divides. Each final hillside has become a unit wash slope, and the relatively large number of such units, because of the more complete drainage net, gives the landscape its characteristic undulatory aspect.

In humid regions the controlling baselevel in virtually all cases would be sea level. This is likewise true of many if not most of the extensive pediments now developing in southwestern United States which are graded to through-flowing, though intermittent, drainage lines. However, many of the African pedimented landscapes described by King (1953) apparently have no graded continuity to the sea, but have developed under local baselevel control.

In this connection it is significant that Davis (1909, ch. 10) regarded the uplands of eastern Montana as a typical peneplain now well started on a second erosion cycle. The surface of that semi-arid region bevels the simple sedimentary structure, with residual "lava-capped" buttes indicating the minimum thickness of strata thus beveled. It was this area that in 1883 confirmed the Davisian peneplain, and the interpretation of its development first enlivened for his students that which hitherto had been a dull and boresome subject.[3] To the present writer the butte scarps in that area are typical gravity slopes and the long wash slopes (the peneplain) are equally typical of graded transportation surfaces controlled by both vegetation and transport of sandy-silty sediment. Therefore it appears that Davis' type peneplain or "ultimate stage in the sequence of a simple cycle of development" (1909, ch. 10) has at least as much in common with the Arizona pediments as with the Harrisburg or any other demonstrable humid-climate peneplain. In accord with priority usage, all these near-plains of subaerial erosion should be included in the broad category of peneplains, though the arid-climate forms may constitute a pediplain class within the larger category.

Relatively resistant rock masses may become residual eminences on a low-relief landscape in any climate, whether by virtue of especially resistant lithology or merely a relative lack of jointing or other structural advantage. Due account must be taken of the fact that in many cases the degree of resistance to weathering depends as much on climate as on rock type. In so far as superior resistance to weathering can be demonstrated, the term monadnock would seem applicable to any conspicuous residual eminence, though the environment may indicate advisability of reference to a bornhardt or other type of monadnock. Further subdivision of the bornhardt type should be made where possible in order to distinguish between those having monad-

[3] Davis, W. M., unpublished communication to V. A. Rigdon.

nock qualities and those which are merely traces of drainage-divide heights (Penck's *Fernlinge*) and without relatively superior resistance to weathering.

CONCLUDING REMARKS

For purposes of topographic description and interpretation of subaerially produced topography, no better framework of reference than the erosion cycle of Davis has yet been devised. Historically it is only natural that humid-climate topography should have been considered first, although areally it is far surpassed by the arid and semi-arid types. However, the shortcomings of the Davis system appear to have been in the realm of detail recognition and correlation rather than in fundamentally limited applicability. A concept so useful cannot and should not be abandoned, though it must be augmented and improved where possible. In particular, the many varieties of stream-developed topography over the Earth should be considered from the standpoint of their similarities and common elements, so that their distinctive differences shall then be seen in proper systematic relationship in a single comprehensive and adequate scheme.

REFERENCES

Bryan, Kirk, 1922, Erosion and sedimentation in the Papago country, Arizona: U. S. Geol. Survey Bull. 730, p. 19-90.

Davis, W. M., 1905, The geographic cycle in an arid region: Jour. Geology, v. 13, p. 381-407.

————, 1909, Geographical essays: Boston, Ginn and Co. Reprinted 1954, New York, Dover Publications.

————, 1930, Rock floors in arid and humid climates: Jour. Geology, v. 38, p. 1-27, 136-158.

————, 1938, Sheetfloods and streamfloods: Geol. Soc. America Bull., v. 49, p. 1337-1416.

Fenneman, N. M., 1908, Some features of erosion by unconcentrated wash: Jour. Geology, v. 16, p. 746-754.

Frye, J. C., 1954, Graded slopes in western Kansas: Kansas Geol. Survey Bull. 109, pt. 6, p. 85-96.

Frye, J. C., and Smith, H. T. U., 1942, Preliminary observations on pediment-like slopes in the central High Plains: Jour. Geomorphology, v. 5, p. 215-221.

Gilluly, James, Waters, A. A., and Woodford, A. O., 1951, Principles of geology: San Francisco, W. H. Freeman Co.

Glock, W. S., 1932, Available relief as a factor in the profile of a land form: Jour. Geology, v. 40, p. 74-83.

Holmes, C. D., 1952, Stream competence and the graded stream profile: Am. Jour. Sci., v. 250, p. 899-906.

King, L. C., 1953, Canons of landscape evolution: Geol. Soc. America Bull., v. 64, p. 721-752.

King, P. B., 1949, The floor of the Shenandoah Valley: Am. Jour. Sci., v. 247, p. 73-93.

Lawson, A. C., 1932, Rain-wash erosion in humid regions: Geol. Soc. America Bull., v. 43, p. 703-724.

Mackin, J. H., 1948, Concept of the graded river: Geol. Soc. America Bull., v. 59, p. 463-511.

McGee, W J, 1897, Sheetflood erosion: Geol. Soc. America Bull., v. 8, p. 87-112.

Meyerhoff, H. A., 1940, Migration of erosion surfaces: Assoc. Am. Geographers Annals, v. 30, p. 247-254.

Penck, Walther, 1924, Die morphologische analyse: Stuttgart. (Translated by Hella Czech and K. C. Boswell, 1953, Morphological analysis of land forms: New York, St. Martin's Press).

Rich, J. L., 1938, Recognition and significance of multiple erosion surfaces: Geol. Soc. America Bull., v. 49, p. 1695-1722.

von Engeln, O. D., 1942, Geomorphology: New York, The Macmillan Co.

Wood, Alan, 1942, The development of hillside slopes: Geologists' Assoc. London Proc., v. 53, p. 128-140.

UNIVERSITY OF MISSOURI
COLUMBIA, MISSOURI

20

Reprinted from *Geol. Soc. Amer. Spec. Paper 84*, pp. 433, 435–446, 448–451 (1965)

Contribution to the Study of the Brazilian Quaternary

João José Bigarella

Dept. of Geology, University of Paraná, Paraná, Brazil

Gilberto Osório de Andrade

Dept. of Geology, University of Recife, Recife, Brazil

Abstract

After a résumé of the morphoclimatic antecedents of the Cenozoic (pediplanes Pd_3 and Pd_2), a classification is presented of the Brazilian Quaternary on the basis of climatic fluctuations, correlating erosion surfaces with distinct, isolated stratigraphic sequences of semiarid sediments in a large area extending from the Rio Plata (Uruguay) to northeastern Brazil.

In a period of positive epeirogenesis four well-defined periods of mechanical morphogenesis are identified, alternating with humid periods that probably correlate with the Pleistocene glacial-eustatic regressions. The first (Pd_1) is the youngest and most extensive, and its correlative sediments are found in many places (Guabirotuba Formation, Graxaim Formation) including the corresponding phase of pedimentation (Alexandra Formation, Nebraskan?). These are followed by successive phases of pedimentation (P_2, Iquererim II Formation, Graxaim II Formation, Kansan?; and P_1, Iquererim I, Graxaim I Formation, Canhanduva-Cachoeira Beds, Illinoisan?), and a detrital pavement (Wisconsin?) may have developed much later, at the transition from Pleistocene to Holocene.

The best records of glacial-eustatic variations are probably limited to post-Wisconsin events; they may be related to Fairbridge's curves, but they also are concerned with some sediments representing Pleistocene emergences and submergences. These records appear to be related to the Pleistocene glaciations, so that the Brazilian sequence can be compared to that found in the Colorado and the Rocky Mountains.

Because in all areas investigated traces of pediments and the correlated deposits signify extensive mechanical morphogenesis during periods of Pleistocene glaciation, it is necessary to make a new approach to the general conception that the glacial phases with lower temperatures correspond to the pluvial periods.

Contents

Introduction

This paper discusses the results of a tentative systematic study of the Brazilian Quaternary. The origin of erosion surfaces and the filling of the basins with sediments are investigated, and the events of the interior are correlated with those of the coastal area.

Although extensive, the Brazilian literature on this subject is neither detailed nor systematic. Study of the Quaternary in Brazil is still only beginning: in northeast Brazil some research has been done by Andrade, in São Paulo by Ab'Sáber, and in Paraná and Santa Catarina by Bigarella.

The discovery of residual pediments at the foot of the Serra do Mar in the present pluvial forested areas completely modified the results of geomorphic studies in southeastern and southern Brazil (Bigarella and others, 1961; Bigarella and Salamuni, 1961; Bigarella and Ab'Sáber, in press; Andrade and others, 1963).

Recent detailed studies in the Quaternary of Brazil suggest major climatic changes and fluctuations. Maack (1947, p. 150) was among the first to call

TABLE 1. QUATERNARY STRATIGRAPHY IN BRAZIL

System	Series	Stage		Tectonics	Climate	Process	Geomorphic feature	Formation			
		Glacial	Interglacial					Rio Grande do Sul	Santa Catarina	Paraná	Pernambuco
Quaternary	Holocene			Positive epeirogenesis (?)	Humid with dry phases	Dissection and sedimentation	Low fluviatile terraces and wave-built terraces	Rudaceous, sandy, and clayey-silty sediments			
	Pleistocene	Wisconsin			Semiarid	Mechanical morphogenesis	Paleopavement	Rudaceous deposits			
			Sangamon	Positive epeirogenesis	Humid	Dissection					
		Illinoisan			Semiarid	Mechanical morphogenesis	Pediment P_1	Graxaim Fm. (1)	Iquererim Fm. (1), Canhanduva beds, Cachoeira beds	not named	
			Yarmouth	Positive epeirogenesis	Humid	Dissection					Riacho Mórno Fm.
		Kansan			Semiarid	Mechanical morphogenesis	Pediment P_2	Graxaim Fm. (II)	Iquererim Fm. (II)	not named	
			Aftonian	Positive epeirogenesis	Humid	Dissection					
		Nebraskan			Semiarid	Mechanical morphogenesis	Pediplane Pd_1	Graxaim Fm. (III)		Alexandra Fm., Guabirotuba Fm.	

attention to semiarid conditions during an undetermined epoch of the Quaternary. Cailleux and Tricart (1957) mentioned the importance of climatic fluctuations in southeastern Brazil. However, none of these authors referred to the occurrence of pediments.

Sedimentary features in many localities document cyclic climatic changes during the Quaternary. On the basis of both these sedimentary features and the erosional surfaces, we shall try to determine the history and chronology of the events of this period. Table 1 shows the Quaternary stratigraphy of Brazil.

Small climatic fluctuations seemingly were present in the humid epochs, as well as in the rough semiarid ones. They played an important role in wearing down the surface. Mortensen (1947) has called attention to this phenomenon. The last climatic fluctuation in the present humid epoch was a dry phase ending at about 450 B.C. (Bigarella, in press).

Acknowledgments

Grants from the Conselho Nacional de Pesquisas (Rio de Janeiro, Brazil), University of Paraná, and University of Recife made this paper possible. The authors are grateful to Edwin D. McKee for his kindness in revising and improving the manuscript.

Methods

Because of the discontinuous character of Quaternary sedimentation, which makes the correlation of several small, generally isolated sedimentary basins difficult, a new but appropriate method has been used in place of the classical stratigraphic approach. Erosional surfaces (pediments and pediplanes)[1], normally coinciding with aggradation surfaces developed at the end of the process, are used as reference planes. This method makes it possible to correlate several isolated stratigraphic sequences over large distances, from the Plata River in Uruguay to northeastern Brazil (Fig. 1). The importance of a rational interpretation of the degradation and aggradation levels is emphasized.

One gains clearer understanding of the phenomena responsible for the development of the landscape after analyzing several well-developed erosion surfaces whose remnants make possible a reconstruction of the post-Cretaceous physiographic history.

---

[1] *Pediment* refers to the sloping erosional plain at the foot of mountains or inside valleys developed under semiarid conditions. The word *pedimentation* is applied to the phenomenon of pediment formation. The erosion surface developed under semiarid climate by the coalescence of pediments here is referred to as a *pediplane*.

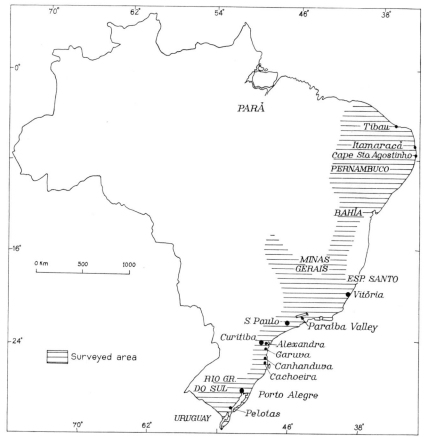

Figure 1.   Distribution of pediments P_1 and P_2 and pediplane Pd_1 in Brazil. Locations are shown for the main sites mentioned in the text.

Detailed sedimentological studies have been made in the Curitiba basin, with which one may analyze the hydrodynamic character of the processes involved in the sedimentation, as well as explain the various types of sediment and their probable origin (Bigarella and Salamuni, 1962).

Cenozoic Antecedents

Extensive marine sedimentation in Brazil ends with deposition of Cretaceous rocks. Cenozoic deposits are mostly continental, except in northeastern Brazil, where some late marine invasions occurred. Climatic conditions ranged from subhumid to moderately semiarid; the latter condition was

311

sometimes very intense. The Cenozoic sediments extended beyond the limits of the Mesozoic basins.

The end of Cretaceous sedimentation coincided with formation of a well-developed pediplane, whose remnants are preserved in many places as summit surfaces. The pediplane was formed under semiarid conditions. This surface has been described under many local names, and it is treated here as pediplane Pd$_3$. Its correlative deposits are found in the upper part of Cretaceous sequences, like the Bauru Series, Santa Tecla Series, and others. According to the literature, this pediplane is considered to be Cretaceous-Eocene.

The Cenozoic Era in Brazil was characterized by intense erosional processes, beginning with pediplane Pd$_3$. The cyclic nature of the erosive phases is unknown, for they lack an adequate stratigraphic record. However, some special, local conditions favored deposition of sediments in certain areas, for example in the basins of São Paulo, Curitiba, Paraiba Valley, Gandarela, and Barreiras. Among these basins only the last one covers a large area, extending along the coast from Pará to Rio de Janeiro and penetrating a considerable distance into the Amazon Valley.

Sediments included in the so-called Barreiras Series are complex and probably represent a varied record of more than one geologic epoch. As yet there is no systematic study on the stratigraphy of the whole Barreiras basin sediments.

The Cenozoic sediments are not widespread, being mostly products of brief deposition and frequently repeated in various climatic phases and separated by irregular erosion surfaces. As they usually represent a fragmentary record of a much discussed and very controversial chronology, they fail to furnish a reasonable picture of Cenozoic events in Brazil.

Pediplane Pd$_3$ was warped during the Cenozoic, probably by deep-seated warping. This was followed by faulting in eastern Brazil, which produced the embryonic organization of the present assemblage of block mountains of the Serra do Mar, later continuously accentuated by tectonic and erosive processes. Part of the great flexure of the Pd$_3$ surface dipped toward the sea, and for the first time in northeastern, southeastern, and southern Brazil the drainage flowed directly to the Atlantic.

New bevelling, possibly of middle Tertiary age, developed an interplateau and peripheral surface through pediplanation. This surface, although very eroded, has notable residual relief (old inselbergs from the period of semiaridity). Correlative deposits of this surface are unknown in southern and southeastern Brazil. However, in northeastern Brazil, the Guararapes Formation (Barreiras Group) on the eastern coast, and possibly the Serra do Martins Formation inland, represent correlative deposits of this erosion surface. It here is referred to as pediplane Pd$_2$.

Morphoclimatic Processes

The Quaternary in Brazil has been characterized by great and extreme climatic changes, *i.e.* long semiarid epochs alternating with humid ones. During semiarid periods, mechanical erosion with lateral planation developed pediment and pediplane surfaces. In the humid epoch chemical weathering was conspicuous, forming thick regoliths. Linear erosion accentuated the thalwegs, pediplanes were rejuvenated, and topography developed into a more rugged relief.

During times of semiarid climate forests were possibly restricted to small refuge areas, where local climatic conditions would allow them to survive. Nothing yet is known about the flora of these desert-forming phases, especially at the time of climax development of the pediments and pediplanes.

In the humid epoch the forest reached its maximum expansion. Such an epoch, however, was not entirely humid; it was marked by climatic fluctuations of short dry phases and extended wet periods. These fluctuations were responsible for the deep dissection of the terrain and were helped by positive epeirogenic movements.

In a forest area subjected to a humid climate, little erosion would be possible in such a short geologic time as a Pleistocene interval. Under such conditions erosion would be more effective in leaching than in the removal of clastics. The existence of climatic fluctuations within the humid epochs resulted in climatic conditions alternating between wet and dry phases. The periods of drought caused the forest to retreat, at the same time increasing areas of grassland, *cerrado*, or *caatinga*.

The thinner vegetative cover exposed the soils to a different rain regimen, probably one of concentrated rainfall, with accelerated dissection of the surface. In the same way, supposedly, humid phases should be found within the semiarid epochs; they would greatly facilitate the development of pediments and pediplanes.

Three well-defined epochs of mechanical morphogenesis have been identified in the Brazilian Quaternary. Corresponding pediplanes and pediments received in our field work the designation of Pd_1, P_2, and P_1, respectively. A remnant of pediment P_3 found at the Serra do Mar and elsewhere corresponded to Pd_1.

Topographically, the pediplane and the pediments represent well-marked levels.

The accumulation of debris is related to epochs of mechanical morphogenesis. Between epochs of mechanical morphogenesis were epochs of humid climate, responsible for stream erosion and the mammillary features of the landscape. Positive epeirogenesis also occurred at this time.

The mechanical morphogenetic epochs seem to relate to a lower sea level. In Rio Grande do Sul the sediments of the Graxaim Formation, deposited during the time of pedimentation, stand today in a great extension below sea level. This criterion correlates pedimentation epochs with times of Quaternary glaciation.

The detrital paleopavements are younger than the last pediment that is recognized throughout Brazil. The pebbly layers of the paleopavement are buried in the subsurface, roughly following the present topography. An epoch of semiarid climate is perhaps responsible for its development.

Pediplane Pd$_1$

Pediplane Pd$_1$ is the most recent and extensive erosion surface in Brazil. Inland it forms interplateau surfaces, and in the coastal area it forms a surface that dips seaward (Figs. 2 and 3). It is well exposed from Rio de Janeiro to the Amazon.

Correlative deposits are present in many places: Graxaim Formation (older sequences) in Rio Grande do Sul, Guabirotuba Formation in Paraná, and the upper part of Barreiras sediments (Riacho Môrno Formation) in eastern and northeastern Brazil. This surface has several local names: "Chãs" and "Tabuleiros" surface in Pernambuco, Neogenic surface in São Paulo, Curitiba surface in Paraná, Campanha surface in Rio Grande do Sul, Montevideo surface in Uruguay.

In Paraná, the Curitiba surface (Pd$_1$) directly cuts into the Alto Iguaçu surface (Pd$_2$). The end of its development coincides with the filling of the Curitiba basin, represented by the Guabirotuba Formation.

The age of pediplane Pd$_1$ is still an unsolved problem. Its chronology, however, is related to the end of deposition of the nonfossiliferous sequences mentioned previously. If one considers the sediments as lower Pleistocene, Pd$_1$ would be Pleistocene in age. Nevertheless, if one prefers to consider them as older, probably Pliocene, the surface would be placed near the boundary between Pliocene and Pleistocene. We believe that this pediplane was completed during the first Pleistocene glaciation (Nebraskan).

Pediments

Pediment remnants from more than one stage of Pleistocene pedimentation are distributed throughout Brazil, from the northeast to the La Plata River in Uruguay.

In southern Brazil, pediments were discovered in Garuva at the foot of the Serra do Mar on the Paraná-Santa Catarina border. Three different levels of pediment remnants have been found. They are designated P$_1$, P$_2$, and P$_3$, in

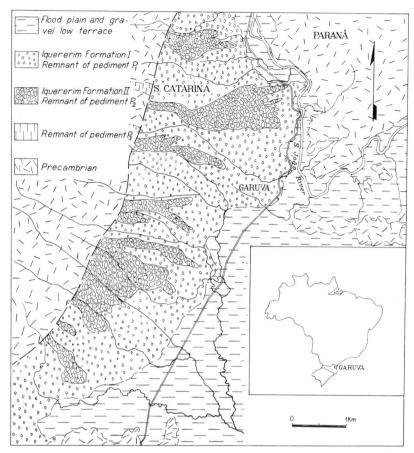

Figure 2. Distribution of pediments and correlative deposits near Garuva at the Paraná–Santa Catarina border, Brazil. A schematic profile of the area is shown in Figure 3B. The Precambrian area is made up of crystalline rocks and is mountainous. The pediments are composed mainly of detrital material, with rocky pediments only in the western part near the mountains.

increasing order of antiquity. Pediment remnants have similarly been found in other places of the Serra do Mar from Santa Catarina to Espírito Santo. Pediment remnants also have been found in other areas in Rio Grande do Sul, Uruguay, São Paulo, Minas Gerais, and Bahia, as well as in semiarid northeastern Brazil, where these features are conspicuous. North from Vittoria (Espírito Santo) the P_3 pediment remnant can be seen to correspond to the pediplane Pd_1.

Between epochs of pedimentation humid conditions prevailed, and epeirogenic movements seemingly were active. We interpret pediments P_1 and P_2 as developments of glaciated periods, probably of the Illinoisan and Kansan, respectively.

Detrital Paleopavement

After the development of pediment P_1 a humid epoch occurred, causing generalized mammillary development of the topography across large areas of Brazil, except in some regions of semiarid northeastern Brazil.

A climatic change, bringing back semiarid conditions, resulted in an extensive detrital pavement almost all over the country above the mammillary areas, as well as on other surfaces. This desert pavement was later covered by the colluvial material introduced during the present humid epoch.

The problem of the paleopavement was recently discussed by Ab'Sáber (1962), who contends that the paleopavement is made up of a pebbly zone of varied thickness, buried by silt and clay, as well as by actual soils. The climatic conditions under which the paleopavement was developed apparently were neither so rigorous nor so extensive as the conditions responsible for the pediments and the pediplane.

Outcrops suggest that more than one phase of paleopavement formation is represented. The present feature is the result of successive reworkings of pebbles by several dry phases following pediment P_1.

The detrital paleopavement was previously dated by Bigarella and Ab'Sáber, who believed that it developed during the transition from Pleistocene to Holocene. However, it seems older, possibly contemporaneous with the last glaciation (Wisconsin). Nevertheless, the problem is not simple. In its development a recurrence of dry phases with successive reworking, which is not stratigraphically distinguishable, must be considered. In this way, the subsurface pebbly horizon may have different ages in different areas.

Sedimentation

The history of Quaternary deposition is mainly related to semiarid climatic phases. The Guabirotuba Formation in Paraná is of uncertain age (Pliocene-Pleistocene), but it may correspond entirely to the Nebraskan glaciation, if the end of sedimentation proves to be related to the aggradation surface of pediplane Pd_1 (Curitiba surface). A detailed study of the textural characteristics of this formation, made by Bigarella and Salamuni (1962), clearly indicates a semiarid climate at the time of its deposition.

Dissection of the Alto Iguaçu surface (Pd_2) under conditions of humid climate resulted in an irregular erosional depression with valleys more than

100 m deep. A large valley occurs along the main drainage (former Iguaçu), flat in relation to its extent, with lateral valleys forming a dendritic complex. This phase of sculpture did not produce the flat surface; it resulted from an interruption in the formation of an irregularly hilly surface.

The previously mentioned erosional phase, which occurred in a humid climate, was interrupted by climatic change through which the climate became semiarid, with concentration of rain in short periods. Therefore the regional aspect of the landscape was modified in various details. As the vegetation cover became sparse, the soil surface was exposed to intensive erosion. The rainfall, concentrated in definite periods, was torrential. Falling on an unprotected soil, the heavy sheets of rains could carry sediment downslope toward the valleys and drainage channels. These sheet floods behaved like muddy rivers of high density, unable either to select sediment according to grain size or to transport its load far. Most of the sediment was deposited in the first break of gradient of the intermittent stream. Such procedure caused the formation of a group of coalescent alluvial fans typical of semiarid climates.

When a stream in a semiarid climate loses the first part of its load mainly because the gradient changes, a large percentage of the clayey-silty particles remain suspended, and transport of coarser grains continues because the fluid is denser. Such sediments are carried to depressions of the terrain, where they are deposited in playa lakes.

During the humid-climate phase that preceded the phase of semiarid climate, a fairly thick regolith was formed, which in textural composition was not unlike that formed today. This material was the source of sediments that filled the basin during the epoch of the Guabirotuba Formation.

The regoliths capping the internal and surrounding hills of the basin and the Alto Iguaçu surface (Pd$_2$) were probably easily eroded and transported. Therefore most of the sediments composing the Guabirotuba Formation were derived from chemical decomposition of the crystalline Precambrian rocks under the influence of a humid climate. They were eroded, transported, and deposited under semiarid climatic conditions. The kaolinitic composition confirms the climatic conditions existing at the time of regolith formation, inasmuch as it was the source of these sediments.

In the Curitiba Basin, Bigarella and Salamuni studied 127 samples of sediments from the Guabirotuba Formation, comparing them with 89 Holocene samples of the floodplain (várzeas) and 37 samples of Precambrian gneisses derived from regolith.

An analysis of the granulometric composition reveals that the Guabirotuba Formation presents properties different from those of floodplain deposits. Median diameter and Inman's sorting coefficient show that the floodplain sediments were deposited under less severe climatic conditions. They present a better sorting of grains, as the median diameter is nearer to 0.18 mm.

This is not the case with deposits of the Guabirotuba Formation, in which no sorting occurs.

Deposition of Quaternary sediments was generally from sheet floods, rather than in channels. Mud torrents were widespread. In many places, mainly in mountain foothills and in rough mountainous areas, they were responsible for the formation of coarse rudaceous deposits.

Under the classical view, detrital pediments are formed with only a thin veneer of rudaceous deposits. However, in occurrences described from the Garuva area, the detrital pediment assumes larger importance because of climatic alternation, epeirogenesis, dissection, and base-level change that occurred during a humid interval.

GUABIROTUBA FORMATION (CURITIBA BASIN): The deposits of this formation present a varied texture and are not well sorted. The formation overlies an irregularly eroded surface, cut in Precambrian metamorphic rocks. It is composed of clayey-silty sediments, arkosic sands, and conglomerates. No stratification is visible. These sediments were deposited by sheet floods in a bajada environment, as alluvial fans, and as playa-lake deposits (Bigarella and Salamuni, 1962).

The end of deposition of the Guabirotuba Formation coincides with the formation of pediplane Pd$_1$ (Curitiba surface).

ALEXANDRA FORMATION: The deposits of this formation overlie an irregular surface cut in Precambrian crystalline rocks, occurring in the littoral of Paraná. They are composed of arkosic clayey-silty sands and conglomerates, deposited in a semiarid environment (bajada). The close of deposition of this formation coincides with formation of pediment P$_3$ of the eastern slope of the Serra do Mar. This pediment is correlated with pediplane Pd$_1$.

IQUERERIM FORMATION: This unit, described by Bigarella and others (1961) as a group of rudaceous deposits in the foothills of the Serra do Mar in Garuva, is situated near the Paraná-Santa Catarina boundary (Fig. 3). It was originally linked to the phases of pedimentation that formed pediments P$_2$ and P$_1$ and are thus named Iquererim Formation (II) or Iquererim Formation (I), separated by an erosional unconformity. Phase II of the Iquererim Formation is composed of boulders and angular to subangular pebbles of heterogeneous composition (gneiss, granite, diabase), with a diameter ranging from a few centimeters to 4 m, in a clayey sandy-silty matrix.

CANHANDUVA-CACHOEIRA BEDS: Most deposits related to times of pedimentation are widespread. They are alike in their characteristics, including deposition from sheet flood, evidence of a semiarid climate, and a distinctive texture. Grains are angular or subangular and rudaceous when near to source areas where morphogenetic processes were operating mainly through mechanical weathering. Where deposited far away in a bajada, they are clayey-silty, with frequent inclusions of sand-sized quartz and feldspar; also they contain granules or small pebbles.

The Canhanduva locality is situated between Itajai and Camboriú (Santa Catarina). It was described by Bigarella and Salamuni (1961) and is composed of two distinct deposits. The basal portion lies over an irregular surface cut in sericite quartzite and sericite schist of the Brusque Series (Precambrian). It is composed of unstratified clayey-silty sediments and is rich in

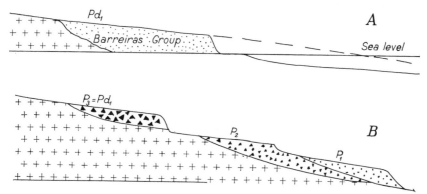

Figure 3.   Schematic stratigraphic relationship between sediment and erosional surfaces or benches (not to scale). A shows pediplane Pd_1 (Chãs and Taboleiro surfaces in Pernambuco); B shows the relationship between the rocky and detrital pediments of Figure 2. The thickness of the detrital pediment results from positive epeirogenesis and deep dissection mainly in a humid climate.

quartz and quartzite pebbles of varied sizes, with a basal rudaceous layer. The source of these beds was a regolith, formed in a humid climate and removed to depressions in the terrain during a semiarid interval during the formation of pediment P_1 by mechanical morphogenesis.

The Cachoeira beds occur about 20 km south of Tijucas (Santa Catarina) in the BR-59 highway cuts. They are composed of sediments deposited in bajada environments and are correlated with the pedimentation P_1 epoch.

GRAXAIM FORMATION: This occurs mainly in highway cuts between Pôrto Alegre and Pelotas. It includes three distinct units of mechanical morphogenesis, separated by deep erosional unconformities developed under a humid climate. The Graxaim Formation (III) was deposited during the development of Pd_1. Its sediments (II and I) were formed respectively during the development of pediments P_2 and P_1. Most of these sediments are today below sea level.

[*Editor's Note:* A short section on wave-built terraces and sea-level fluctuations has been omitted.]

Correlations and Conclusions

In the absence of an appropriate basis for chronology through radiometry, the age of the erosion surfaces (pediments P_2 and P_1 and pediplane Pd_1) and the detrital paleopavement are tentatively correlated with the stages of Pleistocene glaciation. Such surfaces were developed during the Pleistocene, when sea level was much lower than during the present.

The surfaces slope slightly seaward, in some places about 10 m per km. When one of these surfaces, with its correlative deposits, is projected under ideal conditions, it passes below sea level, in places less than 10 km off the coast. In Rio Grande do Sul, continental sediments of the Graxaim Formation document a continental environment down to about 100 m below the present sea level. Samples from the Argentine continental shelf, dated by the C^{14} method, indicate great sea-level displacements during the last glacial period. Fray and Ewing (1963, p. 126) state that sea level was lowered a minimum of 120 m and perhaps as much as 150 m prior to 35,000 years B.P., and a minimum of 110 m about 11,000 to 12,000 years B.P. Possibly a former shore line also developed when sea level was 55 to 73 m below the present level, at a time 15,300 years B.P. World literature frequently refers to other lower sea levels developed during other glacial periods, but in Brazil no reference has yet been made to them, inasmuch as correlative deposits are not known.

We correlate the pediments (P_2 and P_1) and pediplane (Pd_1) with Pleistocene glacial stages, and, for reasons given previously and in a speculative way, we correlate the Brazilian sequence of events here described with that established in the Rocky Mountains of Colorado between Denver and Boulder (Scott, 1960, p. 1543). In that region, Dr. G. Richmond guided one of us (J. J. Bigarella) through the area, defining the most recent pediment as Illinoisan (P_1 according to our field designation), the intermediate (pediment P_2) as Kansan, and the oldest, which is very extensive (pediplane Pd_1), as Nebraskan. A pebbly terrace inlaid in the Illinoisan pediment was referred to the Wisconsin age.

All across Brazil we noticed a similar distribution of pediplane and pediment. Therefore this correlation is suggested by the attitude and sequence of the erosion levels in relation to the sea level and by the continental nature of its deposits.

The interpretation of pediplane Pd₁, subsequent pediments, the more recent detrital paleopavement, and the correlative deposits led us to conclude that the climatic conditions prevailing at these times were semiarid.

In Büdel's opinion (1959, p. 3), during the Pleistocene cold phases, the tropical humid belt was somewhat enlarged by increasing humidity and rainfall, whereas the subtropical belt was drier than today's. Büdel, however, admits exceptions to this principle. On the other hand, it is a general opinion that the glacial stages correspond to pluvial periods. In spite of this, we came to the conclusion that semiarid conditions prevailed from north to south in Brazil during the times of Pleistocene ice.

Evidence gathered in our field work indicates the necessity for a new approach to the problem. In all the surveyed areas, geomorphic processes different from those operating today prevailed in the Pleistocene stages of low temperature, producing a distinctive assemblage of forms. Even in low-latitude areas, like the wet coastal area of northeastern Brazil (2000 mm rainfall), as well as in the middle subtropical humid latitudes of southern Brazil (1400 to more than 3000 mm rainfall), remnants of pediments and correlative sediments laid down in a semiarid environment are present.

We deduced that aridity in the eastern part of South America extended very close to the equator. At present, we do not know what the situation was in other intertropical South American areas, for instance in the Amazon Valley, where information on pediment remnants and other evidences of semiarid climate are lacking.

During the lower-temperature Pleistocene phases, across large surveyed areas of South America from La Plata River to northeastern Brazil (lat. 36° to 3° S.), the climate was semiarid and not pluvial as is generally believed. This phenomenon does not seem to be local, but of widespread character. Tropical humid conditions may have been restricted to smaller areas where topography was favorable for rainfall, and these areas were natural refuges for the flora and fauna.

Pleistocene climatic changes seem to have been universal and synchronous in both hemispheres (Büdel, 1951, p. 1; Fairbridge, 1961). Pediment remnants with strong morphologic resemblances can be seen in Europe, North America, and other continents, in places where the present climate is not favorable to this topographic morphology. Thus a world-wide aridity during the cold Pleistocene phases apparently affected large areas.

We are not yet able to decide which were the semiarid climatic conditions that promoted formations of pediplane and pediments during the Pleistocene glaciations. Also, similar features have been developed in a cyclic manner outside glacial conditions during the Tertiary and even probably before. In

Brazil one notices the Cretaceous-Eocene erosion surface (pediplane Pd$_3$) and the middle Tertiary erosion surface (pediplane Pd$_2$), for the development of which semiarid climate doubtless played an important role. Between these surfaces, several erosion levels (shoulders) may be interpreted as pediment remnants. In Brazil, we have noticed these remnants in the northeast as well as in the south. Between the climatic phases during which the pediplanes and the pediments developed, humid climatic phases occurred.

It is still too speculative to speak of a world-wide cycle of aridity and humidity, but a working hypothesis could be established. Many continental sediments in the geologic column were deposited under semiarid conditions. This may be related to the degree of radiation and so may have an astronomic origin.

Concerning the origin of the Pleistocene glaciations, Fairbridge (1961) emphasizes that they are due to radiation cycles accentuated at a time after the Antarctic continent reached its polar position at the end of the Pliocene Epoch.

Our approach to the problem is new, our conclusions are preliminary, and much more must be done to achieve a broad general view of this geological epoch.

References Cited

AB'SÁBER, A. N., 1962, Revisão dos conhecimentos sôbre o horizonte, subsuperficial de cascalhos inhumanos no Brasil Oriental: Curitiba, Univ. Paraná Bol., Geog. Fis., no. 2, 32 p.

ANDRADE, G. O., 1955, Itamaracá; contribuição para o estudo geomorfológico da costa pernambucana: Recife, Imprensa Oficial do Estado de Pernambuco, 84 p.

ANDRADE, G. O., BIGARELLA, J. J., and LINS, R. C., 1963, Contribuição à geomorfologia e paleoclimatologia do Rio Grande do Sul e do Uruguai: Curitiba, Bol. Paran. Geografia, nos. 8 and 9, p. 123–131

BIGARELLA, J. J., 1964, Variações climáticas no Quaternário e suas implicaçoes no revestimento florístico do Paraná: Curitiba, Bol. Paran. Geografia, no. 10/15, p. 211–231

BIGARELLA, J. J., and AB'SÁBER, A. N., 1964, Palaeogeographische und Palaeoklimatische Aspekte des Kaenozoikums in Suedbrasilien: Zeitschr. für Geomorphologie, v. 8, p. 286–312

BIGARELLA, J. J., and FREIRE, SONIA S., 1960, Nota sobre a ocorrencia de cascalheiro marinho no litoral do Paraná: Curitiba, Univ. Paraná Bol., Geologia 3, 22 p. 3, 22 p.

BIGARELLA, J. J., and SALAMUNI, R., 1961, Ocorrências de sedimentos continentais na região litorânea de Santa Catarina e sua significação paleoclimática: Curitiba, Bol. Paran. Geografia, nos. 4 and 5, p. 179–187

—— 1962, Caracteres texturais dos sedimentos da bacia de Curitiba: Curitiba, Univ. Paraná Bol., Geologia 7, p. 164

BIGARELLA, J. J., MARQUES Fo., P. L., and AB'SÁBER, A. N., 1961, Ocorrência de pedimentos remanescentes nas fraldas da serra do Iquererim (Garuva, SC): Curitiba, Bol. Paran. Geografia, nos. 4 and 5, p. 82–93

BIGARELLA, J. J., SALAMUNI, R., and MARQUES FO., P. L., 1961, Método para avaliação do nível oceânico à época da formação dos terraços de construção marinha: Curitiba, Bol. Paran. Geografia, nos. 4 and 5, p. 111–115

BÜDEL, J., 1959, The "Periglacial"—morphologic effects of the Pleistocene climate over the entire world (contribution to the Geomorphology of the Climatic Zones and Past Climates IX): Internat. Geol. Review, v. 1, p. 1–16 (Trans. by H. E. Wright, Jr., and D. Alt).

CAILLEUX, A., and TRICART, J., 1957, Zones fitogeographiques et morphoclimatiques du Quaternaire du Brésil: Paris, Soc. Biogeog., Comptes Rendus, no. 293, p. 7–13

FAIRBRIDGE, R. W., 1961, Convergence of evidence on climatic change and ice ages: N. Y. Acad. Sci. Annals, v. 95, p. 542–579

—— 1962, World sea-level and climatic changes: Rome, Quaternaria, v. 6, p. 111–134

FRAY, C., and EWING, M., 1963, Wisconsin sea level as indicated in Argentine continental shelf sediments in Pleistocene sedimentation and fauna of the Argentine shelf: Acad. Nat. Sci. Philadelphia Proc., v. 115, p. 113–152

LIMA, A., 1962, A vegetação da ilha de Santo Aleixo: Recife, Cadernos Fac. Filosofia de Pernambuco da Univ. Recife, Imprensa Universitária, no. 7, 6 p. (Dept. Geog., ser. VI-5, no. XXb1)

MAACK, R., 1947, Breves notícias sôbre a geologia dos Estados do Paraná e Santa Catarina: Curitiba, Arq. Biol. Tecn., v. 2, art. 7, p. 66–154

MORTENSEN, H., 1947, Alternierende Abtragung: Göttingen, Akad. Wissensch., Math.-Phys. Klasse Nachrichten, v. 2, p. 3–30

RICHARDS, H. G., and BROECKER, W., 1963, Emerged Holocene South American shore-lines: Science, v. 141, p. 1044–1045

SCOTT, G. R., 1960, Subdivision of the Quaternary alluvium east of the Front Range near Denver, Colorado: Geol. Soc. America Bull., v. 71, p. 1541–1544

Part III

ETCHPLAINS

In 1933, E. J. Wayland (Paper 23) suggested that some planation surfaces in Uganda may have been formed by the *etching* of a previously formed peneplain (or pediplain) under the action of deep chemical weathering and subsequent surface wash. Such plains were called *etched plains*. Since then the idea has been rather warmly accepted and applied to other areas in Africa, Australia, and India.

A similar theme runs through papers by Julius Büdel; he recognizes the role of deep weathering and wash, but does not use the term etched plain, pointing out that *double surfaces of planation (doppelten Einebnungsflächen)* are developed as the *weathering front* descends more deeply into the bedrock while wash tends to remove the upper weathered zones.

In considering the broad relations between etched plains, peneplains, and pediplains, we note:

1. Etched plains seem to require the presence or former existence of a previously developed peneplain or pediplain, or the plains that have been etched.

2. The term now in common use, *etchplain,* connotes a plain formed by etching. Although etching suggests roughening of a surface rather than smoothing, the latter may develop in deeply weathered material with a suitable base level; roughening of the bedrock may not destroy the essential plain character of the pre-etched surface. It may therefore be appropriate to speak of *etchplains* as new plains formed by etching.

3. The deep contemporaneous weathered zone on a peneplain, if stripped away, may reveal a surface that, since it was formed by chemical etching, should be considered an etchplain.

Editor's Comments
on Papers 21 and 22

21   WOOLNOUGH
The Influence of Climate and Topography in the Formation and Distribution of Products of Weathering

22   WAHLSTROM
Pre-fountain and Recent Weathering on Flagstaff Mountain near Boulder, Colorado

EARLY DEEP WEATHERING STUDIES

In 1895, G. F. Becker recognized that deeply weathered, chemically rotted rock that remained in situ over its parent bedrock needed a distinctive label; he termed it *saprolite*. This is essentially the B horizon of an in situ soil. If there is alternatively, or in addition, a mantle of transported soil material, a more encompassing term, *regolith* (Merrill, 1897), should be used. Under a specific sequence of changing climatic regimes this weathered mantle gains a cemented crust near its surface, named by G. W. Lamplugh (1902) according to its chemistry—*ferricrete, silcrete,* or *calcrete*. J. Walther (1924) expressed many of these concepts as applied to a lateritic saprolite with an iron crust (ferricrete) (see the diagram reproduced here as Figure 1). The production of saprolite requires extended conditions of warm, wet climates, whereas the production of the crusts requires relatively brief periods of semiarid conditions.

Collectively, these cemented veneers were dubbed *duricrusts* by W. G. Woolnough in a 1927 paper. In Paper 21 we reproduce his 1930 discussion in which he considers changes in the regolith that might have had enduring effects on the Australian landscape. He discusses the processes and climatic conditions necessary to develop a cemented zone, resistant to weathering and erosion, near the surface of the saprolite. Duricrusts have developed over such widespread areas in Australia that Woolnough suggests that they be given stratigraphic status and called *"the* duricrust."

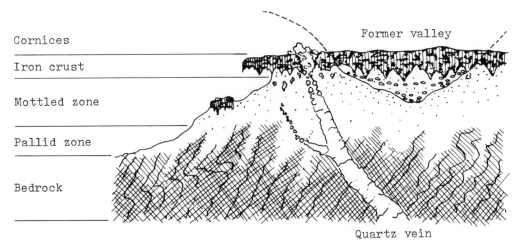

Figure 1. Laterite cover over a folded schist outcrop injected by a quartz vein. The schist was cut by a former valley, through whose gravel filling the laterite continues unchanged. The arrows indicate the supposed land shape before the period of laterization.

When a regime that permits the development of deep saprolites changes to one of dissection, duricrusted areas tend to remain as cap rock on residual hills or as *gibber plains* (*hamada* in Africa) strewn with duricrust fragments.

Although deep weathering is not always accompanied by duricrusting, areas where it has developed tell us something about former climates (alternating wet and dry seasons where now it is either arid or humid). Duricrusts may find use as horizon markers for former planation surfaces now out of equilibrium with the landscape and in the process of destruction or burial. A duricrust is thus a form of *paleosol*.

Woolnough was of the opinion that duricrusts are not forming now in Australia and that *"the* duricrust" is of early Tertiary age. This is a matter for further discussion.

We pass now to Paper 22, which presents an instructive case of deep weathering, laterization, and possible etchplanation of Mississippian–Pennsylvanian age in the tilted beds flanking the Rocky Mountains near Boulder, Colorado.

The author demonstrates that (1) the lateritic weathering proceeded in a climatic regime unlike that of the present, (2) lateritic saprolite of varying thickness was eroded in some areas before burial by the Fountain Formation, (3) at the outcrop shown, little internal disturbance, that is, shear, accompanied the tilting of the laterite beds, and (4) conditions for laterite formation existed in this latitude (40° N) near the close of the Paleozoic. (Paleomagnetic studies suggest that this region was close to the equator at that time.)

Some petrochemical data have been omitted.

REFERENCES

Becker, G. F., 1895. U.S. Geol. Surv. Ann. Rept. 16, Pt. III.

Lamplugh, G. W., 1902. Calcrete. Geol. Mag., 575 (letter).

Merrill, G. P., 1897. A treatise on rocks, rock weathering and soils. Macmillan, New York, 411 p.

Walther, J., 1924. Das Gesetz der Wüstenbildung, 4th ed. Quelle & Meyer, Leipzig, 421 p.

Woolnough, W. G., 1927. The duricrust of Australia. J. Proc. Roy. Soc. New South Wales, 61, p. 1–53.

21

Reprinted from *Geol. Mag.*, **67**(3), 123–132 (1930)

The Influence of Climate and Topography in the Formation and Distribution of Products of Weathering.

By W. G. Woolnough, D.Sc.

FOR many years the writer has been specially interested in the variations apparent in the processes of weathering of rocks under different climatic and physiographic conditions, and has made several attempts to explain such features. Of these published attempts the most detailed took the form of a Presidential Address to the Royal Society of New South Wales in 1927. Certain portions of the argument were repeated in *Economic Geology*, but the regional aspects of the question were omitted.

The reading of a masterly treatise by Storz on the problem of silicification suggests that those regional aspects are of sufficient general importance to warrant wide consideration by geologists. As the medium of their earlier publication may not be readily accessible to some who would find the subject of interest it is thought that a brief résumé of the conclusions arrived at may be welcome to geologists in general.

The paucity of references in this contribution must be put down to the fact that the writer is working far from geological libraries. The views expressed are the results of personal observations throughout the length and breadth of Australia. It is certain that variations and differences will be met with in other lands, but the records of structures in the wide range of latitudes and climates in Australia may suggest comparisons which might otherwise be overlooked.

Varieties of Chemically Formed Surface Rocks in Australia.

Very widely distributed throughout Australia are remnants of what was once, in all probability, a continuous sheet of chemically formed rock material, very variable in composition and thickness. Universally, this rests upon a thick substratum of intensely leached bed rock. In some places, as in parts of Western Australia and Queensland, these deposits are continuous over tens of thousands of square miles of almost level upland. Over still more vast areas, however, the deposits are represented by residuals only, forming the mesas and buttes which are the most characteristic features in the scenery of what is now the arid and semi-arid interior of Australia.

The nature of the deposit varies from a mere infiltration of pre-existing surface rock, to a thick mass of relatively pure chemical precipitate. The two types differ quantitatively rather than qualitatively, and every intermediate variety is found.

The mineral matter deposited from solution falls into three main groups :—

>   (*a*) aluminous and ferruginous,
>   (*b*) siliceous,
>   (*c*) calcareous and magnesian.

When reasonably pure, the precipitates of the first type, varying in composition from somewhat siliceous bauxite to equally siliceous göthite, are usually referred to as " laterite " in Western Australia, where they attain their maximum development.

The more definitely siliceous varieties form quartzite, porcellanite, " desert sandstone ", and opal.

The calcareous and magnesian types constitute the ubiquitous " travertine " crusts which are most extensively developed in South Australia.

A very definite relationship can be traced between the composition of the crust and that of the underlying bed rock, wherever the nature of the latter can be determined. Thus, in Western Australia, the aluminous varieties of " laterite " cap felspathic rocks, low in iron (granites). Ferruginous " laterite " accompanies basic felspathic rocks (greenstones). Both varieties of " laterite " contain considerable percentages of silica, derived, as claimed by Storz, from the decomposition of the silicates.

Argillaceous sediments give rise to opaline crusts of very variable character. Arenaceous sediments produce silica in smaller amounts, and the surface manifestations are chiefly in the form of a " case hardening " of the rock formations : indurated sandstones, " desert sandstones," and " quartzites " more or less ferruginous.

Sediments rich in lime yield " travertine " crusts. Such crusts, however, on account of the greater solubility of their constituent minerals, are in a somewhat different category from the " laterites " and " silicificates ".

That these chemical distinctions are not due to variations in climate brought about by differences of latitude, rainfall, etc., is very clearly proved by the sections obtainable in the Irwin River District of Western Australia (see Fig. 1).

Here, within a very short distance, we have " laterite " resting on Pre-Cambrian granite and greenstone ; porcellanite on argillaceous sediments of Permo-Carboniferous age ; ferruginous silicified sand-stones on coarse, freshwater Jurassic sandstones, and " travertine " on Kainozoic calcareous dune rock.

The Duricrust of Australia.

In every instance examined by me in Australia (and the observations extend over the whole of the continent with the exception of the extreme centre and the extreme north-eastern portion) these chemically formed cappings rest upon a substratum of completely and deeply leached bed rock. That the phenomena have nothing whatever to do with selective transportation by water of material in mechanical suspension is conclusively proved by the occurrence of the remnants of delicate structures, such as quartz veins only a fraction of a millimetre in thickness, and of the incoherent fragments of decomposed micaceous aplite and pegmatite veins, still *in situ*, traversing both cappings and substrata, throughout

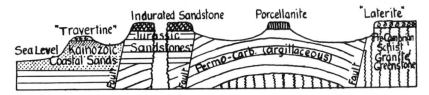

Fig. 1.—Section across Irwin River Basin of Western Australia, somewhat generalized, showing variation in the character of the "duricrust" in sympathy with the nature of the bed rock.

the whole of the areas examined. So universal is this feature, at points separated by a couple of thousands of miles, that it points to a factor of regional importance.

For this reason I have suggested that the widespread chemically formed capping in Australia, resting on a thoroughly leached substratum (which is an *essential* feature), and formed *in situ*, and apparently synchronously in every region, is worthy of formational rank, and the name " Duricrust " is proposed. It is believed that the duricrust was formed throughout Australia during a period of highly perfect peneplanation, about Miocene in age, on a land surface almost devoid of topographic relief, and in a climate marked by the dominantly seasonal character of its rainfall. Although differences were introduced by the *absolute* magnitude of the rainfall, and by temperature effects dependent on latitude, these factors are regarded as completely subordinate to those of *topographic relief* and of *seasonal distribution of rainfall*. In respect of the last-mentioned factor, the length of the " seasons " appears to be of somewhat minor importance. The " seasons " may have been annual in recurrence, or may have been of the nature of occasional wet periods alternating with years of drought.

Typical duricrust appears to have developed alike in humid and semi-arid areas, within the tropics and in the temperate south, in regions marked by annual alternation of dry and wet seasons and in places where wet periods are separated by long years of drought.

It will be seen, then, that a very different emphasis is placed upon the different factors from that postulated by Storz, Walther and others. It is quite likely that, in Australia, such differences may have been of local application only ; but, as it is thought that the postulates *may* be of universal applicability, it is desirable that the processes involved should be stated in some detail.

Perfection of Peneplanation, and its Influence on the Weathering of Rocks.

Ultimate perfection of peneplanation is very rarely attained because of the extreme slowness of the processes of erosion during the later stages, when the land surface has been reduced to the condition of an almost featureless plain very little above sea level (in the general case).

When approximate perfection of peneplanation has been attained, run-off of rainfall is so exceedingly sluggish that mechanical transportation of sediment is reduced to the vanishing point.

Soakage of the land surface by meteoric water is favoured. In humid regions the products of decomposition of organic matter reinforce the chemically active constituents of the rainwater. In such circumstances deep and perfect weathering of silicate and carbonate rocks is brought about.

Removal of soluble products of decomposition (electrolytes and colloids alike) may take place on a large scale, since such a form of transportation is independent of the *velocity* of the moving water. The most soluble products of weathering, the salts of the alkalis, are usually removed completely. Products of moderate solubility such as the carbonates of calcium, magnesium and iron, migrate with considerable freedom. Colloidal suspensions of silica and the oxides of iron and aluminium move almost as freely under most circumstances. Only the most insoluble substances such as crystalline silica and hydrated alumina, and the silicates of aluminium are left behind in the subsoil. From the fact that such delicate original structures as those described above show little or no sign of mechanical disruption, it would appear that the volume of the original rock is very little altered by the processes of decomposition. Increase in volume brought about by oxidation and hydration just about balances the loss of material by removal in solution.

Such a type of action may be postulated for the process of weathering in all types of climate, under conditions of ultimate perfection of peneplanation. The rapidity and extent of the action may vary with differences in temperature and absolute rainfall, but the nature of the process is not notably affected by these factors, except, perhaps, in the case of extreme aridity.

The subsequent fate of the products of weathering is vastly different in different cases, and, in my opinion, the *seasonal incidence* of the rainfall is the most important factor involved.

INFLUENCE OF RAINFALL.

In humid climates with well distributed rainfall the result is the leaching from the land surface of all of the soluble products of weathering. Even under the most perfect conditions of peneplanation there is a movement of the solutions, sluggish it is true, towards the " sink ", which, in the general case, is the sea.

The observations of the officers of the Indian Geological Survey, of Walther and others, suggest that, in these circumstances, temperature plays a not unimportant part. These workers have shown that, under such conditions, tropical " laterite " is produced. Its occurrence shows that the higher temperatures met with in the tropics attain a point which is critical in relation to the retention of iron oxide in the colloidal form. Tropical laterite is produced by leaching, *in situ*, of rocks exposed to per-humid, high temperature conditions or by the precipitation of more or less colloidal material in " sinks " (Storz, fig. 26, p. 52). Such a type of " laterite " must necessarily be confined to tropical countries.

It is unfortunate that the term " laterite " has been employed in Western Australia for a rock type, which, while possessing many points of similarity with the laterite of India and Java, possesses also striking points of difference, and is probably different in mode of origin.

In climates marked by " seasonal " rainfall (independent of the durations of the " seasons "), capillarity brings to the surface during the " dry season " solutions formed in the subsoil during the " wet season ". The physico-chemical conditions in the surface layers during the " dry season " are such that colloidal silica, alumina and iron oxide are precipitated in forms insoluble in the meteoric waters which fall during the " wet season ". These precipitated oxides, and particularly those of iron and aluminium, assume concretionary forms, and give rise, eventually to the pisolitic granules and crusts so typical of the duricrust in Australia. Opaline silica shows less tendency towards this type of aggregation and tends to occur rather as veins and infiltrations.

Carbonates of calcium and magnesium are precipitated in much the same way, but, on account of their higher solubility appear to be capable of ultimate removal in solution. The distribution of " travertine " crusts in Australia suggests that they are less permanent than are those composed of iron, alumina and silica. In this respect the carbonates are intermediate between the colloidal oxides and the alkaline salts.

The precipitation of the colloids is profoundly affected by the salinity and hydrogen ion concentration of the waters, and in this, possibly, may be found at least one cause of the difference between the laterites of per-humid tropical latitudes, and those of semi-arid and sub-humid temperate ones.

In the presidential address cited above the writer did not deal with arid climates, although the weathering processes under such conditions have been extensively studied. At the present time, in Australia, areas partially covered by duricrust are subject to conditions of extreme aridity of climate. In these areas the duricrust is being denuded, and is not in process of formation.

It appears that, in desert conditions, the saturation of the rocks is never complete, even under conditions of peneplanation. In these circumstances, the complete leaching of subjacent formations does not occur, and the small amount of redistribution of mineral matter in solution which takes place results rather in "case hardening" of surface rocks than in the formation of chemical crusts.[1] This is well exemplified in the granites of Western Australia.

It is thought that similar conditions may be very widespread; but that they have not been sufficiently recognized by workers in tropical and sub-tropical areas. In Australia, at least, there is very distinct evidence that the present era was preceded by a pluvial period, perhaps the equivalent of the Glacial Epoch of higher latitudes. The strong probability that formations like the duricrust, in areas which are now arid, may be residual, and not characteristic, should be discussed in each case.

IMPERFECT PENEPLANATION.

When peneplanation is advanced, but still not perfect, (a much more usual condition than perfection), run-off, though relatively feeble as compared with that of an immature land surface, is effective to some extent. Coarse detritus cannot be moved, but finer mineral matter can be carried in mechanical suspension. Consequently, there is a subordinate amount of formation of fine textured sediments in local sinks. With these there are likely to be associated chemically formed deposits of "lateritic" or "tufaceous" character, produced where the slow seepage of subsoil solutions from the insignificant uplands enters the sinks, in which precipitation is brought about by oxidation, and by coagulation of colloids through changes in salinity and hydrogen ion concentration of the waters present in the sinks.

The general impression gained in the course of a single very rapid traverse of the region in the neighbourhood of Les Baux in the south of France suggests that some such process was effective in the formation of the bauxite in that classical locality.

[1] Compare also Walther, (7) in references to literature, pp. 144, 145, 147.

In per-humid tropical climates the formation of " laterite" in the sense used by the Indian Survey officers, by Walther, Storz, and others is favoured by slight departure from perfection of peneplanation. Such conditions are not met with in any part of Australia visited by me. It is claimed that most or all of the Australian " laterite " was formed under conditions in which seasonal distribution of the rainfall and perfection of peneplanation were the dominating factors.

In the more normal climates, namely, temperate climates with well distributed moderate rainfall, conditions of approximate perfection of peneplanation favour ordinary soil formation. Since such conditions prevail in countries containing the large populations of highly cultured people, it is naturally such conditions which are looked upon as " normal ". Comparatively few geologists have opportunities of visiting distant lands where conditions are vastly different, and there is, perhaps, a tendency to over-emphasize the Lyellian doctrine of uniformity. It cannot be assumed that the conditions existing now in the temperate lands of the Old and New Worlds are those which have always existed. It is largely for this reason that the conditions " abnormal " in Europe, but " normal " in Australia are described here. There seems strong presumption that the latter conditions may have been the normal ones in Europe at such periods as the Old Red Sandstone, the Permian and the Triassic, and that application of the theory of perfection of peneplanation, combined with seasonal rainfall, may help to a better understanding of such formations.

In arid regions of low relief, but still not attaining to ultimate perfection of peneplanation, precipitation in sinks becomes the dominating process. Under such conditions the sinks are areas of internal drainage for the most part, and all the soluble products of erosion, including even the hygroscopic salts, are precipitated in the basins. As pointed out above, subsoil solution is very limited in such cases, and the less soluble materials tend to produce case-hardening of the surface rocks.

In Central Australia, where existing conditions approximate to those postulated, it may be stated with a considerable degree of certainty that the formation of chemically formed crusts is a very subordinate process. Destruction of the old duricrust is proceeding apace, with the formation of " gibber-plains " (huge areas littered with close-packed fragments of the duricrust). New precipitation is largely in the form of " desert varnish " covering the individual " gibbers ".

The description given by Storz of the conditions prevailing in the Namib Desert of South West Africa is distinctly suggestive of the prevalence of somewhat similar conditions. Here we have, apparently, the existence of table-topped residuals of chemically formed capping, associated with large areas of waste derived therefrom, together with subordinate (?) formation of chemical precipitates in the lower-lying areas. The description suggests that the chemically formed cappings are thought to be growing, a condition certainly absent in the Central Australian duricrust areas.

REGIONS OF IMMATURE TOPOGRAPHY.

In regions of youthful topography mechanical abrasion almost completely masks chemical weathering, no matter what the incidence of the rainfall may be.

In the tropical regions which I have had the opportunity of examining, such as Fiji and Papua, corrasion is phenomenally rapid, and soil formation is quite subordinate. Fresh and undecomposed rock surfaces are conspicuous everywhere, and coarse-grained alluvial deposits obstruct the stream channels.

With more moderate rainfall the rapidity of corrasion is reduced, but its character is unchanged. There is greater opportunity for rock decomposition and soil formation, but these processes do not proceed very far. The sides of young river valleys consist, to a large extent, of nearly fresh rock.

Owing to the perfection of lateral drainage into neighbouring young valleys, subsoil solutions are disposed of in this way, rather than by capillary attraction to the surfaces of the uplands, even when the incidence of the rainfall is markedly seasonal.

Thus, in the highlands of the temperate, sub-humid coastal portions of Western Australia it appears probable that duricrust is not forming at the present time, even though the climate is characterized by the occurrence of a sharply-defined very wet winter alternating with a very dry summer.

It is possible that, where streams have not yet cut back into areas of level upland (uplifted peneplain), duricrust formation may still be in progress. Round the edges of the central plateau, however, stream corrasion is actively removing the hard crust. Resting as it does on a soft and completely decomposed substratum, the duricrust forms an almost ideally weak structure, physiographically. The "feet of clay" crumble on exposure to the weather, bringing down the solid duricrust in picturesque ruins, and producing some of the most characteristic types of Australian scenery.

In arid regions of high relief the topography is of the "bad land" type. Wind action predominates over water erosion, though, as Walther and others have very rightly insisted, torrential streams play a very important part. The presence of the duricrust in the drier parts of the interior of Australia greatly assists in the development of "bad lands". In the process of denudation the action of marsupials and other animals plays a very noteworthy part. The decomposed substratum, being highly porous, tends to absorb and retain a considerable proportion of the scanty rainfall. Springs and "soaks" supply almost the only water supply for the inhabitants of vast areas. Kangaroos and other animals scoop out literally enormous excavations in search of shade and moisture.

Everywhere, so far as I know, destruction of the duricrust is at the present time, the key-note of sub-aerial activity in those parts of Australia in which the duricrust is developed ; and my experience suggests that the chemical precipitation of silicificates (in the terminology of Storz), or of " lateritic " material, is extremely subordinate.

Conclusion.

It would be impertinence on my part to criticize adversely the observations of German observers like Blanckenhorn, Kaiser, Passarge and Walther, of Englishmen like Hume (quoted by Storz), or the monumental researches of Storz himself. I have not had the opportunity as yet of studying much of the literature dealing with arid and sub-arid weathering in foreign countries ; but, from the summary given by Storz, it is clear that he, and most of his authorities at all events, regard latitude (as the factor controlling atmospheric temperatures) and volume of rainfall as the dominating factors in determining the character and distribution of the products of sub-aerial weathering.

That these are important, and perhaps dominating factors, at the present day, over the greater part of the tropical and temperate parts of the earth's surface is unquestioned. As Storz himself points out, however, in his Introduction, the conditions familiar to the great majority of geologists in their home lands may have had a much more limited extension and scope in the geological past.

A different set of conditions is remarkably well illustrated in what I have called the " Duricrust " of Australia ; and it is believed that consideration of the phenomena there revealed may help geologists in other countries to a better understanding of some of the anomalous and puzzling formations of the geological record, notably those produced during periods of long continued continental emergence, such as the Old Red Sandstone, the Permian and the Triassic.

While recognizing the importance of temperature and rainfall in the formation and distribution of the products of weathering, I hold that the dominating factors are :—

(*a*) The degree of perfection of the process of peneplanation (in other words, *topographic relief*) ;

(*b*) the *seasonal incidence of the rainfall*, be the latter scanty or copious.

REFERENCES TO LITERATURE.

(1) Blanckenhorn, M. "Ägypten," *Handbuch der regionalen Geologie*, Heidelberg, 1921, Bd. 7, H. 93.
(2) Harrassowitz, H. " Laterit, Material und Versuch erdgeschichtlicher Auswertung," *Fortschritte d. Geol. u. Pal.*, Bd. iv, Heft. 14, Gebr. Borntraeger, Berlin, 1926.
(3) Hume, W. F. *Geology of Egypt*, vol. i. The surface features of Egypt, their determining causes and relation to geological structure. Survey of Egypt, Government Press, Cairo, 1925.

(4) KAISER, E. *Die Diamantenwüste Südwestafrikas.* Dietrich Reimer, Berlin, 1926.

(5) PASSARGE, S. *Die Kalahari.* Versuch einer physisch-geographischen Darstellung der Sandfelder des Südafrikanischen Beckens, Dietrich Reimer, Berlin, 1904.

(6) SIMPSON, E. S. "Laterite in Western Australia," GEOL. MAG., 1912, 399–406.

(7) WALTHER, J. *Das Gesetz der Wüstenbildung in Gegenwart und Vorzeit.* Vierte Auflage, Leipzig, Quelle und Meyer, 1914.

(8) —— "Der Laterit in Westaustralien," *Zeits. d. D. Geol. Gesellschaft,* Bd. 67, Monatsber. 4, 1915.

(9) WOOLNOUGH, W. G. "The Physiographic Significance of Laterite in Western Australia," GEOL. MAG., 1918, 385–93.

(10) —— "Presidential Address to the Royal Society of New South Wales," *Journ. Roy. Soc. N.S.W.,* 1927, 61, 1–53.

(11) —— "Origin of White Clays and Bauxite, and Chemical Criteria of Peneplanation," *Econ. Geol.,* 1928, xxiii, 887–94.

(12) STORZ, M. "Die sekundäre authigene Kieselsäure in ihrer petrographisch-geologischen Bedeutung," *Mon. zur Geol. und Pal.,* Herausgegeben von Professor Dr. W. Soergel, Breslau, Serie ii, Heft 4, Berlin, 1928. Gebr. Borntraeger, pp. xi, 137, illust.

(The author will be pleased to send a copy of his presidential address to any geologist who is interested. Address : Department of Home Affairs, Canberra, F.C.T., Australia.)

22

Reprinted from *Bull. Geol. Soc. America,* **59,** 1173–1179, 1180–1181, 1182–1184, 1186–1189 (Dec. 1948)

PRE-FOUNTAIN AND RECENT WEATHERING ON FLAGSTAFF MOUNTAIN NEAR BOULDER, COLORADO

BY ERNEST E. WAHLSTROM

CONTENTS

ILLUSTRATIONS

ABSTRACT

Decomposition of granodiorite underlying basal Fountain (Pennsylvanian) formation on Flagstaff Mountain near Boulder, Colorado, was the result of pre-Fountain, probably post-Madison weathering. Chemical and mineralogical studies indicate that the weathering took place in a humid, warm or hot climate and that laterization was the dominant process.

The results of pre-Fountain and recent weathering are compared and contrasted.

The red color of the Fountain formation is regarded as inherited from a deeply weathered extensive red-stained pre-Fountain regolith.

INTRODUCTION

On Flagstaff Mountain, just west of Boulder, Colorado (Fig. 1), basal Fountain arkose (Pennsylvanian) rests unconformably on a steeply dipping erosion surface

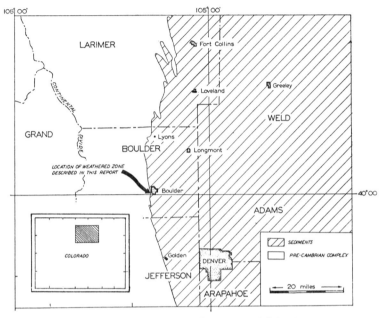

FIGURE 1.—*Outline map of north-central Colorado*
Showing location of weathered rocks described in this report.

which truncates the pre-Cambrian Boulder Creek granodiorite batholith. The contact between the Fountain arkosic sandstone and conglomerate and the granodiorite is well exposed in a road cut near the summit and on the east slope of Flagstaff Mountain, where the granodiorite shows two distinct types of weathering, one of recent origin and the other pre-Fountain. Samples from both fresh and weathered exposures of the granodiorite were analyzed chemically and mineralogically.

ACKNOWLEDGMENTS

The writer gratefully acknowledges the counsel of Professor W. O. Thompson of the Department of Geology at the University of Colorado, whose stimulating discussions and criticism have been of invaluable assistance in the preparation of the manuscript.

FOUNTAIN FORMATION

The Fountain formation in the vicinity of Boulder has been described by Fenneman (1905), Butters (1913), Tieje (1923), and Henderson (1920). It is composed of coarse arkosic sandstone locally grading into quartzose conglomerates, sandstones, and shales. Fresh fragments of microcline are abundant, and mica is widespread, especially in the finer-grained portions. The color ranges from reddish brown through red to pink and is due to the reddish feldspar fragments and a cement of silica and iron oxides. A few beds of arkose and sandstone are nearly white.

Rapid lateral variation in thickness is the rule. In the vicinity of Boulder the average thickness is about 1000 feet. Cross-bedding, lenticular beds, and cut-and-fill structures are conspicuous. Abundant buried stream channels suggest a highland source to the west. Evidence points to a fluvial origin, and the Fountain formation has been interpreted as a series of coalescing alluvial fans deposited along the flanks of a north-south uplift.

McLaughlin (1947) states that in portions of the Colorado Springs quadrangle the Fountain formation rests conformably on the Glen Eyrie formation and elsewhere lies with angular unconformity on older Paleozoic formations or directly on eroded pre-Cambrian rocks. The youngest Paleozoic rocks beneath the unconformity are of Madison age. Fossils from the Glen Eyrie and Fountain formations "indicate that the two formations are equivalent to, or perhaps younger than, some part of the Des Moines series of the Pennsylvanian section of Kansas" (McLaughlin, 1947).

The widespread reddish color of the Fountain formation is inherited. Porous light-pinkish-gray to almost dead white layers are locally present and prove that the iron-oxide stain in the red and pink layers is not secondary.

PRE-CAMBRIAN ROCKS

The pre-Cambrian complex in the core of the Front Range west of Boulder consists of igneous and metamorphic rocks of Archeozoic and Proterozoic age. Tertiary stocks and dikes cut the older rocks. In the vicinity of Boulder three rock assemblages predominate (from oldest to youngest): the gneisses and schists of the Idaho Springs formation, the granodiorite and associated pegmatites of the Boulder Creek granodiorite batholith (by some authors called the "Boulder Creek granite batholith"), and the Silver Plume granite with its satellitic pegmatite bodies. The erosion surface at the base of the Fountain formation truncates each of these rocks at one place or another along the east side of the Front Range.

As the present study is confined to an examination of the contact between the Fountain formation and a portion of the Boulder Creek granodiorite batholith, only this rock will be described. An analysis of the granodiorite is given in Table 1.

In fresh exposures the granodiorite is a medium-gray to pinkish-gray, coarse-grained locally porphyritic rock in which the groundmass minerals average about 5 mm. in diameter.

Phenocrysts of microcline commonly are about 10–15 mm. in average diameter but locally are tabular and attain lengths of 30 to 40 mm. A gneissoid structure is generally apparent and is emphasized by the parallel orientation of flakes of biotite, needles of hornblende, plates of plagioclase, and tabular phenocrysts of microcline. In some exposures numerous schlieren and partly assimilated or recrystallized inclusions of Idaho Springs formation are abundant.

Under the microscope the following minerals, listed in order of decreasing volume percentages, are found: oligoclase (Ab$_{75}$), 35; microcline, 25; quartz, 20; biotite, 16; hornblende, 3; minor accessory minerals including apatite, zircon, sphene, and allanite, 1. Micrometric analyses using the Rosiwal technique do not yield consistent results because the rock is coarse-grained and porphyritic. The oligoclase is tabular,

TABLE 1.—*Granodiorite from Flagstaff Mountain*

Analyst: E. W. Wahlstrom

SiO$_2$	67.92	K$_2$O	4.38
TiO$_2$	0.55	H$_2$O+	0.80
Al$_2$O$_3$	14.70	H$_2$O−	0.36
Fe$_2$O$_3$	0.91	P$_2$O$_5$	0.18
FeO	2.61	CO$_2$	None
CaO	2.94	MnO	0.03
MgO	0.98		
Na$_2$O	3.31		99.67

unzoned, polysynthetically twinned, and in some thin sections shows a myrmekitic intergrowth near the rims of its crystals. The peripheries of many microcline crystals also show quartz-feldspar intergrowths in a micrographic pattern. Quartz is interstitial to microcline and oligoclase.

Biotite and hornblende, intimately associated in clusters of subhedral crystals, closely follow the contacts between quartz and feldspar grains. The mica is pleochroic in olive green and light yellowish brown. 2V is near zero, α = 1.587 ±0.003, γ = 1.621 ±0.003. The hornblende is pleochroic in dark green and yellowish greens. An analysis of the biotite is given in Table 3.

Apatite, zircon, sphene, magnetite, and allanite are associated with the biotite and hornblende.

Numerous pegmatite and aplite dikes ranging from an inch to several feet in width intersect the granodiorite and apparently are genetically related to it.

WEATHERING

GENERAL STATEMENT

The pre-Fountain weathering is related to the surface beneath basal Fountain and is parallel to the Fountain-granodiorite contact (Fig. 2; Pl. 1, fig. 1). Residual masses of fresh granodiorite mark the lower limit of the pre-Fountain weathered mantle.

Recent weathering (Fig. 2) is related to the present land surface. The recent weathering has affected only the fresh granodiorite and has not resulted in any physical or chemical changes in the rocks altered by pre-Fountain weathering.

PRE-FOUNTAIN WEATHERING

General statement.—Pre-Fountain weathering in this area has affected the granodiorite to a depth of approximately 80 feet beneath and normal to the contact. This

FIGURE 1. VIEW LOOKING NORTHEAST AT WEATHERED GRANODIORITE EXPOSURE
In roadcut near the top of Flagstaff Mountain. A — Basal Fountain formation, B — Fountain-granodiorite contact, C — Red-stained rock, D — Bleached horizon, E — Residual boulder near base of bleached horizon, F — Brown-stained rock.

FIGURE 2. FOUNTAIN ARKOSE (RIGHT) RESTING ON RED ALTERED GRANODIORITE (CENTER) WHICH GRADES INTO BLEACHED ALTERED GRANODIORITE (LEFT).

PHOTOGRAPHS OF WEATHERED ROCKS

depth is nearly a maximum for the region. The rocks beneath the Fountain formation were examined in several other exposures along the east side of the Front Range, and nowhere was the weathering as extensive or as conspicuous as on Flagstaff Mountain.

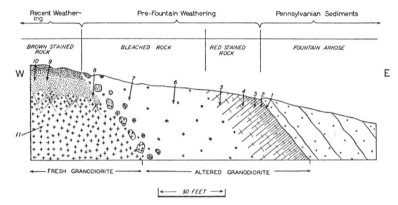

FIGURE 2.—*Cross section at right angles to Fountain-granodiorite contact*
Flagstaff Mountain. Numbers and arrows indicate locations of specimens collected for analysis.

At some localities the Fountain rests on fresh rock. This should be expected, for the Fountain formation was derived from the same rocks as those on which it now rests, and erosion must have stripped the weathered rocks in many places, leaving basement rocks or only vestiges of the weathered mantle. There is no assurance that the uppermost part of the weathered mantle on Flagstaff Mountain has not been stripped away.

The altered granodiorite at the contact (Pl. 1, fig. 2) is a deep brownish red and is enriched in hematite and limonite. The brownish-red rock grades downward into a lighter-red rock not as rich in iron oxides. The red-stained horizon is about 30 feet thick. At the bottom of the red horizon the rock changes within a few feet into a light-gray or pinkish-gray, almost white, slightly mottled rock. The light-colored rock looks bleached and offers a decided color contrast to the medium-gray underlying fresh granodiorite. The bleached horizon is about 50 to 60 feet thick, and its base contains residual boulderlike masses of relatively fresh granodiorite. These masses are residual in the sense that the weathering near the bottom of the bleached material was guided by joint cracks, and the alteration was not so pervasive as it was closer to the surface. Thus, isolated rounded fresh boulders are separated by altered matrix.

Chemical characteristics.—The chemical changes as a result of pre-Fountain weathering were determined by several chemical analyses (Table 2). Uncombined silica was determined by the method outlined by Knopf (1933). The analysis of a biotite-rich schlieren in the granodiorite (No. 12) is included to show the composition of the least silicic portion of the fresh granodiorite mass found in field exposures.

[*Editor's Note:* Table 2 has been omitted. For present purposes, Figure 3 is more useful.]

Figure 3 shows the variation of the oxides in the analyses with depth in the pre-Fountain weathered rocks. In the red rock at the top, immediately beneath the Fountain arkose, Fe_2O_3 and Al_2O_3 are at maxima, and SiO_2 and Na_2O are at minima. K_2O and H_2O reach maxima at the level of the bleached rock as does CO_2, although it is not shown in the graph. FeO is present in small amounts throughout the weathered material. The other oxides do not show significant variations.

Outstanding is the apparent enrichment of the upper portion of the weathered mantle in Fe_2O_3 and Al_2O_3 and the progressive desilication from the bottom of the mantle to its upper limit.

The iron oxide in the red-stained rock is residual. Proof that there has been little or no migration of iron oxide either upward or downward in the weathered mantle is found in altered pegmatite and aplite dikes in the red-stained weathered granodiorite. Weathering in the dikes is similar to that in the granodiorite but, in general, is not as pervasive because of the protective action of abundant quartz.

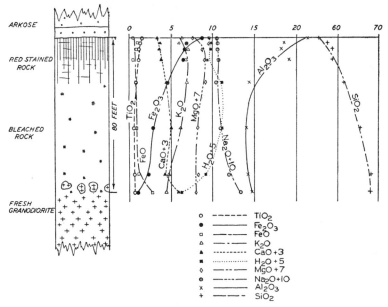

FIGURE 3.—*Variation of weight percentages of oxides with depth*
Pre-Fountain weathered mantle.

The red stain is not present in the dikes, although the decomposed, porous nature of the altered rocks in the dikes would permit easy ingress of iron-bearing solutions.

[*Editor's Note:* For present purposes, a triangular diagram (Figure 4) and a brief discussion have been omitted.]

Mineralogical characteristics.—X-ray and microscopic studies of the samples from the pre-Fountain weathered material yielded significant data. Heavy minerals were separated in bromoform and examined optically. The lighter fractions were studied before and after fine grinding and centrifuging. The X-ray technique was useful in identifying clay minerals in centrifuged portions of the light fractions. An effort was made to determine all significant systematic or nonsystematic mineralogical changes with depth.

Original minerals in the granodiorite are quartz, microcline, oligoclase, biotite, hornblende, and minor accessories. Secondary minerals resulting from pre-Fountain weathering are iron oxides, hydrated mica, clay minerals, and dolomite. Figure 5 shows the relative persistence of each of the abundant minerals and the variation in amounts of each individual mineral with depth.

Quartz, as would be expected, is resistant and persists throughout the weathered mantle. However, in the upper part of the mantle the quartz grains are somewhat rounded and embayed and show the effects of peripheral dissolution. The microcline reacted in much the same way as quartz. It is present in subrounded embayed fragments near the top of the mantle but in unaffected grains in deeper horizons. Although it has been corroded by the weathering solutions, it is as fresh in the most intensively altered portion of the granodiorite as in the original granodiorite. A few grains show a dust of clay minerals. Veins, generally parallel to cleavages and containing fine-grained silica and clay minerals, are present in some grains.

The oligoclase is one of the least stable minerals.   It has been converted to clay minerals and fine-grained silica in samples extending to the very bottom of the altered mantle where the fresh granodiorite first shows the effects of pre-Fountain weathering. However, despite the extreme susceptibility of the oligoclase to weathering the

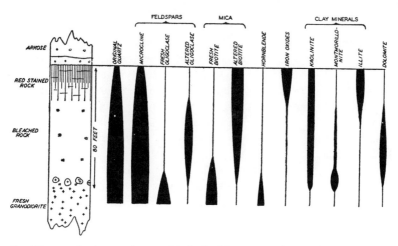

FIGURE 5.—*Diagram showing persistence with depth of important minerals in pre-Fountain mantle*

outlines of its grains can be identified almost all the way through the bleached horizon and up to the bottom of the red-stained horizon.   The clay minerals resulting from the alteration of the oligoclase show a variable distribution with depth.   At the bottom of the weathered mantle the oligoclase has been converted largely to kaolinite and montmorillonite.   In the central and upper portions of the bleached horizon illite, kaolinite, and dolomite aggregates mark the position of altered oligoclase crystals.   No vestige of oligoclase remains in the red-stained horizon.

Inasmuch as micaceous minerals persist in abundance throughout the weathered rock, the writer separated and analyzed micas from each horizon to determine how the chemical and optical properties vary with depth.   For several reasons the results were of indifferent value.   The biotite weathered erratically giving at any horizon in the weathered mantle plates with separate alternating layers of considerably different composition.   Moreover, iron oxide resulting from decomposition was deposited unevenly in a very fine state of subdivision through the mica making it virtually impossible to obtain uncontaminated material for analysis.

[*Editor's Note:* Detailed material on biotite weathering is omitted.]

The hornblende altered easily, and it is not present a short distance above the bottom of the weathered mantle.   However, its former presence is indicated by aggregates of iron oxides, clay minerals, carbonate, and a micaceous substance. No trace of hornblende or its alteration products is found in the upper part of the bleached horizon or in the red-stained horizon.

X-ray examination of centrifuged portions of the lighter fraction of the altered rock shows that kaolinite is persistent throughout the weathered mass, montmorillonite is present only in the bleached horizon, and illite is abundant in the red-stained horizon. Intensive alteration in the red-stained rock seems to have favored the development of illite at the expense of both kaolinite and altered mica.

Dolomite is the only mineral that gives positive evidence of having been deposited from solutions. It is confined to the bleached rock and is generally found in altered aggregates resulting from the decomposition of oligoclase. It is also present in veinlets cutting other minerals. Several fragments of quartz and microcline contain intersecting cracks filled with pure dolomite. Probably the dolomite in the veinlets was derived from adjacent altered minerals and not from solutions moving downward from the original land surface.

The minor accessory minerals of the fresh granodiorite—chiefly apatite, sphene, magnetite, and allanite—are not visible in any of the thin sections of the altered rocks and presumably were altered or leached beyond recognition.

The order in which the minerals of the granodiorite decomposed is in general agreement with the order of decomposition given by Goldich (1938) in his mineral stability series.

Volume changes.—The rocks in the bleached horizon preserve the original texture of the granodiorite although they are much more porous than the fresh rock. In the red-stained horizon the leaching has been extreme. The minerals in the red horizon are chiefly illitic mica, hematite, quartz, and microcline; under the microscope, these minerals form a clastic mass resembling in many respects the overlying arkosic sediments of the Fountain formation. The micaceous material forms jumbled aggregates surrounding rounded and subrounded grains of quartz and microcline. The texture is the result of removal of large amounts of the original and secondary minerals by leaching with concomitant collapse and compaction of the residual material.

The exact amount of bulk volume change due to compaction could not be determined by any means available to the writer, because too many unknown factors must enter into any computation that might be made.

RECENT WEATHERING

Recent weathering has converted the fresh granodiorite into a loosely coherent, crumbly, brown-stained rock. This type of weathering is seen in exposed crystalline rocks over most of the foothills belt of the Front Range and is characteristic of the arid to semiarid climate that prevails over much of the region. Recent weathering has not affected the rocks in the pre-Fountain weathered mantle because the decomposition of these rocks has proceeded to a stage far beyond that resulting from recent weathering.

Recent weathering has stained the exposed residual boulders near the base of the pre-Fountain mantle a characteristic brown and in this respect has simulated an effect similar to one in northern Colorado described by Thompson (1931) who found brown, limonite-stained residual granite boulders embedded in a light-gray to pinkish-gray altered rock below the Fountain formation.   Thompson interpreted the brown alteration as the result of recent weathering acting on exposed residual boulders of fresh granite that originally were present near the base of a pre-Fountain regolith.

The recently weathered rocks are easily identified inasmuch as they are related to the present land surface.   In contradistinction, the pre-Fountain weathered mantle parallels the contact between the granodiorite and the Fountain formation.

[*Editor's Note:* Data on recent weathering are omitted.]

349

PRE-FOUNTAIN CLIMATE

A typical soil profile is characterized by three horizons: the top or "A" horizon, in which leaching is evident; the intermediate or "B" horizon, to which substances presumably derived from "A" have been added; and the bottom or "C" horizon, the parent material. Some soil scientists regard the "C" zone as fresh rock, others regard it as consisting of the altered material below the "B" horizon and extending downward into unaltered rock. It is in the latter sense that the present writer uses the term "C" horizon.

On Flagstaff Mountain the pre-Fountain weathered rock shows the "C" horizon and all or part of the "B" horizon. The "A" zone appears to be absent, presumably removed by erosion. Unfortunately, most of the literature in which soil characteristics are related to climate deals only with the "A" zone and the upper part of the "B" zone.

According to Reiche (1945) three distinct associations of soil-forming processes depend on climate:

(1) Podsolization is normal for temperate humid climates and forest vegetation and results in concentration of iron or iron and aluminum in the "B" zone. Characteristic are acid soils in which silica removal and rather complete removal of alkalis and alkaline earths are noteworthy. Clay minerals are the normal end products of this type of weathering. Below the "A" zone the podsolic soils are brown to gray brown, grading into yellows and reds in their southern development where the influence of laterization is felt.

(2) Calcification results in accumulation of notable amounts of calcium and magnesium carbonates in the "B" horizon. It is distinguished by the presence of "caliche" at the depth of average penetration of surface water. Calcification is normal to dry climates and to areas characterized by brush or grass vegetation. The clay minerals may be bentonitic. The "C" zone of regions characterized by calcified soils is generally only a few feet thick.

(3) Laterization is characteristic of tropical weathering. The term "laterite" was originally applied to a material from the "B" zone in tropical humid regions. Containing an unusual concentration of iron and aluminum sesquioxides, laterite is typically brownish red or yellowish. Evidence for the removal of considerable silica is generally present. Laterites derived from silica-deficient source rocks may contain hydrated aluminum oxides. Laterites derived from siliceous source rocks generally contain abundant clay minerals, notably kaolinite. Tropical humid weathering commonly affects rocks to considerable depths.

Features in the pre-Fountain weathered mantle on Flagstaff Mountain that affect the interpretation of the climate at the time of the weathering are:

(1) A notable thickness of preserved weathered mantle—approximately 80 feet.

(2) A "C" zone about 50 feet thick grading upward into an iron-oxide-rich zone about 30 feet thick which may represent all or part of the "B" zone.

(3) Carbonates in the "C" zone which are interpreted as resulting from the weathering of ferromagnesian minerals and oligoclase in place, with some solution and redeposition near the point of origin. The dolomite probably was not derived from rocks in higher horizons.

(4) Chemical trends which show increasing concentration of iron oxides and a notable decrease in silica from the bottom of the weathered mantle to its present top. If the trends continued upward into a portion of the mantle removed by erosion prior to the deposition of the Fountain arkose the land surface prior to erosion locally may have exhibited an iron-oxide crust.

(5) Presence of abundant clay minerals with a characteristic distribution through the weathered mantle; montmorillonite is present near the base; illite is abundant in the iron-enriched rock; and kaolinite is abundant throughout the whole weathered mass.

(6) The sharp contrast in extent, color, and zonal characteristics when compared to the rocks altered by recent weathering.

Considering all factors, three conclusions emerge: (1) the climate was humid; (2) the average temperature was moderate to high; (3) if a different type of rock had been present, a typical laterite would have formed.

STRATIGRAPHIC RELATIONSHIPS

Some notion of the age of the pre-Fountain weathered mantle and the prevailing climate may be obtained by a study of the relationships of the basal sedimentary rocks along the east side of the Front Range. The Fountain formation of Pennsylvanian (Des Moines or post-Des Moines) age rests on the pre-Cambrian rocks for a distance of at least 70 miles north of Boulder as far as the Colorado-Wyoming State line where it has been identified as the lower part of the Casper formation. In northern Colorado near the State line the basal portion of the Fountain formation contains red-stained chert nodules containing Mississippian fossils derived from the Madison limestone. South of Boulder the Fountain arkose is the basal sedimentary formation at least as far as an unknown point between Deer Creek and Perry Park— a distance somewhere between 40 and 60 miles. At Perry Park the Fountain rests unconformably on beds at least as young as the Leadville limestone of Madison age.

The fossiliferous chert pebbles in the lower portion of the Fountain in northern Colorado apparently are residual in the sense that they were originally present in the Madison limestone and were left behind on the land surface after the erosion and solution of the limestone. The brownish-red color is the result of introduction of iron oxides into the pebbles as the limestone was being eroded, and while the pebbles lay exposed on the land surface. Chert pebbles in unweathered Madison are normally white to light gray. The red color was not introduced into the pebbles after incorporation into the Fountain formation because boulders of a variety of other rock materials in the Fountain have not been similarly affected. The red color is the same as that in the "B" zone of the weathered mantle on Flagstaff Mountain. The stratigraphic position of the chert pebbles and their characteristic coloration indicate that they were exposed to the same climatic and other conditions as the rocks beneath the Fountain contact on Flagstaff Mountain.

At Perry Park limestones beneath the unconformity at the base of the Fountain formation are red-stained to a depth of several feet. The amount of staining is a function of the distance below the unconformity and is most intense at and immediately below the unconformity. Moreover, small residual hematite deposits are common on the buried land surface that marks the unconformity and probably formed in the same manner as similar deposits of hematite in undisputed laterites in other parts of the world.

The above observations lead to the following conclusions: (1) The weathered mantle on Flagstaff Mountain is pre-Fountain and post-Madison; (2) it was formed while the unconformity between the Fountain arkose and Leadville limestone at Perry Park was in process of development; (3) the weathered rocks on Flagstaff Mountain stratigraphically are equivalent to laterized rocks elsewhere in the region.

The chert pebbles in northern Colorado indicate that Madison limestone was once present there or in immediately adjacent regions. Thompson (1931) found evidence for pre-Fountain weathering in a pinkish-white altered pre-Cambrian rock beneath the Fountain formation. The weathering Thompson describes probably was synchronous with that on Flagstaff Mountain and developed largely during and after the removal of the Madison limestone.

The evidence is clear that after the deposition of limestones of Madison age of the northern part of the Front Range was gently uplifted and a broad surface of low relief, but with sufficient drainage to permit deep weathering, was exposed to the corrosive action of the atmosphere and ground water. After a period of weathering, during which the Madison limestone and older formations were locally removed by erosion and a thick, reddish regolith was developed, the uplift was intensified, and the uplifted area began rapidly to shed the Fountain sediments. Some geologists refer to the uplift that lay slightly to the west of the present hogback exposures of the Fountain formation as the "ancestral Rockies."

Thus the red color of the Fountain formation is inherited from a pre-Fountain regolith which was conditioned by pre-Fountain and post-Madison weathering and was rendered susceptible to rapid denudation during and after uplift. The red color of the Lyons sandstone and the Lykins shales and sandstones which overlie the Fountain formation in the Boulder district probably has a similar ultimate origin. The climate was conducive to the development of laterite.

REFERENCES CITED

Butters, R. M. (1913) *Permian or "Permo-Carboniferous" of the eastern foothills of the Rocky Mountains in Colorado*, Colo. Geol. Survey, Bull. 5.

Fenneman, N. M. (1905) *Geology of the Boulder District, Colorado*, U. S. Geol. Survey, Bull. 265, p. 22–23.

Goldich, Samuel S. (1938) *A study in rock weathering*, Jour. Geol., vol. 46, p. 17–58.

Gruner, J. W. (1934) *The structures of vermiculites and their collapse by dehydration*, Am. Mineral., vol. 19, p. 557–575.

Henderson, Junius (1920) *The foothills formations of north-central Colorado*, Colo. Geol. Survey, Bull. 19, p. 58–92.

Knopf, Adolph (1933) *The quantitative determination of quartz in dusts*, U. S. Public Health Service, Repts., vol. 48, p. 183–190.

McLaughlin, K. P. (1947) *Pennsylvanian stratigraphy of Colorado Springs quadrangle, Colorado*, Am. Assoc. Petrol. Geol., Bull., vol. 31, p. 1936–1981.

Reiche, Parry (1945) *A survey of weathering processes and products*, Univ. N. Mex., Pubs., no. 1, Univ. N. Mex. Press, 87 pp.

Thompson, W. O. (1931) *Notes on the contact of the Fountain with pre-Fountain rocks along the northern Front Range of Colorado* (Abstract), Colo.-Wyo. Acad. Sci., Jour., vol. 1, p. 28.

Tieje, A. J. (1923) *The red beds of the Front Range in Colorado: A study in sedimentation*, Jour. Geol., vol. 31, p. 192–207.

UNIVERSITY OF COLORADO, BOULDER, COLORADO.
MANUSCRIPT RECEIVED BY THE SECRETARY OF THE SOCIETY, DECEMBER 23, 1947.

Editor's Comments
on Paper 23

23   WAYLAND
Peneplains and Some Other Erosional Platforms

THE CONCEPT

In Uganda, Wayland recognizes the highest undulating planation surface as a peneplain below which at least one later surface
has been produced by "etching."

There is possible confusion between the nature of the present "etched" plain and the general "etching" process. Wayland
describes the etched lower surfaces as "characteristically blanketed" (by lateritic ironstone) and as being "amazingly flat"
(paragraph 367). From this description it would appear that there
is little fresh bedrock showing and that the flat surface is to be
equated to the partial wash of the weathered material leaving
large patches of "lateritic ironstone" as a crust (Paper 21).

In discussing the process of etching, and after describing the
nature of weathering in a seasonal climate that would produce a
saprolite blanket "tens of feet" thick (paragraph 376), Wayland
says, "this zone of rotted rock is largely removed by denudation. . . ." Such removal would tend to expose the fresh bedrock
and present a somewhat rough etched surface and not the even
plateau described above. Since the surface cannot be smooth and
irregular at the same time, perhaps there are two possible kinds
of etched surfaces, one on the bedrock and another cut into the
saprolite. Wayland's type example seems to this observer to be
the latter.

23

Reprinted from *Ann. Rept. Bull., Protectorate of Uganda Geol. Surv. Dept. of Mines*, Note 1, pp. 77–79 (1933)

PENEPLAINS AND SOME OTHER EROSIONAL PLATFORMS

E. J. Wayland

There has recently been a discussion in the *Geographical Journal* concerning the peneplains of Eastern Central Africa and some clarification appears necessary. Many geographers and geologists will have misgivings with regard to a succession of peneplains more or less analogous to a series of river terraces or berms,[1] and for that reason it is thought that the present note might be helpful.

At one time we recognised but a single peneplain in Uganda, later it became evident that there were three, still later it was apparent that Peneplain II. could be any one of three erosional surfaces (other than Peneplains I. and III.) locally expressed at the expense of the other two.[2] Moreover, Peneplain I. appears to differ from the others in that (a) it has but local patches of lateritic ironstone upon its surface, while the lower erosion levels are characteristically blanketed by it, and (b) in the possession of an uneven or undulating surface in contradistinction to the amazingly flat plateaux that form the lower levels.

It might be argued that these differences are consequent upon the greater amount of denudation that Peneplain I. has undergone; but the present writer, while not wishing to deny that the oldest peneplain must have suffered most from erosion, nor that its undulations in question are other than erosional in origin,[3] is strongly inclined to regard the undulating unblanketed surface of that plateau as, in large measure, an original feature.

It seems implicit in various definitions that the base to which a true peneplain must be eroded is that provided by sea-level, and in respect of this few geologists will question that the earliest extreme old age erosion surface of which some very imposing remnants are left in Eastern Central Africa (Peneplain I.) is a true peneplain. It might be thought, however, that the later and lower widespread plateaux of Uganda, that are morphologically peneplains, have been graded to that vast sheet of water Lake Victoria, but geological evidence discredits the suggestion, though it cannot be denied that denudation to that level is slowly proceeding at the present day, but it is more than likely that the future planation, thus apparently foreshadowed, will never be anything like complete.

The latest plateau surface (originally called Peneplain III.) is almost certainly of Pliocene age, and it attains a maximum altitude of about 4,000 feet above sea level. It would appear, therefore, that Uganda, and parts of the adjoining territories,

---

[1] Berm: a term, given morphological significance by Bascom, to denote an erosion surface consequent upon dissection following upon elevation of land (*Bull. Geog. Soc. America*, Vol. 44, No. 3, footnote p. 555).
[2] *Geog. Jour.* Jan. 1934, p. 79.
[3] These undulations are distinct from regional warping of which there is evidence.

have suffered elevation to at least half that amount since Pliocene times; but, neglecting the faulted upthrust block of Ruwenzori which reaches nearly 17,000 feet, the total uplift is at least two thousand feet greater, for the maximum altitude of Peneplain I. is considerably more than 6,000 feet above M.S.L.

This stupendous movement, which has taken place during post-Cretaceous time and is probably still proceeding, though very widespread has not been similar in degree throughout the area of its expression. It has been culminative about an east-central, roughly N.—S., line, as evidenced by the original but now greatly modified divide between the drainage systems to the Indian Ocean, on the one hand, and to the Atlantic on the other. But the movement has been discontinuous and recrudescent thus producing a series of successively lower erosion planes, and relies (often of great extent) of preceding levels.

It has been held in certain quarters that if these lower levels are true peneplains no remnant of Peneplain I. should now be left, for, given the neccessary tectonic stability, the forces of denudation must act on a land-mass during an extremely long period in order to produce a peneplain as flat and as definite as those of Uganda that post-date Peneplain I.

This argument appears to rest upon an assumption, namely that, granting the time required for "complete" peneplanation of a country is, say, $r \times 10^7$ years, another and lower erosion plain in the same country, and of no less perfection, would require a similar period for its production.

But would it? We have reason to believe that Peneplain I. was forming at some time during, and perhaps throughout most, if not all, of Jurassic and Cretaceous time (a span of 36×10^7 years) while the succeeding and lower levels, of which there are four, were all completed in Tertiary time (some 9×10^7 years). But it may be that Peneplain I. had reached a stage of considerable perfection long before its disturbance in late pre- or early-Tertiary days.

The conversion of a mountainous country to a vast plain, although an imperfect one, must inevitably be a very lengthy affair, but granting the same intensity to the agents of denudation and the tectonic uplift necessary for their effective operation, the production of a plain from a plain would seem to involve much less time. Fourteen years ago[1] I attempted to provide some account of the mechanism of this process and of the formation of Inselberg as residuals during its action, but this was better done the other day by Bailey Willis who aptly names this erosive operation *etching*. Peneplain I. may be a true peneplain produced by the base-levelling of a diversified topography, or it may not, but it clearly preceded the other erosion planes, now manifested as extensive plateaux, and thereby influenced their development from the start. They are younger plains (or planes) etched in the surface of older ones, and the manner of etching appears to be this:—

Absence of any marked surface relief, a flat gradient and a seasonal climate lead to vertical rather than horizontal movements of ground waters and the consequent rotting of all but chemically resistant rocks, such as certain quartzites, to a depth of tens of feet. This zone of rotted rock, or saprolite, is largely removed by denudation if and when land elevation supervenes, and the process may be repeated

[1] *Ann. Rep. Geol. Surv. Uganda,* for the year ended 31st March, 1920, p. 40.

again and again as the country rises slowly and discontinuously. Thus the surface (or large parts of it whose areas are determined by drainage basins) is lowered and kept at or near base level by superficial removal against elevation from below.

The orthodox view is that peneplanation results from the erosion of a tectonically stable land-mass. If that be accepted, then it would seem to the writer that the accepted nomenclature in this regard is well chosen, for the resulting land-form could never be more than an imperfect plain, it would at best be undulating and could never present an exceedingly flat surface such as those displayed by the erosion levels below Peneplain I. in Uganda. Peneplain I. is likely enough a true peneplain by the terms of orthodox definition, but in my view the other erosion levels are not true peneplains (although we have used that term for them) they are etched plains, and as such are indicative not of tectonic stability and quiescence but of instability and upward movement.

The fact that the etched plains are individualized by marked altitude differences is evidence of strong, relatively rapid vertical movements punctuating the slow discontinuous rise, and in respect of this the testimony of the sedimentation in the Albertine depression is corroborative.

The altitude of the country increases toward the rift and reaches a maximum about twenty-three miles from the escarpment.[1] The elevation of the country fringing the rift and the depression of its bottom are associated phenomena. As the sides have risen so the bottom has been depressed, slowly for a considerable period, then rapidly, and the process has been a recurrent one. The Kisegi beds, a thick accumulation of shallow-water sediments in the rift, are separated by an unconformity from the overlying Kaiso deposits,[2] and the latter have been faulted and displaced to the extent of a thousand feet with the consequent formation of an imposing scarp. It is true that the movements of which these facts are indicative post-date the latest of the great erosion surfaces (that which we used to call Peneplain III.), but they are events in the elevation of Eastern Central Africa and, as such, are doubtless representative.

In the writer's opinion, then, Peneplain I. is in all probability a true peneplain, formed during a long period of secular quiescence, but the erosion levels beneath it are etched platforms formed during periods of slow elevation.

[1] It is along this line parallel with the scarp that modern swamp-divides are situated in ancient erosion valleys.
[2] The Kaiso deposits are early middle-Pleistocene in age.

Editor's Comments
on Papers 24 Through 27

DEEP WEATHERING AND ETCHING

Büdel, in Paper 24, considers the development of planation
surfaces in the humid tropics with monsoon and savanna climatic
characteristics. The two-page summary presented here includes
Büdel's essential thoughts. These revolve around the general
concept that, in tropical regions where thick soil or saprolite has
developed, there are two concurrently active leveling surfaces
(double surfaces of leveling): an exposed wash surface at the top
of the deep saprolite and a buried basal surface of weathering at
the bottom in contact with unweathered bedrock. The processes
of surface denudation and subsurface weathering may be in
equilibrium if slow uplift is in progress, which would permit sub-
surface weathering to keep pace with the rate of surface wash,
thus continuously maintaining a thick soil cover. With the exception of narrow stepped-rim pediments near highlands and around
outlying inselbergs, Büdel sees the frequent occurrence of broad
wash plains in between. In denying that broad rock pediments
are widely developed, Büdel could be thinking of etchplains, although he does not use the term. Büdel allows for almost.com-
plete denudation of the weathered cover and exposure of the
irregular weathered surface, which may have considerable relief

(the *Grundhöckerrelief* or basement topography). Later pedimentation may destroy or modify this relief.

Büdel plays down the importance of the work of rivers in his double-development surfaces. He also points out that basement high relief and rim pediments may be recognized in fossil (or defunct) form in regions outside the tropics, for example, in Central Germany.

In Paper 25 Büdel applies the idea of double surfaces of leveling to a broad, almost monolithic granite terrain in the Precambrian shield of India, where the influence of climate and "epirovariability" emerges the more clearly. He emphasizes that some shallow depressions that carry a linear flow are hydrologic rather than morphologic units, and that deeper valleys which show some side cutting do not corrade but merely pick up small-sized particles from the weathered surface. Other regions of high soil moisture content are found at the base of inselbergs; according to Büdel, these account for the undermining of the inselberg base and thereby ensure the continued existence of a steep slope.

Büdel describes two well-developed planation surfaces at different heights separated by a scarp. The higher and older of these is dated as pre-Miocene, when tropical weathering was more extensive than now, and is referred to as a "fossilized rock peneplain." It is fossilized only in the sense that it is preserved, but not that it has been buried and exhumed. Instead it could represent the lower member of a pair of double planation surfaces exposed by uplift and deep washing.

The topographically lower planation surface is now graded with respect to sea level. This is an example of upper-surface leveling in deep soil, which presumably developed under late Tertiary to present monsoon conditions. Büdel considers this an active "peneplain." Perhaps it would be more precise to call it an etchplain.

(*Note:* This English summary uses "duplicate planation surfaces" as a translation of *doppelten Einebnungsflächen*, which is better rendered as double, meaning a pair, not identical twins.)

Paper 26, by Ollier, on the inselbergs of Uganda is concerned with showing that deep weathering may well be the essential prelude to the formation of inselbergs as well as the broad areas of relatively smooth regolith between their bases. This is essentially an etchplain situation, with the emphasis on showing that pediments, inselbergs, and even pediplains may be the result of deep weathering preceding surface wash or stream erosion. Ollier sees only one long period of deep weathering, whereas Büdel favors repeated and rapid episodes of weathering.

The thoughtful paper by Thomas (Paper 27) examines the validity of the evidence for deep weathering in west central tropical Africa. The author submits maps and other data to show that weathering is indeed deep in many places, but that the weathering front is quite irregular; he cites cases of streams superposed from the regolith onto rocky "highs" along the weathering front.

The author is critical of many of the pediment interpretations of L. C. King, preferring instead the etchplain idea. As to the initial condition of the weathered plain, Thomas merely refers to "some earlier well-planed land surface." The type of original plain is left indeterminate.

24

Reprinted from *Zeit. Geomorph.*, **1**(2), 223–225 and Figs. 5–12 (1957)

DOUBLE SURFACES OF LEVELING IN THE HUMID TROPICS

Julius Büdel

Summary

In the tropics, especially in the region of the intermittent-humid savanna and monsoon climates, areas of little to moderate tectonic uplifting are occupied by wide characteristic plains. Scattered inselbergs and the higher bordering mountains rise sharply above these plains for the formation of which the process of "Flächenspülung" (sheetwash) plays an essential part. In details the mechanism of formation of this very striking sub-aerial typ of forms (ref. fig. 1–4) is only little revealed; especially unexplained is the contradiction that these plains according to their function represent the most important e r o s i o n a l s u r f a c e s of the earth, whereas they correspond in many features to d e p o s i t i o n a l s u r f a c e s outside the tropics.

The theory of the "Doppelte Einebnungsflächen" (double surfaces of levelling) gives a solution of the contradiction and clears some more questions of the genesis of these plains.

From the familiar relations of the regions outside the tropics one assumed that also on these plains the phenomena of weathering and denudation t o g e - t h e r took place within the (in the regions outside the tropics often only 0,5–1,5 m thick) soil-cover, that is to say on one and the same level (ref. fig. 5/1). In fact the soil-cover (caolin-rich red clays and related clays rich of clay-minerals) thickens up to 60 m, on the average 30 m, on such tropic plains. Thus t w o surfaces of levelling exist exhibiting a perfect partition of the processes of weathering and denudation ("Doppelte Einebnungsfläche", ref. fig. 5/2). The lower "weathering base-level" is the front of attack of the chemical decomposition progressing rapidly to depth with the constant high temperature and moisture (also in more resistant rocks; even boulders coming from the slopes of the inselbergs or the rapids of rivers into this mill of sorting and decomposition disintegrate to the common clay fine-sand mixture). Just so exclusively the upper "Spül-Oberfläche" (wash-surface) ist the place of denudation, the transportation of the very movable final products of weathering. This happens together — simultaneously and rather homogeneously — by the effects of the common washfloods of small rills a n d the high water of the main rivers at rainy season. The fine-material will be suspended becoming so a part of the high water wash-floods. It is not a slowly moving upper bottom-layer (like solifluction) but the running water itself that effects the material transportation of the erosion. That is the

[*Editor's Note:* The German text has been omitted, owing to limitations of space. The original German title is: "Die doppelten Einebnungs-flächen in den feuchten Tropen."]

reason, why the "fastebene" (nearly even) slope of these plains conforms so far to the gradient-curve of the streams; otherwise formulated, according to the conceptions outside the tropics, it means that the streams run "nearly perfectly here in the niveau of the plains". By these factors "Spülpedimente" (wash-pediments) will be formed at the rims of such plains to the steeper slopes of the insel-bergs and the bordering mountains.

If the bed-rock is nearly homogeneous and the tectonic uplift takes place as a repeated intermittent slow rising both levels will be lowered parallel to one another. At the rim of the mountain-fronts a Treppe (succession of step like benches) of raised Rand-Spülpedimenten (rim-wash-pediments) (ref. fig. 5, 3) develops. These pediments, however, are not the remnants or stumps of plains which previously as Fels-Rumpfflächen (ancient erosion surfaces with exposed bed-rocks) covered the intermediate area. S u c h F e l s - R u m p f f l ä c h e n h a v e n e v e r e x i s t e d . The connection between the rimpediments of two such mountain-fronts was ever only effected by wash-surfaces over a thick wea-thering-profil. By that the "Rumpftreppen", too, are better to be interpreted on the flanks of individual Middle-European mountain-ranges which have never been connected by Felsfußflächen (pediments) of so narrow vertical distances but so far horizontal extension (ref. fig. 6).

Other forms develop if the underlying rocks are considerable different in hardness in the area of such a slowly uplifted system. At last residual hills of the sculptured underground bed-rock surface rise to the surface (ref. fig. 7) where they will be levelled by wash processes to "isolated wash-pediments" or even grow up as "shield-inselbergs" over the surface (ref. fig. 8). In contrast to the larger "Auslieger-Inselbergs" in front of the mountains rims these shield-insel-bergs have no relations to the genesis of the neighbouring mountains. On their tops no relicts of higher lying ancient erosion surfaces (Altflächen) exist. All these forms can be observed in fossil state in Middle-Europe, too, e. g. on the Muschelkalk-Gäufläche in Franconia (ref. fig. 9).

Some cases exist where the thick soil-cover of these plains suddenly will be removed for wide distances so that the uneven, rugged relief of the bed-rock surface is exposed and appears now as a strange complex of forms not to be explained by sub-aerial processes. At the one hand this happens with a rapid rising of crustal blocks at the other with a sudden change from a tropic inter-mittent-humid climate to a tropic arid one. For both examples from the tropics and — fossil — the European middle latitudes will be cited: for the first case — very rapid rising — from Abyssinia and the Swabian Kuppenalb (ref. fig. 12a). Widespread areas are controlled by such an emerged "Grundhöckerrelief" (expo-sed hilly relief of the bed-rock surface) with wide fields of hilly shield-inselbergs in the southern Sahara (ref. fig. 10 and 11) where in the Quarternary repeated rapid changes took place between an intermittent-humid climate and an arid one as it can be proved.

If the uplifting is moderate the isolated-pediments and rim-pediments coalesce step by to an almost uniform r e a l F e l s f l ä c h e (coalescing pedi-ment surface) with a thin soil cover. Such forms, overtopped by shield-inselbergs, are often spread on the heights of mountain ranges in and outside the tropics. In the tropics, too, the further development of such Rumpfflächen (surfaces) highly raised over the base of denudation finally ceases and by degrees starting from the rims they will be cut by valleys, destroyed and removed. The quickness of this process of destruction only varies greatly in the different climatic-morphological regions of the earth. In the humid tropics such raised Rumpfflächen remain much longer preserved than outside the tropics. The down cutting of valleys on raised ancient erosion surfaces (Altflächen) of petrographical homogeneous rocks is especially quick in the tropic- and sub-tropic-arid regions, but just so the destruction of a Rumpffläche supported by a competent bed is especially greatly delayed in areas with nearly horizontal, alternating beds of competent and incompetent rocks (ref. fig. 12b).

At the border zones of the tropics the climatic suppositions for the formation of such plains gradually go out: here in the lowland, too, the "zone of plain formation" finally grades to a "zone of plain preservation", untill ultimately in the regions outside the tropics such (inherited from prior tropic climatic conditions) plains notwithstanding favourable tectonic conditions are no more fit for preservation. The reasons for this change will be described in context at another place.

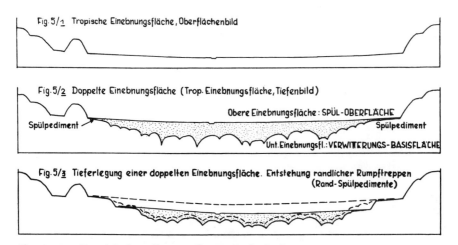

Figure 5. Double leveling surfaces and piedmont steps.
5/1: Tropical leveling surface (surface profile).
5/2: Double leveling surface (tropical leveling surface, cross section).
Obere Einebnungsfläche: SPÜL-OBERFLÄCHE—upper leveling surface: WASH SURFACE; *Spülpediment:* wash pediment; *Unt. Einebnungsfl.: VERWITTERUNGS-BASISFLÄCHE*—lower leveling surface: BASAL WEATHERING SURFACE.
5/3: Lowering of a double leveling surface: development of rock steps (Rumpftreppen). *Rand-Spülpedimente:* rim wash pediments.

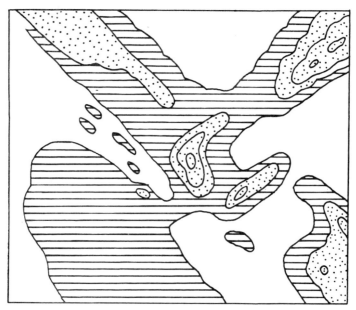

Figure 6. Higher bordering rock surfaces on a single mountain (rim wash pediments) (Examples: Frankish Alps, Fichtelgebirge, and Erzgebirge). Horizontally lined: Upper Miocene surface areas; dotted: Lower Pliocene surface areas. Areas are only approximately indicated.

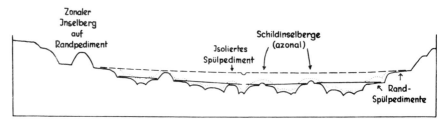

Figure 7. Tropical level surfaces with rim wash pediments, zonal and azonal (shield) inselbergs. (Examples: Northern Nigeria and Southern Somaliland.) Labels from left to right: zonal inselberg on rim pediment; isolated wash pediment; shield inselberg (azonal); rim wash pediments.

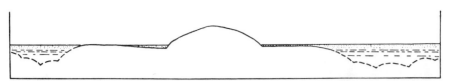

Figure 8a. Modern shield inselberg with wash pediment NW of ARAK (central Sahara).

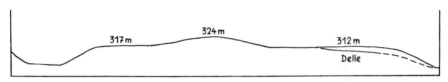

Figure 8b. Fossil shield inselberg with pediment (Geroldhauser upland, southern Würzburg regional surface).

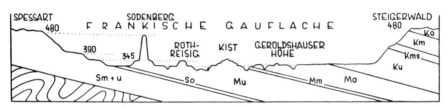

Figure 9. The Frankish regional surface (Gäufläche), a fossil tropical leveling surface with isolated and rim wash pediments, outlying and shield inselbergs (strongly surmounted).

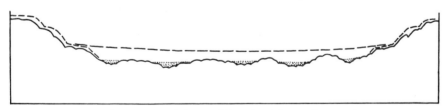

Figure 10. Development of dome-like shield inselberg plain from rapid deep weathering of a tropical surface.

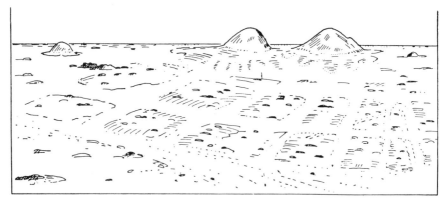

Figure 11. Outlying inselbergs, dome-like shield inselberg plain, flat rock pediment (tectonic structural lineaments shown), and the drifting sand plain in the southern Hoggar foreland (southern Sahara).

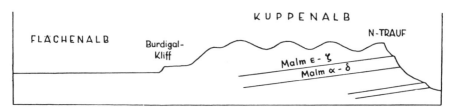

Figure 12a. S-N-Profile of the Swabian Alps. *Flächenalb:* surface alb; *Burdigal Kliff:* Miocene cliff; *Kuppenalb:* Kuppenalb; *N-Trauf:* N-trough; *Malm $\epsilon - \zeta$:* Upper Jurassic $\epsilon - \zeta$; *Malm $\alpha - \delta$:* Upper Jurassic $\alpha - \delta$.

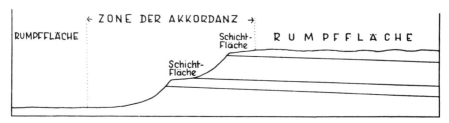

Figure 12b. Modification of beds in peneplain on Plateau de Tademeit (northern Sahara). *Rumpffläche:* peneplain; *Zone der Akkordanz:* accordant zone; *Schichtfläche:* bed-controlled surfaces.

25

Reprinted from *Colloquium Geographicum*, **8**, 25, 33, 93–95 (1965)

THE RELIEF TYPES OF THE SHEETWASH ZONE OF SOUTHERN INDIA ON THE EASTERN SLOPE OF THE DECCAN HIGHLANDS TOWARD MADRAS

Julius Büdel

XVI. Summary

The eastern slope of the Deccan Highlands towards the Coromandel Coast in the region between Bangalore and Madras offers a model example of a Rumpftreppe. Essentially it consists of two large plains. Only in the north of the field under consideration exists a third plain which is included as an intermediate form.

The lower of these plains, the Tamilnad Plain, broadly follows the coastline and reaches inland an average of 100 km and in places to 200 km. It rises to an altitude of 200 m and on occasion exceeds 500 m. It is then succeeded by a steep step some 100 m in height which leads to the higher Bangalore Plain (750—900 m) overtopped, however, by inselberge, the highest of which reach almost 1500 m. But these occur less at the centre than on the fringe areas of the Bangalore Plain (cf. Fig. 5). There are two other relief types in the region of decreasing gradation between both the large plains. Firstly a "Tropisches Gebirgsrelief" (tropical mountain relief) with narrow, and in the long profile much graded true erosion valleys, immediately joins the abrupt upper edge of the lower lying Tamilnad Plain. A "Tropisches Rückenrelief" (tropical ridge relief) follows, beginning above the border zone so much dissected by valleys and continuing as far as the lower edge of the higher Bangalore Plain. The shallow trough-shaped valleys — "Spültäler" (rills) — which produce the ridge relief, do not owe their existence to river erosion but to a line-like concentration of both the process of weathering in the lower strata and the process of surface run-off. This complex of processes is described as "Linienspülung" (linear run-off).

Here the functional explanation of the four examples of relief formation in the tropical zone of alternating humidity has been made especially easy by the fact that the entire region consists uniformly of granite with only very small extents of sedimentary flakes and penetrated by only a few dolerite veins. Thus the influence of climate and epirovariability emerge the more clearly.

The Tamilnad Plain is a model of a Rumpffläche in the process of active formation. Its entire slope reaches no more than 2 ‰ or 0.6°; the gradation of the

[*Editor's Note:* The German text has been omitted, owing to limitations of space. The original German title is: "Die Relieftypen der Flächenspülzone Süd-Indiens am Ostabfall Dekans gegen Madras."]

hydrological network operating on the plain — that is between the "Spülschei-
den" (run-off divides) and the "Spülmulden" (run-off troughs) which collect
the water here — show in detail a mean value of only 10 ‰, with extremes of
20 to 30 ‰ (= 1.2 to 1.7°). Only in the area on the immediate front of the
upper edge of the plain or the high inselberge ("Auslieger-Inselberge") perched
on top far away from rivers and in markedly localized situations does the gradient of these areas rise to 3.5 to 4°. These never more than a few hundred
metres-wide and somewhat steeper margins of the plain which begin at the
sharp break on the upper edge — i. e. at the foot of the inselberge or their
upper step edge — we term "Spülpedimente" (run-off pediments).

The formation of the Tamilnad Plain has occured and continues to occur as
a result of the "Mechanismus der doppelten Einebnungsflächen" (mechanism
of duplicate planation surfaces). A heavy sheet of red loam of an average thickness of 4 to 10 metres covers the entire area and is the product of a chemical
decomposition above the granite at the base of the red loam cover. Along
fissures in the granite the decomposition reaches the lower levels more speedily.
"Grundhöcker" (basement humps) persist between them. Thus a close-up of
the basic plain of decomposition presents an uneven relief. With hard granite
measures spreading out or fissures diverging widely, the basement humps rise
above the soil surface to form "shield inselberge".

The entire profile of the red loam is without motion and lacks any denudation
process on these plains: it is nothing but a product of the first chemical decomposition. Denudation only occurs on the surface of the red loam cover the "Spül-
fläche" (run-off plain). It occurs by fits and starts, continuing step by step, during the rainy season by way of small grooves evenly distributed over the entire
plain. These coalesce to make up bigger streams and finally to form rivers. The
smallest brooks and the biggest rivers alike, however, only transport material
of the same grain size which has to be provided by decomposition. It is a clay
rich in kaolin on the one hand and fine sand of 50 to 200 µ on the other. Larger
rivers too, flowing in the hyper-shallow "Spültäler" are not able to acquire
bigger material from the underground by their own erosive activity: they do
not erode but are merely drawn passively into the general process of superficial
run-off which governs and shapes these entire surfaces.

In paragraphs VII—X it has been proved in detail that these rivers are inca-
pable of depth erosion as they are incapable of lateral and regressive erosion.
The run-off troughs they use are not morphological but only hydrological units.
Along their margins all signs of working edges are absent.

A morphological strict unit defined by distinct working edges at their upper
brink and at the foot of each "outlier inselberg" can only be seen in the area as
a whole. It continues to extend further at these working edges by the process
of sub-cutaneous lateral denudation: it is an intensified form of the mechanism
of the duplicate planation surfaces generally prevalent on these plains. The
intensification is brought about by the fact that precipitation falling during the
rainy season is joined at this brink by the water running off the steep and often
rocky slopes of the inselberge or running down the break-of-slope from above.
Thus there is a zone of particularly high moisture saturation of the red loam
cover and of especially intense decomposition in depths extending along the

foot of such slopes (Fig. 2 & Photo 2). Such slopes can even be actually under-mined and thus kept steep. In this way an active peneplain continually extends at this working edge (SB-Akk, Fig. 2) at the expense of the terrain above.

As long as the erosion base-level of the sea remains constant the mechanism of duplicate planation surfaces will lower this area as a whole parallel to itself and become flatter and flatter in every direction. The formation of such pene-plains is conditioned by the alternately humid tropical climate of the savanna with 1 to 10 humid months besides distinct dry seasons; it occurs on continents with only little or no tectonic uplift and in connection with a fixed base-level of erosion — the sea or large inland waters — which does not allow further incisions of erosion lines (valleys). The higher plain of Bangalore once formed before the Miocene in conjunction with a marine erosion-base which persisted over a long period. The vaulting of the Deccan followed in the Miocene, accom-panied by the uplifting of the summit region of Bangalore almost en bloc. This uplift declines in wave-like fashion towards the coast. At this point the younger plain was vertically let into the body of the upper one and horizontally extended. The upper plain of Bangalore is able to extend its upper edge against its "per-ched" inselberge by virtue of the same mechanism of the dual levelling. But now this plain is being dissolved from below by valley formation which is connected with the zone marked by steps. Thus the red loam cover decreases here and the basic plain of decomposition comes more and more to the surface in the form of extensive shield inselberge. In this way the active plain of run-off gradually turns into a more and more fossilized rock-peneplain. In the Tertiary, the period of the "old tropical earth", variants of the tropical savanna climate favouring the formation of peneplains extended as far as the subpolar regions in the Eocene and later, in the Miocene and older Pliocene, as far as Central Europe. It is from this age that nearly everywhere on earth the oldest relief generations of our present range of forms have been preserved in the shape of fossil peneplains (old plains) or remains of such. This applies to long consolidated continents as well as to plateaux formed of flat strata deposits and to young regions with alpidian or saxon tectonic structures.

At the present time the old plains are to be found at various altitudes and in various latitudes: pg. XIII—XIV shows that these flat forms are endowed with an exceptional ability to survive even extreme climatic conditions. For long periods even floating ice and the sub-polar climate of the „excessive valley for-mation" do not succeed in subduing them for the break-down processes effective there and thus they manage to persist in remnant fragments.

Lastly, paragraph XIV presents the methods which enable us to distinguish genuine remnant fragments of fossil peneplains form the period of "old tropical earth" in the central latitudes from ridges formed by recent breakdown pro-cesses and working stages which are accidentally situated at the same height.

Figure 1. Surface development in the extra-tropical valley-forming zone (I) and in the tropical surface-forming zone (II).

I: Fluvial surface formation through lateral erosion in the extra-tropical zone.

Model of the Munich graded plain or of the Vienna–Neustadt stone plain (without late and post-glacial cutting). *Altfläche:* Old surface.

 1 Stream transported Wurm glacial gravel, covering but not modifying the planate floor. The gravel surface is slightly arched in the middle.

 2 Contiguous basement (Molasse). Old surfaces (Altflächen) preserved under fossil alluvial deposits containing remnants of older glacial gravels.

3 The working edges (Akk, concave; Akv, convex), produced by lateral erosion, bound the surface.

II: Wash-surface formation through double surface leveling in the monsoon tropics.

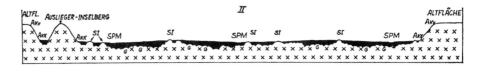

Model of the Tamilnad surface in southern India. *Altfläche:* old surface; *Auslieger Inselberg:* outlying inselberg.

 1 Tropical red loam (laterite) formed by weathering in place while the surface is completely unaffected. Later this becomes a wash surface.

 2 Underlying granite

 3 Basement humps of the basal weathering surface.

 4 Corresponding working edges (Akk, concave; Akv, convex), produced through subcutaneous lateral denudation (Figure 2), which bound the surface. The old surfaces are higher-lying planation surfaces in granite. SPM, wash troughs; SI, shield inselbergs.

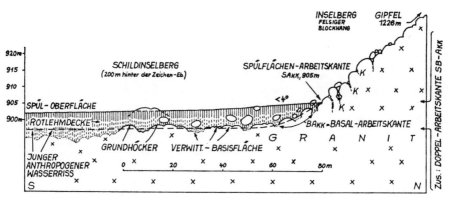

Figure 2. Undercutting of an inselberg near Kolar through subcutaneous lateral denudation of the double leveling surface at the two working edges, SAkk and BAkk. K: joint-weathering which this lateral erosion produces. *Inselberg, Felsiger Blockhang:* Inselberg, rock slope; *Gipfel:* peak; *Schildinselberg (200 m hinter der Zeichen-Eb.):* shield inselberg (200 m beyond section); *Spülflächen-Arbeitskante:* wash surface working edge; *Spül-Oberfläche:* wash surface; *Rotlehmdecke:* laterite layer; *BAkk-Basal-Arbeitskante:* BAkk basal working edge; *Grundhöcker:* basement humps; *Verwitt.-Basisfläche:* weathered lower surface; *junger anthropogener Wasserriss:* later man-made water level; *Doppel-Arbeitskante SB-A$_{kk}$:* double working edge of SB-A$_{kk}$.

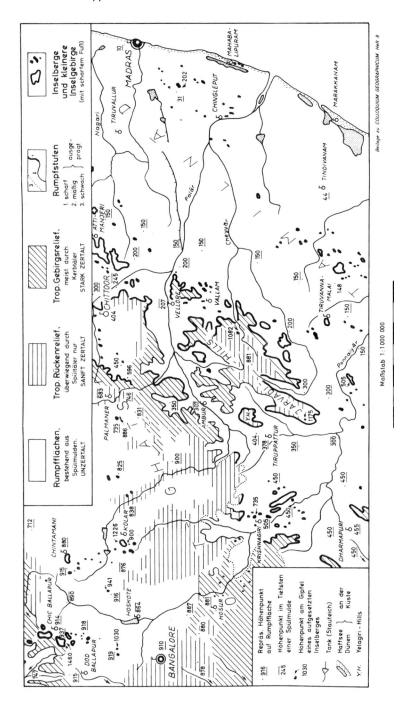

Figure 5. The tropical relief types on the eastern border of the Deccan Highlands (Southern India). *Rumpfflächen:* planation surfaces with wash troughs prominent, no valleys; *Trop. Rückenrelief:* tropical upland relief mainly through wash valleys, sparse valleys; *Trop. Gebirgsrelief:* tropical mountain relief mostly by notched valleys, many valleys; *Rumpfstufen:* planation stages: (1) sharp, (2) medium, and (3) faint relief; *Inselberge:* inselbergs and smaller inselbergs with sharp bases.

26

Reprinted from *Zeit. Geomorph.*, **4**(1), 43–52 (1960)

The Inselbergs of Uganda

By

C. D. OLLIER-Melbourne

with 3 figures and 2 photos

Introduction

The observations and ideas presented in this paper were accumulated over a period of two and a half years field work while making a provisional soil survey of the Northern and Eastern Provinces of Uganda. During that time every motorable road in the two provinces and much of the land between was traversed, and as many profiles as possible recorded and studied. All other districts in Uganda were visited but not studied in such detail.

In the course of this work it became clear that many of the soils were formed not on fresh rock as a parent material, but on a pre-weathered regolith. This was nowhere more apparent than in inselberg landscapes. Following up this work it was realized that many theories of inselberg formation were quite incapable of accounting for the geomorphic features of Uganda inselbergs. Others provided fairly satisfactory explanations, but as they had been evolved in other parts of the world they did not account for all the characteristic features of Uganda inselbergs.

In this paper the features of inselbergs and their settings will be described, and a hypothesis of their development proposed, which will later be compared and contrasted with other hypotheses and theories.

The Inselbergs and their Setting

The main features of the inselbergs of Uganda which must be taken into account in any theory of their origin will now be described. Many of the features are common to inselbergs of other areas, and have frequently been recorded before.

They rise very abruptly from the surrounding land, often on all sides, and there are the usual exfoliation surfaces, jointing, perched boulders and so on. Most of them occur at the top of ridges of very subdued relief, but some occur in more rugged areas, and there are others which occur on lower slopes. A whole group of inselbergs often occur together, and isolated ones, although they are found, are rare. They range in height from a few tens of feet to a few hundred feet, and for their size they cover a relatively small area. Large hills which occupy several square miles and rise for several hundreds or thousands of feet are not true inselbergs.

Granite is most commonly an inselberg producing rock, but gneisses may also be etched into inselbergs. Amphibolite and volcanic rocks never give rise to them and quartzite bands always form ridges – never true inselbergs.

The inselbergs are often of irregular shape, although simple dome shaped ones are sometimes found. They are frequently asymmetrical, with markedly different slopes on different sides. They usually have rounded tops, and although there may be some accordance of summit levels there are no remnants of previous erosion levels on the inselbergs proper. Where remnants of old erosion surfaces are found they are always higher than the inselberg summits.

In more rugged areas, as in Mubende district, there is a complete range from true inselbergs to small tors, but all seem to share a common origin.

The most remarkable feature of the inselbergs is that they are separated from each other by wide areas of very weathered rock. Many profiles in road cuttings, wells, gullies, quarries and specially dug soil profile pits show extensive and deep weathering in all areas, and yet the inselbergs themselves are of fresh rock. Bore-holes on the plains between inselbergs frequently go through over 200 ft. of weathered rock before reaching fresh rock.

This deep weathering is the most significant feature of inselberg country, and the main features of the regolith will be briefly described. A very common type of profile shows a variable thickness of soil overlying a stone line, which in turn overlies weathered rock. Quartz bands often traverse the rotted rock, and they provide the material which forms the stone line. The top part of the profile is a more or less uniform soil which is probably formed by resorting of the weathered rock by termites, leaving the coarse material at the base to form a stone line. Where there are no quartz bands to provide material for a stone line there is no clear division between soil and rotted rock. On amphibolite the weathered material goes down for many tens of feet with very little change. On granite there is small angular quartz where the weathered rock is more or less in situ but no other obvious difference from the soil in the upper part of the profile. Metamorphic rocks consist of bands of rock of different weatherability and therefore give rise to a wider variety of profiles, to irregular topography and to fewer tors or inselbergs than does the more uniform granite. A more elaborate account of the profiles formed on the regolith is given elsewhere (OLLIER [1959]) and a study of a selected catena (RADWANSKI & OLLIER [1959] provides detailed pedological and mineralogical information.

Such a thick regolith has been often described, as for instance by GREENE (1945), NYE (1954), ANDERSON (1957) and BRÜCKNER (1955), but it is usual to regard it as formed in the present cycle, and its significance to geomorphic development has often been neglected. LINTON (1955), however, has postulated pre-weathering in his theory of tors, and BÜDEL's account of inselbergs also puts emphasis on the formation and destruction of the regolith.

Between the fresh rock of the inselbergs and the rotted rock of the surrounding areas there is a remarkably abrupt junction. This is present in weathered granite areas in other parts of the world and has been called the "basal surface of weathering" by RUXTON & BERRY (1959), who describe it in detail and discuss many of the implications involved. LINTON (1955), in his account of Dartmoor, described it as a "basal platform", and BÜDEL calls it the "base level of weathering". Many authors maintain that the fresh rock merges into rotted rock, which may well happen in some localities, but the existence of abrupt junctions in other areas seems well documented.

Some fresh rock may be cut off from the main mass through the attack by weathering from all sides. These become rounded corestones, and also have abrupt junctions between fresh and weathered rock. RUXTON & BERRY (1957) have described such corestones from Hong Kong, and BRÜCKNER (1955) figures epidiorite corestones from the Gold Coast with a sharp separation from weathered rock. Just why the junction is so abrupt is not well known.

The evidence in Uganda indicates that the basal surface is very irregular, and where it crosses the ground surface inselbergs emerge. Because of the irregularity RUXTON & BERRY's term "basal surface" is preferred to the other names which have been proposed.

It is only possible to photograph the basal surface where it is close to the surface, and it is usually difficult to get a good picture because excavations stop when fresh rock is reached. However, Fig. 1 shows a basal surface junction and a corestone in granite.

Fig. 1a: Photo of the basal surface of weathering and a corestone. Kakumiro, Mubende, Uganda

The setting of the inselbergs in relation to the known geomorphic development of Uganda will now be summarised.

In the south of Uganda there are the remains of an old erosion surface,
mostly preserved on quartzite ridges, and below it there is a younger surface which
is extensively developed further north. There is considerable confusion about the
naming of these surfaces, but the upper surface may be called the Gondwana sur-

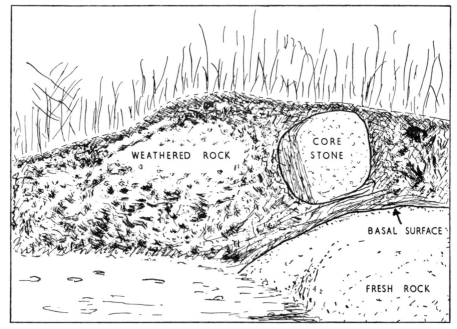

Fig. 1b: Explanatory diagram of photo above

face and the lower one the African surface. The Gondwana surface is represented
by flat topped hills, usually capped by massive laterite. The younger African surface shows all the stages of development from small valley side pediments in the
south to mature, very flat plains further north. The African surface is studded
with inselbergs, not found on the Gondwana surface. Inselbergs tend to be more
numerous and larger away from the remains of the Gondwana surface – that is
where erosion of the African cycle is more advanced.

The arrangement of the two surfaces and the inselsbergs is in accord with
KING's statement (1948) that inselbergs only occur where there is more than one
cycle of erosion.

A Hypothesis of Inselberg Formation

Landforms having the features described above may be explained by a composite hypothesis which is derived from the tor hypothesis of LINTON (1955), the
shield inselberg hypothesis of BÜDEL (1957), the bornhardt theory of KING

(1948) and the landscape development views of PALLISTER (1956), but there are differences from all previous hypotheses.

As a starting point we will take the Gondwana surface where it is on granite, Fig. 2a. We can ignore the laterite layer as it does not play an important part in the story to be outlined. The Gondwana surface was a very senile, very flat erosion level and probably existed for a very long time. During an extended period of weathering erosion was very slight and a great depth of rotted rock accumulated. There were probably several zones of weathering, but there was an abrupt junction between fresh rock and rotted rock. Different rock types would weather to different degrees and such features as jointing and texture would also influence the intensity of the weathering, so the basal surface of weathering would be irregular. This stage is shown in Fig 2b.

Eventually a new period of erosion was initiated. A drainage system developed and valleys began to cut through the weathered rock. This stage is shown in Fig. 2c. The rivers would soon become adjusted to structure and would tend to pick out the most weathered and softest rock, which would also be where the basal surface of weathering was deepest. Thus although the streams may be situated anywhere on the Gondwana surface at the start, at a later stage the main valleys would be those between incipient inselbergs as shown in Fig. 2c. Tributary valleys too would tend to erode along lines of most intense weathering, and so the less weathered rock, where the basal surface is highest, tends to be between valleys.

Valley widening later became the main feature of erosion and downcutting came to an end. The slope development and formation of a lower surface which took place in this stage have been described by PALLISTER (1956). There was parallel retreat of slopes and formation of pediments, which would eventually coalesce to form a lower erosion surface – a pediplain. The new surface was cut across rotted rock. This stage is shown in Fig. 2d. PALLISTER followed the process as far as the complete removal of the old surface, after which there is lowering of the convex summits, but this is not the end of the story.

PALLLISTER's work concerned an area on the watershed between Kyoga and Lake Victoria drainage systems, where there are many remnants of the Gondwana surface and erosion has not gone very deep. This area is not typical of much of Uganda and in the north and central parts erosion has reached a more advanced stage.

With further erosion the basal surface of weathering is reached and fresh rock is exposed at the ground surface. Rock might be exposed on any part of a slope, but it is usually on the tops of interfluves that it outcrops, and then as inselbergs. The sharp angle between inselbergs and the surrounding land is not so difficult to understand when it is realised that it marks the junction between fresh and very weathered rock. The final stage of the process is shown in Fig. 2e.

Other Hypotheses of Inselberg Formation

There are many older theories of inselberg formation which have been summarised and discussed by COTTON (1942), but most of these do not account satisfactorily for the inselbergs of Uganda, and will not be discussed here. The main modern hypotheses to be contrasted with the hypothesis given in the previous section are those of KING (1948), LINTON (1955) supported by WATERS (1957), and BÜDEL (1957).

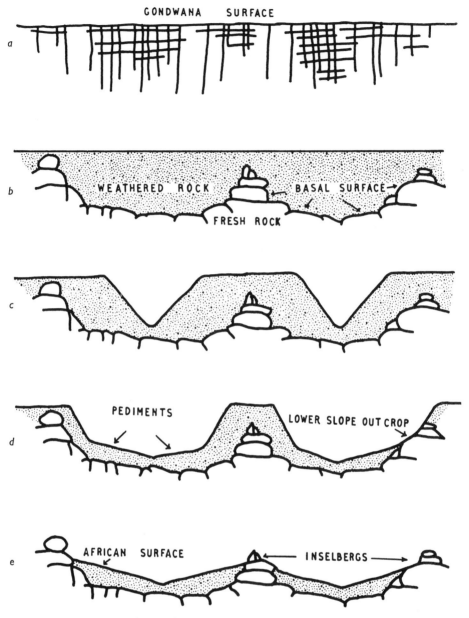

Fig. 2a—e. Diagram showing the evolution of inselbergs in Uganda

LINTON's solution to the problem of tors includes deep weathering and the removal in a later cycle of the weathered material. This only differs from the present account in that it is supposed that almost all the rotted rock is removed and that the flat areas between tors are also on the "basal platform", where fresh rock is close to the surface. The account of Dartmoor topography by WATERS indicates that this not so, and the inter-tor areas are largely cut across weathered rock. The formation of tors in humid temperate regions seems to have much in common with inselberg formation of drier areas.

KING, in his original theory of bornhardts (1948), explained them as residuals left after extensive erosion on Penckian lines had cut great pediplains across solid rock. Without denying that this could happen, it is quite certain that this is not the mode of origin of the Uganda inselbergs. The pediments and pediplains in Uganda are cut across a regolith and not across fresh rock. They do not have symmetrical slopes such as would result from simple backwearing, and if KING's theory were accepted it would be necessary to invoke weathering of the residual to give the present rounded form, perched blocks and so on. The evidence seems to indicate that present day weathering is working to break down the inselbergs by cracking, and that it was subsurface weathering which gave them their predominantly rounded form. It is interesting that KING recorded that wells are often sunk in the weathered rock around inselbergs. This happens around many Uganda inselbergs too.

In his criticism of LINTON's theory, KING (1958) postulates two sorts of tors – skyline and subskyline tors; the former his original type and the latter a special case formed as in LINTON's theory. Many Uganda inselbergs occupy the highest parts of the landcape, rising above the African surface, but they are not surrounded by aprons of planed bedrock. Just occasionally fresh rock might surround

Fig. 3. Subskyline tors. Kungu, 14 miles north of Kampala, Mengo, Uganda. The flat topped hill in the background is a remnant of the Gondwana surface, and is higher than the large tor on the left

an inselberg where the basal surface was flat as shown in the central inselberg in Fig. 2e.

LINTON's "basal platform" comes in for criticism from KING, being considered too smooth and regular, and contrasting too much with the irregular tors, but such criticisms would not apply to the present hypothesis as the plains are pediplains cut by all the usual processes, differing from KING's plains only in being cut across pre-weathered rock. The basal surface of weathering is supposed, in the present hypothesis, to be suitably irregular everywhere.

KING's subskyline tors are said to be formed after subsurface weathering, as in LINTON's theory. Such tors can be found in Uganda where there are remains of the Gondwana surface to form a skyline, and an example is shown in Fig. 3. There is no reason to suppose that the two types of tor in Uganda have any difference in mode of origin. The relationship is shown diagrammatically in Fig. 4, which

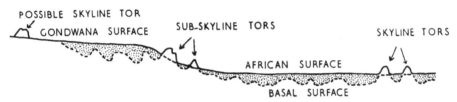

Fig. 4. Diagram illustrating the relationship of skyline and subskyline tors in Uganda

might be regarded as a diagrammatic profile from south (the left side) to north Uganda. If there are any hills rising above the Gondwana surface they might be skyline tors according to Professor KING's definition, but the writer has failed to find any.

BÜDEL's "doppelten Einebnungsflächen" (1957) also involves removal of pre-weathered rock and exposure of the irregular basal surface of weathering to form shield inselbergs To that extent the present account adds support to BÜDEL's hypothesis. BÜDEL, however, thinks that the weathering proceeds rapidly to depth, and is repeated in each cycle. In the present account it is considered that a thick regolith requires a very long time to develop, and that only one major period of regolith formation has occurred – that in between the formation of the Gondwana and the African surfaces.

Two of the older theories seem to be on the same lines as the present hypothesis. COTTON (1942) states that PASSARGE believed that in many cases a rock boundary was situated at the base line of an inselberg. This seems to be so, but the fresh rock and rotted rock on opposite sides of this boundary may have been originally of the same petrology, differing only in fissility. As COTTON points out, in this case the inselberg has not suffered reduction by backwearing but is a landform due to downwearing of the surrounding plains.

WILLIS (1936) believed in a difference in weatherability between the rock of the inselbergs and the surrounding plains. He postulated that a series of cycles of denudation had emphasised this difference. In each cycle the surrounding rock is weathered and eroded, increasing the relative relief of the inselbergs, until very large hills were formed. This is very like the present hypothesis, but makes the process cyclic.

Zusammenfassung

Die Inselberge Ugandas werden zusammen mit den wichtigsten Merkmalen ihrer geomorphologischen Umrahmung beschrieben. Die Inselberge selbst bestehen aus frischem Gestein, aber die umrahmenden Ebenen sind in einen tiefgründig verwitterten Regolith eingeschnitten. Es wird eine Theorie der Inselbergbildung vorgetragen, die die folgenden Entwicklungsphasen einschließt: 1. Einebnung der Gondwana-Rumpffläche. 2. Verwitterung zu einem tiefgründig zersetzten Regolith, der plötzlich und unregelmäßig gegen das liegende frische Gestein absetzt. 3. Abtragung während des „African cycle" mit der Ausbildung von Pedimenten im Regolith. Wo frisches Gestein angeschnitten wird, erscheint es häufig als Inselberg.

Die Inselberge sind also ausgebildet worden durch Tieferlegung des umgebenden Regoliths bei fehlender Zurückschneidung des frischen Gesteins, so daß der Vorgang bei der Bildung der Ebenen im zersetzten Gestein als Pediplanation anzusehen ist.

Résumé

L'auteur décrit les inselbergs de l'Uganda avec les caractères les plus importants de leur encadrement géomorphologique. Les inselbergs eux-mêmes se composent de roche fraîche, mais les plaines encadrantes sont entaillées dans un régolite profondément décomposé. L'auteur propose une théorie de formation des inselbergs, comprenant les phases suivantes:

1) Pénéplanation de la surface de Gondwana.
2) Actions météoriques produisant un régolite profondément décomposé, qui s'interrompt brusquement et irrégulièrement contre la roche fraîche sous-jacente.
3) Déblaiment pendant le «cycle africain» avec formation de pédiments dans le régolite. La roche fraîche apparait fréquemment en inselbergs, là ou elle est entaillée.

Les inselbergs ont donc été formés par abaissement (downwearing) du régolite entourant et sans érosion régréssive (backwearing) de la roche fraîche, de sorte que le processus qui intervient pour former les plaines dans la roche décomposée doit être considéré comme pédiplanation.

References

ANDERSON, B.: A survey of soils in the Kongwa and Nachingwea Districts of Tanganyika. University of Reading, 1957.

BRÜCKNER, W.: The mantle rock ("laterite") of the Gold Coast and its origin. Geol. Rundschau 43, 307—327, 1955.

BÜDEL, J.: Die „Doppelten Einebnungsflächen" in den feuchten Tropen. Zeitsch. für Geomorphologie. NF. 1, 201—228, 1957.

COTTON, C. A.: Climatic Accidents. Whitcomb & Tombs Ltd., Wellington, 1942.

GREENE, H.: Classification and use of tropical soils. Proc. Soil Sci. Soc. Amer. 10, 392—396, 1945.

KING, L. C.: A theory of bornhardts. Geog. Journ. 112, 83—87, 1948.

—: The problem of tors. Geog. Journ. 124, 289, 1958.

LINTON, D. L.: The problem of tors. Geog. Journ. 121, 470—486, 1955.

NYE, P. H.: Some soil-forming processes in the humid tropics. I. A field study of a catena in the Western African forest. Journ. Soil Sci. 5, 7—21, 1954.

Ollier, C. D.: A two cycle theory of tropical pedology. Journ. Soil Sci. In press, 1959.
Pallister, J. W.: Slope development in Buganda. Geog. Journ. **122**, 80—87, 1956.
Radwanski, S. A., & Ollier, C. D.: A study of an East African catena. Journ. Soil Sci.. In press, 1959.
Ruxton, B. P., & Berry, L.: The weathering of granite and associated erosional features in Hongkong. Bull. Geol. Soc. Amer. **68**, 1263—1292, 1957.
— & —: The basal rock surface on weathered granitic rocks. Proc. Geol. Assoc. In press.
Waters, R. S.: Differential weathering and erosion of oldlands. Geog. Journ. **123**, 503—509, 1957
Willis, B.: East African plateaus and rift valleys. Carnegie Inst. Wash. Publ. **470**, 1936.

27

Reprinted from *Inst. Brit. Geogr. Trans.*, **40**, 173–193 (1966)

Some Geomorphological Implications of Deep Weathering Patterns in Crystalline Rocks in Nigeria

M. F. THOMAS, M.A.

(*Lecturer in Geography, St. Salvator's College, University of St. Andrews*)

Revised MS. received 24 February 1966

THE PHENOMENON of deep weathering has received frequent discussion in papers concerned with the geomorphology of terrains developed on crystalline rocks, yet few of these studies show actual depths of weathering except for occasional, isolated locations. This absence of quantitative information obviously weakens arguments based upon the premise that deep weathering occurs widely within a given area or displays particular relationships with surface relief, drainage patterns or individual landforms. This deficiency is explained by the paucity of existing records and the difficulty of obtaining additional data where weathering depths commonly exceed fifty feet. Drilling and portable seismographs are both commonly used by civil engineers[1], and the information used in this paper is derived from these sources.

It is recognized that such data have certain inherent inadequacies. First the sites are chosen because they appear from surface considerations to be suitable for the ultimate purpose of the project. Most surveys of this kind in Nigeria have been undertaken to determine the suitability of sites for dam foundations. As a result the sites selected are mostly well-defined valleys which usually display frequent fresh rock outcrops. They are not therefore located in areas where weathering depths would be expected to be great. In spite of this limitation there is sufficient variety of site to make the profiles representative of more than one type of terrain.

The second difficulty is the lack of uniformity in the terms used to describe the various zones of the weathering profile. Furthermore, subdivision of the profile on the basis of seismic information is not often very accurate. For most of the profiles presented in this study emphasis has therefore been placed only on the recognition of the sound rock line or basal surface of weathering, but there are difficulties even here, for the transition from weathered to fresh rock varies greatly in its characteristics from one borehole to another.

There is also the possibility that the upper part of the profile may contain transported material, either sandy wash similar to that recognized for instance by R. V. Ruhe[2] or wind blown dust that is well known from Northern Nigeria. Both J. D. Falconer[3] and more recently J. Dresch[4] have called attention to the role of alluviation in depressions on the interior plains of West Africa. However, it is not thought likely that any serious errors, resulting from the inclusion of transported materials under the heading of *in situ* weathering products, have been incorporated into the profiles to be described. Seismic traverse information which may be ambiguous in this respect, has been included only from the humid south of Nigeria, where alluvial infilling of the kind envisaged above does not appear to have taken place, and from the Niger valley near Bussa, where a large amount of confirmatory drilling information is available. Other profiles have been constructed from borehole logs in which it is usually possible to distinguish transported from other materials. Only in the case of the Jos Plateau was any considerable thickness of alluvial material recognized and this has been shown on the sections (Fig. 4a and 4b). Bearing all these limitations in mind it is still possible to present fairly accurate data on depths of weathering for selected sites within Nigeria.

Published Records of Deep Weathering Patterns

Deep weathering has been recorded for a very wide range of present-day climatic environments, from Fort Trinquet in Mauritania where the rainfall is two inches and decomposition of granite to depths of sixty-five feet has been recorded by J. Archambault,[5] to northern Scotland, where mean annual temperatures are less than 50° F., and where E. A. Fitzpatrick[6] has recently shown widespread rotting of granite to depths of forty feet. The evidence for deep rotting in both semi-arid and temperate environments is in fact very widespread, and the implication that these profiles are relics of former humid tropical weathering conditions is generally accepted. D. L. Linton's[7] study of the Dartmoor tors and the recent study of similar phenomena in the Bohemian Highlands by J. Demek[8] are examples of this reasoning. In the semi-arid and sub-humid interior of Australia both J. A. Mabbutt[9] and R. L. White[10] have shown the importance of deep weathering of both crystalline and sedimentary rocks, commonly to depths of fifty to seventy-five feet. These profiles are also regarded as relics of a Cretaceous or Tertiary plain.

Within the humid tropics, a characteristic description for Uganda is given by C. B. Bisset:[11] 'although outcrops may locally be frequent, large, and conspicuous, the rock is weathered, disintegrated, broken or well fissured to depths varying from several feet to several hundred feet from the surface over most of the basement areas. The general depth of the weathered zone may be given as about 100 to 150 feet.' In the Serra do Navio district of Brazil, where Pre-Cambrian metasediments, mainly amphibolites, are weathered to depths of over 300 feet, R. H. Nagell[12] records that 'no rocks crop out on the ridge crests and are only sporadically present in the beds of small streams tributary to the rivers. Most exposures are in the main river beds, while beneath the ridge crests weathering extends to depths of more than 100 metres, or nearly to local base level.' Very deep profiles of this order have been recorded from Hong Kong by L. Berry and B. P. Ruxton[13] who also noted that beneath convexo-concave hills the thickest weathering profiles often occurred below the hilltops.

Opinions concerning the age relationships of deep weathering profiles in the tropics differ. C. D. Ollier[14] regards the deep weathering profiles beneath the Gondwana surface in Uganda as having developed during or before the middle Tertiary, that is before the formation of the African surface which is seen to truncate the profiles. In this view most of the deep weathering in tropical Africa is regarded as antecedent to recent cycles of erosion.

Evidence of such relic weathering profiles was demonstrated in Northern Nigeria by Falconer in 1911.[15] At Kano two mesa-shaped hills of weathered crystalline rock rise 150 feet above the surrounding plains, and Falconer commented that they 'bear witness, however, not only to the extensive decomposition in situ to which the crystalline rocks were once exposed, but also to the intensity of subsequent erosion which has left them as solitary relics of the earlier surface of the plain. It is possible indeed that the formation of such a sheet of weathered rock possessed a regional character'. The relationship of these hills to the plains that surround them is clearly similar to that observed in Uganda. But it must be emphasized that these later plains are also deeply weathered in places, demonstrating the continued effectiveness of the tropical weathering processes. The importance of such continued deep weathering is underlined by the observations of Nagell[16] who considers that the ridges of his area are preserved by the lateritic duricrusts, and that the very deep weathering profiles have been formed later, concomitant with the development of the present stream valleys.

Taken together with some of the profiles presented later in this paper such observations make it necessary to qualify assumptions regarding the antiquity of deep weathering profiles in Nigeria. The weathering processes may be assumed to have continued throughout a considerable period of geological time, and their effects will be most advanced beneath older land-surfaces. There is within Nigeria evidence both for deep weathering beneath late Tertiary or even Pleistocene surfaces, and also for the progressive stripping of the basal surface of weathering during more recent cycles of erosion.[17]

The Character of Weathering Profiles

There is no single type of profile illustrating the transition from thoroughly weathered material to fresh and unweathered rock that remains valid for all crystalline rocks type or all locations. F. W. Roe[18] gives three common types of transition found in Malayan granites. The rock may weather '(i) to form core boulders, (ii) to form clayey soil and soft rotted rock, which is separated by a well-defined line of contact from hard fresh granite, (iii) to give a gradual transition from clayey soil to fresh rock'. Although appertaining to granites these three possibilities apply with variations to most if not all weathering profiles in crystalline rocks. Berry and Ruxton and many other authors[19] have described in detail granite weathering profiles of type (i) above.

J. W. Barnes[20] gives a clear description of a typical weathering profile from the Basement Complex gneisses of Uganda: '(e) soil, (d) laterite, sand or clay, (c) highly weathered and incoherent rock appearing as sand and clay, (b) moderately weathered coherent rock, (a) partially weathered to fresh rock.' The same author records that the base of (c) is seldom found below 100 feet, while fresh rock is usually encountered before 160 feet. This profile which is of type (iii) above clearly presents certain problems to the geomorphologist. The question must be asked whether the recognition and mapping of an indistinct basal surface has any real significance, for if the transition from regolith to fresh rock is gradual through several tens of feet, then there is no real basal *surface* of weathering, only a zone of transition. In profiles of this type stripping of the basal surface of weathering to produce tors and domes could be regarded as an unlikely process, because there exists no recognizable surface to be stripped.

However, in most recorded cases, a marked change in mechanical and refractive properties of the rock materials can be detected at a depth corresponding with the occurrence of fresh or little altered rock, and it is concluded that, in general, it is possible to refer to a basal surface of weathering or weathering front. Furthermore, where weathering profiles are relatively thin, the basal surface is usually observed to be fairly sharply defined in a wide variety of rock types. On the other hand many drilling records reveal that, where weathering depths are great, the basal surface is not always well defined. However, there does not appear to be any correlation between total depth of weathering and the abruptness of the transition from weathered to unaltered rock. Examination of approximately 150 boreholes revealed that 41 per cent of the borings encountered a sharp transition from altered to fresh rock, while only 17 per cent of the logs indicated a thick transitional zone of ten feet or more of partially weathered but coherent rock.[21] Even in these cases it is not necessarily arguable that the basal surface would be unrecognizable.

It seems likely that the variations in the character of weathering profiles result less from variations in rock type and structure than from differences in the conditions of ground-water weathering. This introduces the question of duration of weathering attack at certain levels within the rock, and also the relationship between the progress of weathering and the development of a water-table. Berry and Ruxton[22] have pointed out that initially the concept of a water-table is not necessarily involved, but once developed it constitutes an effective base level of weathering. The same authors regard the distinctly zoned profile with the development of core-stones, as a mature profile resulting from a prolonged period of weathering.

However, prolonged weathering attack on the zones containing the core-stones must eventually lead to their disappearance, and if further weathering penetration is prevented,

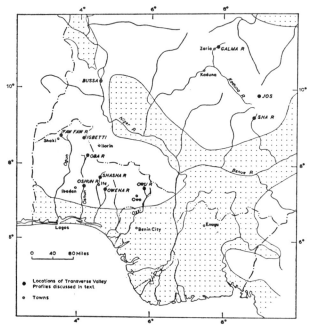

FIGURE 1—Location map of part of Nigeria to show positions of sites discussed in the text. Areas of Basement Complex and intrusive rocks are unshaded; areas of sedimentary rocks stippled.

either by conditions of rock resistance or the development of a permanent water-table, few if any new core-stones will be produced. Instead it seems likely that a sharply defined basal surface will be developed, except in the deeper troughs of weathering, where the formation of a permanent water-table would lead to the cessation of certain weathering processes and the slowing down of others, resulting in partial or arrested decomposition of the rock minerals, and therefore to a poorly defined basal surface of weathering.

It can therefore be argued that the typical granite weathering profile containing core-stones is an immature profile undergoing active development (See Plates I and II), and long maintained stable conditions will ultimately lead to the type of profile described above. But in any case soil water movement will be accelerated and aeration will reach new levels during phases of stream incision, and it is possible that profiles which formerly displayed an indefinite

PLATES I and II—Core-stones within shallow weathering profiles in porphyritic Older Granite on the Jos Plateau, Northern Nigeria. These profiles occur in the dissected zone on the plateau edge, and while core-stone development is active at depth, truncation of the profiles producing small tors in places (see upper Plate) continues as a result of surface erosion. These are regarded as immature profiles.

facing p. 176

PLATE III—A sharp basal surface of weathering in a Pre-Cambrian banded gneiss, exposed in a road cutting near Abuja, Northern Nigeria. Although the more resistant mineral bands persist within the regolith, a definite basal surface transgresses the foliation of the rock.

basal surface will become more clearly zoned, so that as the basal surface nears the land surface, it becomes sharpened and therefore more easily stripped of superincumbent weathered material. The deeper troughs of weathering are in any case seldom stripped of their regolith,[23] so that conditions of profile development within these do not affect arguments concerning the effectiveness of the stripping process in the production of bare rock landforms.

Differences in rock type are also influential in the development of varying types of weathering profiles. Thus the well-known granite profiles displaying well-marked core-stones, according to the zonation described by Ruxton and Berry,[24] are not found in gneissic rocks, nor for that matter are they found in all granites. Tors and core-stones appear to be most characteristic of the coarse, porphyritic biotite granites, but also depend upon the jointing frequencies. Core-stones do not survive where jointing is very close and are not produced so often where open joints are very widely spaced. However, in the latter case, sub-surface domes are formed and exfoliation fractures parallel to the buried dome faces split and are weathered subsequently into core-boulders. In the basement gneisses core-stones are less common and tors rare. Two characteristics of these rocks may be jointly responsible for this phenomenon. First, as F. Dixey[25] has observed, horizontal jointing is seldom present in the metamorphic rocks. Second, the banding and foliation of these rocks leads to differential weathering penetration along linear planes, leading to the early disintegration of any spheroidal boulders previously formed from jointed blocks. However, the bands of particularly susceptible minerals such as biotite decompose most rapidly, and expose the intervening bands of more resistant felspathic and quartzitic minerals to weathering attack from the sides as well as from above. As a result they too are soon weathered and a recognizable basal surface of weathering is produced that transgresses the mineralogical banding of the rock (Plate 3).

It therefore appears possible to retain the term basal surface of weathering with some meaning. Moreover, where crystalline rocks are overlain by a mantle of weathered material, this debris will be differentially stripped from the unaltered rock by any phase of erosion, during which surface removal of material exceeds the rate of deep weathering for long enough. to bring the agents of surface denudation to the level of the basal surface of weathering.

Patterns of Deep Weathering in Nigeria

There are few places in Nigeria where closely spaced measurements can be used to show patterns of deep weathering, and because the basal surface may be highly irregular in profile, any reconstruction of its configuration based on widely spaced measurements is likely to be misleading. Consequently profiles of the basal surface have been constructed only from bore-holes or seismic readings taken at intervals of 200 feet or less. Extrapolation of sub-surface contours has been avoided wherever borehole information is more widely scattered.

Figure 1 shows that the sites discussed are spread over four degrees of latitude embracing the humid forest country in the south, where mean annual rainfall exceeds sixty inches and the dry season lasts barely four months, and also the sub-humid savanna lands of northern Nigeria, where rainfall amounts may be less than thirty-five inches and the dry season as long as seven months. The possible significance of such climatic variations to the development of deep weathering profiles will not be considered, as little evidence can be adduced from the profiles. Furthermore, known Pleistocene variations in climate largely obscure differences in the current rates of deep weathering between one place and another. Variations in susceptibility of the rocks is another factor that is imperfectly understood.

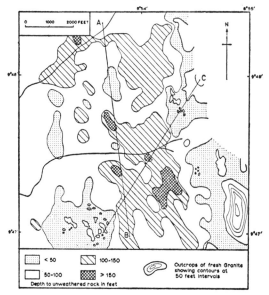

FIGURE 2—Deep weathering patterns near Jos, Northern Nigeria. The map shows depth to unweathered rock rather than true depths of weathering, because the *in situ* weathering profiles are overlain by varying thicknesses of alluvium (see Fig. 4). The rock is a fine-grained biotite granite of Jurassic age.

The greatest depth of weathering in Nigeria known to the author is 184 feet, a figure that is not exceptional for tropical areas. It is found in one of the deep basins of weathering on the Jos Plateau (Fig. 2). However, it is perhaps more significant to emphasize that in none of the sections is the maximum depth of weathering less than fifty feet, in spite of the selection of many of the sites for the frequency of outcropping, unweathered rock.

Reference to the individual profiles (Figs. 2 to 10) facilitates certain general observations. In the first place, marked irregularity of the basal surface of weathering is characteristic of most of the profiles. This irregularity can be shown to consist of a series of basins and domical rises in the rock surface both in the area near Jos (Fig. 2) over biotite granite, and also in basement gneisses around the Owu River in western Nigeria (Fig. 7). In the

former case a fair correspondence exists between the orientation of the basins of weathering and the dominant joint directions in the area, as comparison of Figures 2 and 3 reveals. On the other hand transverse profiles across the Oshun and Oba Rivers (Figs. 8 and 10D), and also the weathering on the right bank of the River Niger (Fig. 6), suggest that an increase of weathering depth irregularly away from the river channel may be a common occurrence; if so it may have some general geomorphological significance. This observation can be corroborated by frequent reference to deep weathering beneath laterite caps which tend to survive as interfluve cappings well away from streams. This may be a likely relationship on general theoretical grounds.

However, although this is the situation in some cases, other profiles exhibit different characteristics, and a comparison of

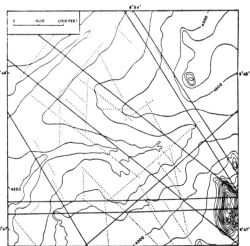

FIGURE 3—Jointing and weathering patterns near Jos, northern Nigeria. This map attempts to show the correspondence between major joint directions and lineaments in the weathering patterns in the same area as Figure 2. Joints have been mapped from air photographs and their bearings extrapolated across the deeply weathered area. They are shown by ruled lines. Lineaments in the deep weathering pattern are derived from Figure 2, and are indicated by dotted lines. Contour interval 20 feet.

profiles A to F in Figure 10 indicates that drainage lines must in some way become super-imposed on the underlying bedrock. The corollary of this observation is that the level and position of the stream channels do not necessarily control the depth or patterns of weathering. In the case of the Galma River near Zaria (Fig. 10A) weathering below the channel of the stream itself is demonstrated. A similar pattern of weathering is displayed by the small left bank tributaries of the River Niger near Bussa (Fig. 5), where depths of decomposition may be as great beneath the stream channels as elsewhere. These examples may be compared with the profile across the Faw Faw River (Fig. 9) which drains from the edge of the granite

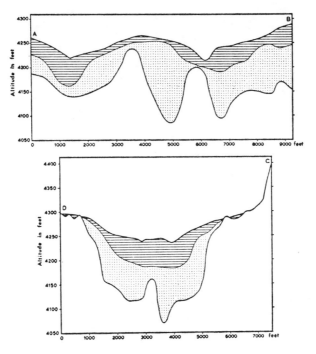

FIGURE 4—(*a*) (upper) Cross section along the line A–B on Figure 2. Ruled shading represents alluvial material; stippled shading the *in situ* weathering products. (*b*) (lower) Cross section along the line C–D on Figure 2.

plateau near Shaki in the Western Region. This plateau stands 600 feet above the adjacent plains and the Faw Faw River is dissecting the deeply weathered margins of the plateau, effecting the removal of the weathered material from the valley floor. A similar pattern of weathering is revealed by the Sha River (Fig. 10C) which drains the southern edge of the Jos Plateau in the Northern Region.

The features displayed by the deep weathering patterns near Jos itself (Fig. 2) are of considerable interest. The map shows the depths to unaltered bedrock within a small area due south of Jos on the main plateau surface at about 4250 feet. It lies within the Jos-Bukuru Ring Complex[26] and the weathering is within a closely jointed fine-grained biotite granite. The weathering pattern displays pronounced basining, and domical rises between the basins break

the surface as outcrops in places. Mapping of the principal joints observable from air photographs for the area reveals a striking conformity between joint directions and weathering
patterns (Fig. 3). The major joints show a dominance of bearing between 120 and 150 degrees
that accords well with the elongation and arrangement of the deeper basins of weathering. A
complementary set of joints between fifty and seventy degrees is less prominent but probably
influential. A further set of joints at about ninety degrees is locally present. Since it is scarcely
possible to observe jointing directions within the weathered area the apparent correspondence
between jointing and weathering patterns must rest on circumstantial evidence alone.

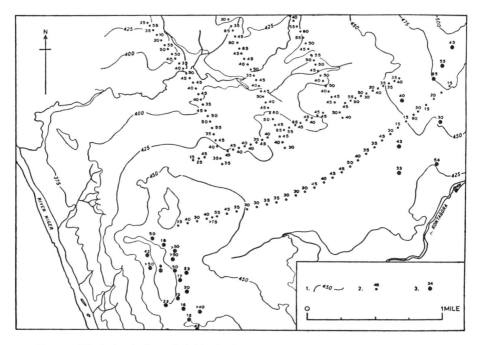

FIGURE 5—Weathering depths on the left bank of the River Niger at the Kainji Dam site, near Bussa,
Northern Nigeria.
    Key: 1. Surface contours in feet; 2. weathering depths from seismic traverses; 3. weathering depths
from diamond drilling. On this map the weathering data are too widely spaced to draw contours of the
basal surface except for small areas, and they have therefore been omitted.

Sections A–B and C–D (Fig. 4) show that the depth to unweathered rock does not accurately
define the depth of the surviving weathering profile because of the presence of considerable
depths of alluvium. Nevertheless the simplification of this picture in Figure 2 probably does
not distort the weathering patterns very much since it may be supposed that the buried river
channels were eroded into a surface of low relief similar to that of the present time. Moreover,
the deepest basins of weathering often lie beneath only a superficial veneer of alluvial material.
    These and other profiles demonstrate the absence of consistent relationships between
stream lines and patterns of weathering, but of equal interest are the relationships between
outcrops and patterns of weathering, for this is relevant to the problem of subsurface development of domes, and to the pediment problem. From the Jos map (Fig. 2) it can be shown that

around many of the outcrops there is a platform of shallow weathering, but the depth of weathering does not increase gradually or regularly with distance from outcrops; beyond the shelf feature the basal surface plunges, often very steeply into the deeper basins (section c–d, Fig. 4*b*). In some places the shelf or platform is absent altogether. This steep plunge of the basal surface around outcrops is well illustrated from the Owu River area (Fig. 7). The weathering here is in Pre-Cambrian banded gneiss, and the outcrops form smooth little jointed domes

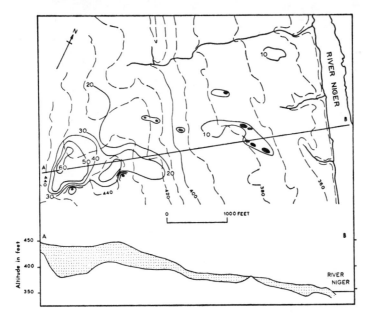

FIGURE 6—Weathering patterns on the right bank of the River Niger at the Kainji Dam site, near Bussa, Northern Nigeria.
    Key: broken lines—surface contours; continuous lines—depths of weathering. The latter are lines joining points of equal depth of weathering and the figures are in feet. It should be realized that these are not identical with bedrock contours and the pattern is partly dependent upon the surface relief. On the section a–b the stippled zone represents the weathering profile; outcrops are in black. These conventions have been adopted for subsequent figures. The rock is predominantly Pre-Cambrian gneiss.

mostly of low relief. The geomorphological implications of these profiles will be considered under four headings.

    1. *The Progress of Weathering and the Development of Deep Weathering Patterns.* No attempt will be made here to discuss the chemistry of rock weathering, but certain deductions with regard to the progress of weathering may be made. The attack of ground water upon crystalline rocks is thought to be most rapid within a zone above the permanant water-table where periodic or seasonal aeration of the partially weathered material takes place. The water-table and the movement of ground water are usually regarded as being controlled ultimately by the level of stream channels and the disposition of stream lines. But these concepts are inadequate as a guide to the controls over weathering penetration into crystalline rocks, particularly under tropical conditions.

As a consequence of their crystalline structure these rocks have a very low porosity, and the deep penetration of ground water can only be achieved by percolation within open tension joints and faults, although as P. Birot[27] points out a measurable porosity exists in these rocks as a result of 'microfissures' representing fractures on a microscopic scale between crystal faces. But important though these may be in admitting water into the surface layers of the rock, they scarcely allow any general percolation or circulation of water within the rock mass. The depth of penetration of ground water will be very variable, but according to Dixey[28] it is not likely greatly to exceed 300 feet, because at greater depths the confining pressures cause joints either to die out altogether or to remain tightly closed. In unweathered crystalline rocks it is therefore necessary to discard the concept of a water-table. Water masses will be small and discrete, having no necessary connection with adjacent stream channels. It seems likely that movement of water towards stream lines will only become important in areas of high relief, and of course within any permeable regolith above the fresh rock. Weathering will follow ground-water penetration and its attack will be concentrated on the faces and vertices of joint blocks. Where jointing is close textured the approach of the weathering from from adjacent joints will lead to the complete breakdown of intervening rock masses. As the rock decomposes so a water-table comes into existence, and this fluctuates in level with the seasonality of climate. Below the permanent water-table weathering is generally thought to be very slow. The characteristic patterns of jointing seem to favour the development of basins of weathering, which may initially be deep and narrow and which are likely to enlarge with time.[29] These remarks assume a homogeneous rock mass, but of course the existence of rocks particularly sensitive to chemical weathering, such as amphibolites or biotite schists will favour deep weathering following the outcrop patterns of such weaker materials.

In advancing an evolutionary scheme for the development of characteristic weathering patterns an important theoretical problem is encountered. It is not possible to assume an initial rock surface, little weathered and of negligible relief, from which the present forms have evolved. Any extensive erosional plain will already exhibit deep weathering patterns, while the absence of known or deducible tectonic relief over wide areas of the African shield makes it impossible to define any initial landforms from which the present patterns have been inherited. Even exhumed surfaces that may exist on the basement rocks around the margins of sedimentary basins cannot be assumed to have emerged unweathered from their former sedimentary cover. Weathering of the basement is recorded beneath the sediments in southern Nigeria,[30] and exploratory work for oil in the Niger Delta suggests that the basement surface beneath the sediments is both weathered and irregular.

The progress of weathering today is influenced not only by the internal structures of the rocks but also by the relief. Thus the rapid run-off from the bare rock domes of the inselberg landscape concentrates water available for sub-surface seepage into intervening and deeply weathered depressions, leading to arid conditions on the outcrops, and to prolonged contact between moisture and rock below the plains. It might be argued therefore that, once formed, the pattern of domes and weathered plains is likely to persist. Furthermore, while it may be argued that deep weathering is favoured by the gentle relief and slow pace of surface erosion over senile plains, it is not known whether much of the development takes place because of changes in ground-water movement consequent upon gradual encroachment of erosional escarpments. Thus the deep weathering found within upstanding granite massifs may in part be explicable in terms of the effect upon ground-water movement, and consequently upon rates of weathering, of their relative relief above adjacent plains.

An increase in erosional activity in streams will incise such profiles along lines that may not be directly related to the underlying structures that have determined the weathering pattern. It is clear from most of the maps and profiles that the streams are not always aligned with the troughs in the basal surface. On the other hand, within rocky areas where pre-weathered material has been largely stripped and streams flow over unaltered bedrock, structural guidance is very pronounced. Indeed the stream plan for many basement areas in Nigeria is markedly rectilinear suggesting the influence of underlying fracture systems. Once stream erosion has been accelerated by changes of climate or base-level, the river channels will increasingly influence the movement of ground water and therefore the further progress of weathering. This question is associated with the problem of slope and valley evolution.

2. *The Evolution of Stream Valleys and Valley-side Slopes.* Valleys of the smaller streams, particularly within the savanna regions of Nigeria are often poorly developed, and even where the relative relief is fifty to one hundred feet the stream channels are frequently no more than concentrated zones of slope wash. Some though not all of these valleys recall the 'pure saucer-shaped valleys' ('Reine Flachmulden-täler') observed by H. Louis[31] in Tanzania, and may be accounted for by the vigorous action of slope wash combined with weak lineal erosion. When this factor is combined with gentle gradient and seasonal flow, conditions for deep weathering beneath valley sides and beneath the stream channel itself are favoured. Conditions of low gradient and seasonal flow are illustrated by the Galma River (Fig. 10A) near Zaria which flows over the senile plains of a mid-Tertiary erosion surface. Weathering of the bedrock beneath the channel is favoured by a long dry season, during the latter part of which the river is reduced to a series

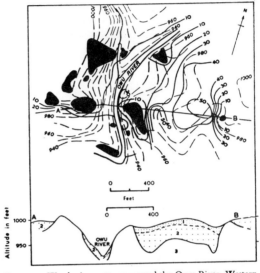

FIGURE 7—Weathering patterns around the Owu River, Western Nigeria.

Key as for Figure 6. On the section: 1. represents the lateritic soil horizon. 2. decomposed rock, 3. unaltered rock. The rock is a Pre-Cambrian gneiss with pegmatite sheets.

of pools. On the other hand vigorous erosion during the flood season is avoided because of the quantity and fineness of its load. The predominance of fine sand and silt in tropical rivers, is often regarded as a function of the speed and thoroughness of tropical weathering, and has been discussed at length by Birot and J. Büdel.[32] Nevertheless vertical incision in response to changes in climate or base-level will take place within the poorly consolidated weathered rock which may occur beneath the stream channels and is likely to extend to greater depths beneath the valley flanks. The efficacy of slope wash during downcutting may be reduced by the existence of duricrust, resulting from the exposure of lateritic horizons, formed perhaps within the weathering profiles of the gently sloping valleys on the original plain. In this way the basal surface may become exposed in the stream channel, while leaving the

N

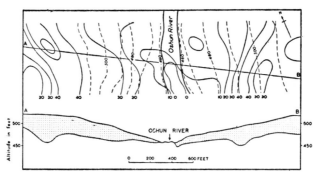

FIGURE 8—Weathering patterns around the Oshun River at the Asejire Dam site, Western Nigeria. Key as for Figure 6. The rock is a Pre-Cambrian gneiss with pegmatite sheets.

weathered mantle protected by the laterite over the valley flanks (Fig. 11B). Because of the irregularities of the basal surface it will be fortuitous if the stream reaches the sound rock at its lowest point. In fact the streams are *superimposed* upon the basal surface; the streams will then migrate along the interface between the weathered and unaltered rock and will become aligned with the troughs in the basal surface. This is indicated by the arrows in Figure 11B, and may be compared with uniclinal shifting in sedimentary rocks. In certain cases the streams may become superimposed upon subsurface domes as is suggested by the section across the Shasha River in western Nigeria (Fig. 10E). The exposure of bornhardt domes in this way can be seen often to result in progressive outward and downward erosion from the exposed dome summit which thus forms a slip-off slope, while under-cutting proceeds into the regolith, or along the curved exfoliation joints down-slope.[33]

The subsequent evolution of valleys and of the inselberg landscape as a whole may proceed in different ways according to the rock structures and the form of the basal surface. Two hypothetical sequences of evolution are suggested in Figure 11 (C, D, E, and C′, D′, E′). In the first case it is assumed that the underlying rocks are relatively uniform and easily weathered.

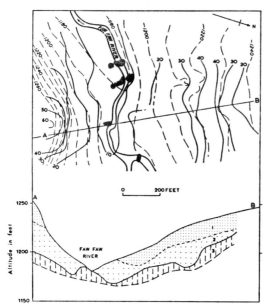

FIGURE 9—Weathering patterns around the Faw Faw River near Shaki, Western Nigeria.

Key as for Figure 6. On the section A–B: 1. represents lateritic soil horizon. 2. decomposed rock. 3. fractured and fissured rock. 4. massive and unaltered rock.

The rock is a well foliated Pre-Cambrian Older Granite containing semi-conformable pegmatite sheets. It is well jointed, and large core-stones were encountered in some of the profiles, the largest being seventeen feet in diameter. The slope between A and the river forms a talus slope beneath rock outcrops beyond the line of section.

Thus while the stream is competent to remove weathered material along its channel, slope processes do not strip the basal surface on the valley flanks. Because of the ground-slope there will be a fairly strong lateral flow of ground water towards the stream channel. Ground water is therefore being continually renewed near the interfluve, and will remove material in solution and perhaps in suspension from the upper slopes of the valley. This combination of lateral movement and replenishment of ground water may lead to more rapid decomposition of the rock beneath the upper slopes of the valley than near the stream channel. This argument gains support from observations made by Berry and Ruxton in a similar context.[34] The result of this greater rate of weathering beneath the valley flanks must be to reduce the relief of the basal surface of weathering at a rate equal to or greater than that by which the land-surface is lowered. As the phase of incision comes to an end, valley-side slopes will be reduced once the laterite caps have been destroyed, and the rates of lateral flow and replenishment of ground water will be correspondingly reduced. Stream erosion will become less efficient and, particularly in channels experiencing seasonal flow, weathering of rock beneath the channels will become possible. Gradually conditions similar to those at the beginning of the phase of erosion will be re-established. Throughout this cyclical scheme the bed-rock surface is actively reduced by ground-water weathering, and the effects of surface erosion are confined to the removal of the regolith materials. This type of land-surface and this mode of evolution accords well with the concept of the etchplain put forward by E. J. Wayland.[35]

The alternative hypothesis (Fig. 11c', d', e') supposes that the incision of the stream not only superimposes the channel on the basal surface of erosion, but also leads, because of the effects of wash processes and creep on the valley slopes, to the stripping of domical exposures of more resistant rocks along the valley flanks. This will occur where there are marked variations in the underlying rocks or their structures. Once exposed the domes become 'permanent' features of the landscape, and increase in size as the surrounding regolith materials are eroded. Between the domes the pockets of basins of weathering may become still deeper, because it is into these that the run-off from the dome surfaces will be concentrated, and they will also be protected from further stripping by the rim of unweathered rock (Fig. 7). This kind of evolutionary scheme produces the characteristic inselberg landscape, and is of course similar to schemes adduced by Büdel and Ollier.[36]

The varying relationships between the rates of deep weathering and stripping, and the effects of base-level and climatic changes, have been discussed elsewhere,[37] and a descriptive scheme devised for the nomenclature of the types of etchplain that may be formed. The other implications of such hypotheses of valley evolution concern the further evolution of the domical exposures of the basal surface that undergoes stripping.

3. *The Formation and Evolution of Bare Rock Landforms.* In common with many other authors the writer has favoured the hypothesis that bornhardt domes are in some sense exhumed from their superincumbent regolith.[38] But all these studies lack detailed evidence of the relationship between the deep weathering patterns and the rock domes on the one hand and of both to fracture systems in the bedrock on the other. The data presented in this paper are in no way capable of proving the general hypothesis, but do allow certain deductions to be made which support its application to specific cases.

The morphology of the basal surface weathering in the Jos granites (Fig. 2) and in the gneisses of the Western Region, for instance at the Owu River site (Fig. 8) and the Shasha River site (Fig. 11E) give unmistakable evidence for the sub-surface origin of the smaller domes, the relative relief of which does not exceed the depth of weathering. In the case of the Jos granites the association of weathering patterns with local joint systems (Fig. 3) gives supporting evidence for the structural guidance of deep weathering. On the other hand the relationship between the ground plan of individual domes and joint patterns is not everywhere established. Berry and Ruxton and C. R. Twidale[39] have all shown close relationships between jointing and dome faces but L. C. King[40] for instance has expressed a contrary opinion. Interpretation of aerial photographs and also ground survey work in Nigeria demonstrate that although

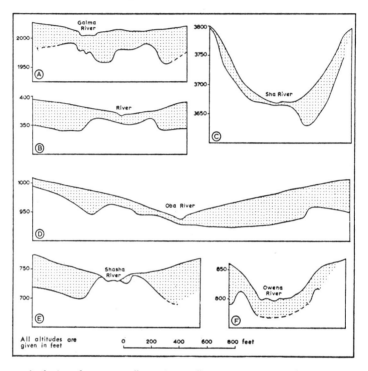

FIGURE 10—A selection of transverse valley sections to illustrate contrasting weathering patterns. A. The Galma River near Zaria, Northern Nigeria. The rock is predominantly a Pre-Cambrian quartz schist. B. The Igbetti Dam site, Western Nigeria, Pre-Cambrian Older Granite. C. The Sha River near the southern edge of the Jos Plateau, Northern Nigeria, Jurassic biotite granite. D. The Oba River near Ogbomosho, Western Nigeria, Pre-Cambrian gneiss with pegmatite sheets. E. The Shasha River near Ife, Western Nigeria, Pre-Cambrian gneiss. F. The Owena River near Ife, Western Nigeria, Pre-Cambrian gneiss.

details of some dome outlines may diverge from observable structural lineaments, many dome margins are in fact joint controlled.

The distribution of domes also gives additional evidence in favour of their origin *as domes* and not their formation by the reduction of larger rock masses by slope retreat, in the manner of L. C. King and J. C. Pugh.[41] The formation of domes by the retreat of slopes within a substantially unweathered and homogeneous rock mass should imply that they are forms of old age in the landscape, in the sense that they will be expected to occur on interfluve sites, and surrounded by more or less extensive pediments falling away towards adjacent stream lines.

Processes operating in a horizontal plane are regarded as being more important in their formation than those operating vertically. All authors would admit the strong control of structural compartmentation in the evolution of the landscape, but emphasis in these schemes is laid upon the guidance of stream courses by fracture patterns, and the extension of pediments from

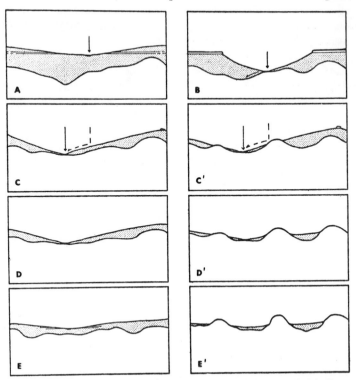

FIGURE 11—Hypothetical transverse valley profiles to illustrate two alternative modes of valley evolution on crystalline rocks in the tropics. A. The 'initial' stage: a deeply weathered and lateritized etchplain within which rivers exhibit senile characteristics. B. Stream incision in a new cycle superimposes the river channel on the irregular basal surface of weathering, indicated by the vertical arrow. A second arrow indicates the future migration of the stream to a neighbouring trough in the basal surface. C, D, E, indicate subsequent evolution of the valley following stream incision and migration, assuming a rate of ground-water weathering greater than the rate of surface erosion. After a prolonged period of denudation a return to the conditions of the 'initial' stage is envisaged (compare A and E). C', D', E', follow the same period of evolution but assume marked inequalities in bedrock resistance to chemical weathering, that give rise to exposures of the basal surface as small domes. Any return to 'initial' conditions will be long delayed, because of the persistence of the domes as landforms.

these to give rise to individual bornhardts. Many different lines of physiographic evidence are adverse to such an hypothesis. These can only be summarized in this study.

The dissection of granite plateaux frequently gives rise to closely grouped domes which may emerge from the regolith in their characteristic form at a very early stage in dissection. Further dissection of such masses results in more numerous bornhardts extending into the heart of the granite intrusion, but these domes are not smaller or more widely spaced at this stage.

Such distributional patterns demonstrate the dominance of vertical weathering and erosion in the formation of bornhardts, an observation corroborated by the smooth, near-vertical, convex faces of many very high domes.[42] Where escarpments are found, emergent half-domes can often be seen along the scarp face,[43] and in less extreme cases it is common to find bornhardts on valley flanks with the foot-slopes inclined *towards* the dome on the up-slope side and away from it towards the river. Notwithstanding the fact that some bornhardts are indeed to be found standing above wide plains, it is relevant to recall Birot's remark 'nous n'oublions pas que les pains de sucre bresiliens sont des formes de jeunesse du cycle, et non des inselberg'.[44] It is not possible fully to explore this problem here, but the evidence of the weathering patterns, taken together with the morphology and distribution of bornhardts argue powerfully against the formation of these hills by the extension of pediments. For even the examples of isolated domes show such features as pronounced asymmetry, associated core-stones and tors, and deeply weathered surrounding plains that can be construed to support formation by exhumation, even if they have become subsequently modified by sub-aerial denudation. However, the pediment landform itself requires comment in this context.

4. *The Formation of Pediments Sensu Stricto in Inselberg Landscapes.* The apparently widespread occurrence of rock-cut pediments in landscapes of this type has been seriously questioned by several writers including the author.[45] Nevertheless it must be admitted that with this as with other morphological problems in the tropics, surficial conditions may often be misleading, and the presence or absence of exposed or thinly veneered rock surfaces is not necessarily an adequate criterion on which to base an assessment of the pediment problem. It is obvious that within an environment where rapid chemical weathering is a feature, rock-cut pediments might undergo weathering following the parallel retreat of the hill-slope. It is therefore necessary to seek independent evidence for the back-wearing of the hill-slope as a rock-cut feature. It might be argued that the depth of weathering should increase with distance from the hill-slope, but because of structural inequalities in the bedrock this increase in the depth of weathering may not occur as a regular progression from the hill-foot. Furthermore one might reason that the run-off and seepage from the slope itself would, by keeping the hill-foot areas moist during any dry season, bring about both a greater rate and consequently a greater depth of weathering in the vicinity of the hill-foot, that would obscure the expected pattern.[46] Using this kind of reasoning it is therefore difficult to refute the theory of pediplanation.

Nevertheless the theory becomes difficult to apply to multicyclic land-surfaces. It is difficult to postulate initial landforms in the basement areas, so that it is possible only to assume some early well planed land-surface as the precursor of the present complex of landforms. The Gondwana surface might be taken for this purpose, and if, following King,[47] it is assumed that this was formed by pediplanation and that it became deeply weathered over wide areas following its formation, as postulated by Ollier,[48] then the first effects of renewed erosion upon such a surface must be the stripping of much of the deep weathering profile to produce a complex rocky inselberg landscape in some areas and widespread weathered plains in others, according to local conditions of rock resistance. Subsequent evolution will largely destroy the rocky relief and lead to further deep weathering over wide areas. A return to a deeply weathered plain with a few residual hills is inevitable, as is a repetition of these events during subsequent phases of erosion. Thus although slope retreat may affect specific landforms, pediplains *sensu stricto* cannot be formed, and it is necessary to observe that the Gondwana surface itself may not have been formed by pediplanation, but in the same manner from some earlier deeply weathered plain. All these land-surfaces are produced by a combination of etching, stripping by incision and slope retreat within the weathering profile, together with a complex of processes leading to the destruction of the rock domes and tors. Such plains are better described as etchplains.

The possibility of parallel slope retreat in the manner of L. C. King[49] becomes relevant in two circumstances. First in the modification of the bare rock landforms produced by stripping, and second along erosional escarpments generated by tilting, warping, or even faulting of the basement rocks in such a way as to produce a relative relief considerably in excess of the depth of the weathering profiles.

Parallel retreat should be evident in some of the profiles discussed in this paper, but such indications are less impressive than the evidences for deep weathering and stripping. In several instances the basal surface of weathering plunges steeply around outcrops as in the areas of the Owu River (Fig. 7) and the Faw Faw River where deep weathering has occurred beneath the scree slopes below adjacent domes (Fig. 9). The Jos Plateau area (Fig. 2) provides more interesting data, for it can be seen that a platform of shallow weathering surrounds certain of the more important outcrops. This is well illustrated along the section C–D (Fig. 4b). But it is equally clear that beyond this shelf zone the depth of weathering increases very rapidly. These facts can be construed to support the notion that the deep weathering pattern is the *primary* feature of the granite terrain, and that the present outcrops are the result of stripping to a certain level. With the cessation of further downcutting, lateral retreat has reduced the area of the residual hills, producing the platforms or shelves of shallow weathering around the present outcrops.

The situation on the Jos Plateau is not a simple one, however. There is evidence of important changes of climate during the Pleistocene period, and furthermore the hills are not true bornhardt domes, but are closely jointed rocky features that will undergo modification by slope retreat far more readily than massive domes. The factor of climatic change is probably relevant to the question why such closely jointed granite should have become exposed to form a bold rocky relief. The extensive stripping evident over many of the residual hill masses of the Plateau area is probably a function of its high internal relative relief together with the effects of a semi-arid or even arid climate on a deeply weathered Tertiary or Cretaceous land-surface. Once exposed to a seasonally humid climate these highly jointed forms become subject to basal sapping, leading to the slope retreat evidenced by the profiles (Figs. 2, 4b). Basal sapping is probably brought about by the retention of moisture between the joint blocks towards the foot of the hill-slope, and the periodic flushing of the material that becomes weathered as a result of the prolonged contact between water and rock. The slope is thus continually undermined and a form of parallel retreat results.

Elsewhere in Nigeria, highly jointed rock forms are less common, probably because they are seldom exposed in the first place. The Older Granites that occur more commonly in other parts of Nigeria are usually massive; moreover few other areas have experienced the same combination of high relief and alternating phases of arid and humid climate. Much of the Northern Region is dominated by senile plains, while the more varied relief of the Western Region lies in a latitudinal belt that may have escaped the extremities of the interpluvial dry phases in West Africa.

The author therefore feels justified in maintaining that parallel retreat of slopes as affecting bare rock forms in Nigeria is not a generally important process. It operates where the rock is well jointed, but for the reasons discussed such areas of rock usually become deeply weathered and do not normally form extensive outcrops. Massiveness is a feature of most bornhardts and there is little evidence to support a general theory of parallel retreat for bare, massive rock-slopes.

Erosional escarpments pose additional problems in relation to slope retreat that are not necessarily encountered when considering individual landforms. An escarpment cannot be considered as a single slope of uniform characteristics. Along the escarpment face or zone of dissection, the agents of erosion will be working variously on regolith and rock materials. Laterite capped regoliths have been found to undergo parallel retreat,[50] but the evolution of the rock forms will depend upon the completeness of stripping and on the characteristics, especially jointing, of the rock mass. Slope retreat, most commonly as a result of basal sapping, will affect many outcrops, but other styles of evolution may occur including outward collapse.[51]

A different situation will arise, however, if a resistant granite mass is considered in relation to a less resistant basement of metamorphic rocks. In many cases within Nigeria the granite mass stands more than 1000 feet above the adjacent basement surface, and a bounding escarpment in the form of an almost continuous wall of fresh rock may be formed. This may be explained as the result of a more rapid lowering by deep weathering and surface erosion of the basement surface than by deep weathering of the granite mass, so that although the upper part of the escarpment may be characterized by tors, domes and kopjes, the lower slopes are dominated by contiguous half-domes which present an unbroken barrier of resistant and massive granite. In such cases the retreat of the escarpment may be negligible, as around the Kagoro Hills in northern Nigeria.

It is not correct therefore to regard the foot-slopes of erosional escarpments as pediments *sensu stricto*. They are pedimented in the sense that they are bordered by gentle wash-slopes with a concave profile, but rock-cut surfaces will occupy a comparatively small area within these slopes.[52]

Conclusions

1. The form of the basal surface of weathering may be highly irregular, and usually exhibits a pattern of discrete basins and rises. This observation accords well with that made by J. F. Enslin[53] in South Africa, where, in connection with ground-water resources, he noted that the weathered rock 'tends to vary in depth and lateral extent, and to form a large number of isolated basin-shaped or trough-like ground-water compartments. These areas of deep decomposition show practically no surface indications but can be differentiated by geophysical methods.'

2. Such weathering patterns appear to be aligned with the fracture systems observable in surrounding rock outcrops, though this does not permit any predictions as to the depth or accurate alignment of weathering troughs, because both the frequency and bearing of joints may vary rapidly from place to place in a manner that cannot be readily explained.

3. Weathering beneath stream channels is restricted in the main to small seasonal streams with gentle gradients, flowing over senile plains; in other cases the streams flow predominantly over sound rock, or over alluvium. In certain instances the depth of weathering increases irregularly away from the stream, towards the adjacent interfluve. This is characteristic of several profiles and may represent a general relationship wherever the landscape has reached a certain stage of development, and where there are no marked inequalities in the resistance of the rocks. Wherever massive rock occurs it is likely to form a rise in the basal surface which may break the land surface to form domical outcrops of varying size.

4. Surface indications are seldom an accurate guide to weathering depths. The presence of deep weathering profiles beneath the Older (Tertiary) Laterite[54] in northern Nigeria has been

commonly observed, and parallels are found in other continents, but this is certainly not true of all laterites or in all places. It may be explained in terms of the protection afforded by the laterite to the less coherent zones of the weathering profile beneath, which allows the progress of deep weathering to continue below a land-surface that may be little affected by erosion over lengthy periods of geological time. Deep weathering may occur quite close to outcrops of fresh rock, but no generally valid statement on this relationship can yet be attempted. It is, however, relevant to comment that the presence of frequent rock outcrops does not necessarily imply the absence of deep troughs of weathering in between.

5. The patterns of weathering and the morphology of bornhardt domes supports the hypothesis of sub-surface origin of the domes as well as of the tors.

6. The relationship of dome foot-slopes with major joints and the steep plunge of the basal surface around outcrop margins suggest that domes are not formed by slope retreat within unweathered rock: that their modification by this process is often insignificant, and in the observed cases where it has occurred the close jointing of the rock or a surviving regolith cover appears to be responsible.

7. The importance of deep weathering in areas of frequently outcropping rock and the alternation of outcrop and weathered basin or trough argue against the use of the term pediplain for inselberg landscape of low relief. The dominant processes are etching and stripping and a terminology utilizing these terms in varieties of etchplain seems more appropriate.[55]

8. Finally it is necessary to recognize that the data which we have on weathering profiles and patterns are as yet inadequate to provide a basis for many general conclusions. Only systematic geophysical surveys carried out in areas of contrasting geology and morphology can reveal general relationships, if these exist, between weathering patterns and other landform distributions. It seems likely that different kinds of weathering pattern will emerge from close study of etchplains of varying age and character, and that degrees of stripping related not only to geological outcrop patterns but also to regional geomorphological history would emerge from such systematic work. This could be of the greatest theoretical and practical value in that distinctive weathering patterns may yet prove to be associated with certain kinds of surface morphology.

ACKNOWLEDGMENTS

The data on weathering patterns could not have been obtained without the help and co-operation of a large number of people and of the commercial firms and government departments in Nigeria concerned with the practical problems of civil engineering projects. An indication is given below of the sources of information for the various sites quoted in this study. My thanks are due to the Amalgamated Tin Mines of Nigeria Limited for data for the Jos Plateau (Fig. 2), to Balfour, Beatty & Company Limited for permission to use their records for the Niger Dam and Sha Falls sites (Figs. 5, 6 and 10C), to Foundation Engineering (Nigeria) Limited for data on the Asejire (Oshun River) and Ogbomosho (Oba River) dam sites (Figs. 8, and 10D), to Scott and Wilson, Kirkpatrick & Partners Limited for details of the Galma River site near Zaria (Fig. 10A), and to Tahal (Water Planning) Limited for provision of detailed geophysical data and other information on the Asejire, Faw Faw River, Igbetti, and Owu River dam sites (Figs. 7, 8, 9, and 10B). I also acknowledge the permission given to use the data contained in the Niger Dams Report by the Niger Dams Authority of the Republic of Nigeria. I also thank the following people for much personal help in obtaining access to necessary data: Dr. C. S. Hitchen, Mr. J. A. Meehan, Mr. O. Nagler, Mr. K. Paulo and Mr. G. Parsons.

I am also grateful to the University of Ibadan, Nigeria, for provision of research funds to cover the cost of field work, to Mr. T. K. P. Amachree who drew many of the diagrams, and to the University of St. Andrews for a grant towards the cost of illustrations.

NOTES

[1] The use of this type of equipment and results obtained are given in the following reports: *Geological Supplement, Niger Dams Report, Federal Government of Nigeria,* Lagos (1961), and *Geophysical Survey of Proposed Dam Sites, Report to the Government of Western Nigeria,* Tahal (*Water Planning*) Ltd., Tel-Aviv (1963). Portable seismographs give accurate results to 70 or 100 feet.

[2] R. V. RUHE, 'Landscape evolution in the High Ituri, Belgian Congo', *Pub. Inst. Nat. pour Étl'ude agron. Congo Belge (I.N.E.A.C.), Sér. Sci.*, 66 (1956).

[3] J. D. FALCONER, *The Geography and Geology of Northern Nigeria* (1911).

[4] J. DRESCH, 'Plaines Soudanaises', *Rev. Géomorph. dynamique* (1953), 39–44.

[5] J. ARCHAMBAULT, 'Les Eaux Souterraines de L'Afrique Occidentale', *Service de l'Hydraulique A.O.F.* (1960).

[6] E. A. FITZPATRICK, 'Deeply weathered rock in Scotland, its occurrence, age and contribution to the soils', *J. soil Sci.*, 14 (1963), 32–43.

[7] D. L. LINTON, 'The problem of tors', *geogr. J.*, 121 (1955), 470–87.

[8] J. DEMEK, 'Castle koppies and tors in the Bohemian Highland (Czechoslovakia)', *Biul. Peryglacjalny*, 14 (1964), 195–216.

[9] J. A. MABBUTT, 'A stripped land surface in Western Australia', *Trans. Inst. Br. Geogr.*, 29 (1961), 101–14; 'The weathered land surface of Central Australia', *Z. Geomorph.*, 9 (1965), 82–114.

[10] R. L. WRIGHT, 'Deep weathering and erosion surfaces in the Daly River basin, Northern Territory', *J. geol. Soc. Australia*, 10 (1963), 151–63.

[11] C. B. BISSETT, 'Water boring in Uganda 1920–1940', *geol. Surv. Uganda*, Water Supply Paper No. 1 (1941).

[12] R. H. NAGELL, 'Geology of the Serra do Navio manganese district, Brazil', *econ. Geol.*, 57 (1962), 481–98.

[13] B. P. RUXTON and L. BERRY, 'Weathering of granite and associated erosional features in Hong Kong', *Bull. geol. Soc. Am.*, 68 (1957), 1263–92, and L. BERRY and B. P. RUXTON, 'Notes on weathering zones and soils on granitic rocks in two tropical regions', *J. soil Sci.*, 10 (1959), 54–63.

[14] C. D. OLLIER, 'A two cycle theory of tropical pedology', *J. soil Sci.*, 10 (1959), 137–49.

[15] J. D. FALCONER, op. cit.

[16] R. H. NAGELL, op. cit.

[17] M. F. THOMAS, 'An approach to some problems of landform analysis in tropical environments', in J. B. WHITTOW and P. D. WOOD (eds.), *Essays in Geography for Austin Miller* (1965), 118–44.

[18] F. W. ROE, 'The geology and mineral resources of the neighbourhood of Kuala Selangor and Rasa Selangor, Federation of Malaya', *geol. Surv. Rep. Fed. of Malaya*, Mem. No. 7 (1953), 53.

[19] B. P. RUXTON and L. BERRY, op. cit. (1957), and also for example, H. WILHELMY, *Klimamorphologie der Massengesteine, Brunswick* (1958), and D. BRUNSDEN, 'The origin of decomposed granite on Dartmoor', in I. G. SIMMONS (ed.), *Dartmoor Essays, Devonshire Ass. Advmt. Sci. Lit. Art* (1964), 97–116.

[20] J. W. BARNES (ed.), 'The mineral resources of Uganda', *geol. Surv. Uganda*, Bull. No. 4 (1961), 48.

[21] Because of differences in terminology used to describe the profiles from different sites no further analysis of this kind of data has been attempted.

[22] B. P. RUXTON and L. BERRY, op. cit. (1957).

[23] This point is referred to by J. A. MABBUTT, op. cit. (1961), and is further discussed by the writer in M. F. THOMAS, op. cit. (1965a).

[24] B. P. BUXTON and L. BERRY, op. cit. (1957).

[25] F. DIXEY, *A Practical Handbook of Water Supply* (1931).

[26] R. R. E. JACOBSON, W. N. MACLEOD, and R. BLACK, 'Ring complexes in the younger granite province of Northern Nigeria', *geol. Soc. London*, Mem. No. 1 (1958).

[27] P. BIROT, 'Les dômes crystallins" *Centre Nat. Recherche Sci.*, Mem. Documents, 6 (1958), 8–34; 'La mésure de la porosité des roches crystallines', *Z. Geomorph. Suppl.*, 5 (1964), 41–52.

[28] F. DIXEY, op. cit. (1931).

[29] This is suggested by observations made in another context by R. S. WATERS, 'Differential weathering and erosion on oldlands', *Geogr. J.* 123 (1957). 503–9.

[30] Information obtained from water supply borehole records (unpublished) of the Geological Survey Department, Nigeria.

[31] H. LOUIS, 'Über rumpfflächen und talbildung in den wechselfeuchten tropen besonders nach studien in Tanganyika', *Z. Geomorph.*, 8, Sonderheft (1964), 43–70.

[32] P. BIROT, 'Le cycle d'érosion en climat tropical humide', in *Le Cycle D'Erosion Sous Les Differents Climats*, Univ. Brasil, Rio De Janeiro (1960), part 11, 75–95, and J. BÜDEL, 'Die "Doppleten Einebnungsflächen" in den feuchten Tropen', *Z. Geomorph.*, NF 1 (1957), 201–25. The writer has also discussed this question in M. F. THOMAS, op. cit. (1965a)

[33] M. F. THOMAS, op. cit. (1965a), 'Some aspects of the geomorphology of tors and domes in Nigeria', *Z. Geomorph.*, 9 (1965), 63–81.

[34] B. P. RUXTON and L. BERRY, 'Weathering profiles and geomorphic position of granite in two tropical regions', *Rev. Géomorph. Dynamique*, 12 (1961), 16–31.

[35] E. J. WAYLAND, 'Peneplains and some other erosional platforms', *Annual Rep. Bull. geol. Surv. Uganda* (1933), Notes 1, 74, 366.

[36] J. BÜDEL, op. cit. (1957); C. D. OLLIER, 'The inselbergs of Uganda', *Z. Geomorph.*, 4 (1960), 43–52; C. D. OLLIER, 'Some features of granite weathering in Australia', ibid., 9 (1965), 285–304.

[37] M. F. THOMAS, op. cit. (1965a). Compare Figure 28 in that study and Figure 11 in the present study.

[38] Writers advancing this general hypothesis include J. D. FALCONER, op. cit. (1911); J. D. FALCONER, 'The origins

of kopjes and inselberge', *Br. Ass. Advmt. Sci.*, Section C (1912), 476; E. BAILEY WILLIS, *Studies in comparative seismology: East African plateaux and rift valleys*, Washington (1936); G. ROUGERIE, 'Un mode de dégagement probable de certains dômes granitiques', *C. r. Acad. Sci.*, 240 (1955), 327–9; J. BÜDEL, op. cit. (1957); P. BIROT, op. cit. (1958); C. D. OLLIER, op. cit. (1960); R. A. G. SAVIGEAR, 'Slopes and hills in West Africa', *Z. Geomorph.*, Supp., 1 (1960) 156–71; C. R. TWI-DALE, 'A contribution to the general theory of domes inselbergs', *Trans. Inst. Br. Geogr.*, 34 (1964), 91–113; and M. F. THOMAS, op. cit. (1965b).

[39] B. P. RUXTON and L. BERRY, 'Notes on facetted slopes, rock fans and domes on granite in the east central Sudan', *Am. J. Sci.*, 259 (1961), 194–206 (see Figure 2), and C. R. TWIDALE, 'Steepened margins of inselbergs from north-western Eyre Peninsula', *Z. Geomorph.*, 6 (1962), 51–69 (see Figure 2).

[40] L. C. KING, 'A theory of bornhardts', *Geogr. J.*, 112 (1948), 83–7.

[41] *Idem* and J. C. PUGH, 'Fringing pediments and marginal depressions in the inselberg landscape of Nigeria', *Trans. Inst. Br. Geogr.*, 22 (1956), 15–31.

[42] M. F. THOMAS, op. cit. (1965b), especially plate 6.

[43] This kind of morphology was described by R. W. CLAYTON, 'Linear depressions (bergfussneiderungen) in savanna landscapes', *Geogr. Studies*, 3 (1956), 102–26, and is also referred to in M. F. THOMAS, op. cit. (1965a).

[44] P. BIROT, op. cit. (1958), 24.

[45] For example see J. R. F. HANDLEY, 'The geomorphology of the Nzega area of Tanganyika with specia lreference to the formation of granite tors', *Congr. géol. Int. Algiers* (1952), Fascicule XXI, 201–10; J. W. PALLISTER, 'Slope development in Buganda', *Geogr. J.*, 122 (1956), 80–7; J. A. MABBUTT, op. cit. (1961), and M. F. THOMAS, op. cit. (1965b).

[46] This phenomenon and further discussion of its possible implications has been considered by R. W. CLAYTON, op. cit. (1956) and J. A. MABBUTT, op. cit. (1965).

[47] L. C. KING, 'The study of the World's Plainlands', *Q. J. geol. Soc.*, 106 (1950), 101–31.

[48] C. D. OLLIER, op. cit. (1959).

[49] L. C. KING, 'Canons of landscape evolution', *Bull. geol. Soc. Am.*, 64 (1953), 721–52; 'The uniformitarian nature of hillslopes', *Trans. Edinburgh geol. Soc.*, 17 (1957), 81–102.

[50] For discussion of this point see J. W. PALLISTER, op. cit.

[51] M. F. THOMAS, op. cit. (1965b).

[52] M. F. THOMAS, 'On the approach to landform studies in Nigeria', *Nigerian geogr. J.*, 5 (1962), 87–101.

[53] J. F. ENSLIN, 'Secondary aquifers in South Africa and the scientific selection of boring sites in them', *Inter-African Conference on Hydrology, C.C.T.A.*, Publication No. 66, Nairobi (1961), Section IV, 'Groundwater Hydrology', 379.

[54] The use of the term 'Older Laterite' derives from the studies of A. M. J. DE SWARDT, 'The recent erosional history of the Kaduna valley', *geol. Surv. Nigeria, Annual Rep.*, 1946, 39–45; 'Recent erosion surfaces on the Jos Plateau', *Proc. Third Int. West African Conference* (1949), 180–6; and 'The geology of the country around Ilesha', *geol. Surv. Nigeria, Bull.* 23 (1953).

[55] The writer's tentative classification may be found in M. F. THOMAS, op. cit. (1965a).

Editor's Comments
on Papers 28 Through 32

ETCHPLAINS, PENEPLAINS, AND PEDIPLAINS

Now we shift from tropical Africa to tropical Australia. W. L. Wright describes in Paper 28 the nature of the erosion surfaces in a part of northern Australia where the terrain is largely sedimentary and the relation of these surfaces to deep weathering. This is one of a number of studies of Australian planation surfaces and represents in part an expansion of Woolnough's pioneering work on weathering and the development of duricrusts. These more recent studies benefit from the realization that Davisian peneplains are not the only possible kind of planation surface.

Perhaps one of the most important points made in the present paper is that each of two lower surfaces is made by stripping deep lateritic soils from the next higher surface to form two etchplains. The highest, oldest, smoothest, and most deeply weathered surface is not assigned a specific origin. It could be either a peneplain or a weathered pediplain.

Another inherent concept is that a kind of etching may be set up in the duricrust itself by differential wash, and that a lower etchplain may be developed in the soft parts of the regolith as well as in or near the weathering front at the base of the saprolite.

In Paper 29 Mabbutt presents a form line map showing the surface subjected to weathering in a part of what is now arid central Australia. This map indicates that the topography was formed of two elements: a pre-Cretaceous summit surface on high-lying resistant rocks and a weathered and duricrusted post-Cretaceous plain, with silicification of sedimentary rocks to the south and laterization on crystalline rocks to the north of the region studied.

The weathering profiles display evidence of more humid former climates and a well-developed stream network (Mulcahy and Bettenay, Paper 31). Silicification may have preceded laterization, which suggests passage from a seasonally humid climate to one of greater aridity, but without notable erosion during the humid periods. Laterization was only moderate, and developed almost exclusively in a quasi-humid climate and on crystalline rocks.

The weathering front was guided by topographic conditions, and has favored in turn the survival of characteristic relief traits.

Space limitations have restricted us to Mabbutt's conclusions, two plates, and a map.

Aside from the subject matter, Paper 30 invites attention because the authors, representing different sides of a controversy, joined forces to reexamine evidence in the field. The resulting report resolved many, if not all, of their original differences. Unfortunately, this interesting procedure is all too rare for many well-known reasons.

The substantive content revolves around distinctions between deep-weathered zones, duricrusts, laterites, and silcretes in southern Queensland. The distinctions are based on the observed features and relations among these materials in their geomorphic setting. The ages of the materials in question are determined from datable weathered sediments and valley fill basalts, the latter standing forth as inverted topography.

The authors agree that there is evidence of one long period of humid weathering, extending from Upper Cretaceous to Upper Oligocene, during which a deep saprolite zone was formed beneath a preexisting Mesozoic pediplain. The weathered zone does not show a good duricrust. Silcrete, which is present in many places, has become dissociated from its original weathered zone by later processes, such as creep. It is not here considered as a duricrust in the geomorphic sense. The authors would only consider silcrete as a duricrust if it overlay a deep weathered zone with which it is genetically associated.

In this part of Australia, the broad upland surfaces are judged to be pediplains rather than peneplains or etchplains because of their general smoothness and because the lower ones consist of pediments flanking present stream valleys. This does not rule out the possibility of other types of contemporary planation surfaces elsewhere on the continent.

Paper 31 is an excellent point of departure from our subject. It is a kind of summary of the discussion of planation surfaces as they have been studied to date.

The authors set up a defense of the Davisian peneplain on the basis of field evidence from Western Australia and the deductions of Davis himself. They point out that, as Davis supposed, in the case of peneplains there is deep weathering under low divides and, as recently discovered by deep borings, also under shallow valleys. These valleys are broad-floored and of extremely low gradient, harboring many ephemeral salt lakes, which may be connected by sluggish streams during periods of heavy rainfall. The drainage on this low relief surface is clearly mappable. The overall relief is generally less than 100 m and slopes are generally low.

Absolute elevations on the plain rise to 600 m in 500 km, suggesting that some uplift has occurred since peneplanation, and rise in such a way as to initiate a cycle of incision starting at the coast, while preserving virtually intact the uplifted peneplain inland. In one sense the peneplain may be in process of becoming still more level; for it is at present subject to "leveling without base leveling" by recent slope wash and deflation grading to the sluggish old stream channels, where the thalweg gradients are too gentle to facilitate removal of the material delivered to them.

At the same time, the present erosional wave appears to be producing a pediplain whose present inner terminus is a narrow zone, the Meckering Line, where it meets the high peneplain. Thus, peneplains and pediplains may "lie down together," though not peacefully in this case, where the pediplain is the aggressor.

This may be the place to review time relationships. The encroaching pediplain must, of course, be considered Holocene since it is an active surface today; when it was initiated is not clear. The age of the ancient peneplain also presents some difficulty. In one sense, if it is still largely intact and undergoing leveling, it, too, is Holocene. But this surface was a peneplain near the end of a cycle that for all practical purposes ended by general emergence and with the beginning of coastal dissection and pedimentation. The peneplain *was there* at the time of the last big emergence, widely thought to be a Miocene event. We must therefore conclude that its inception was more ancient, possibly extending back into the Paleozoic in view of the truncated Precambrian platform. Parenthetically, since there is no surface indication of a Tertiary cover in this area nor any convenient lava flows, there is a general lack of stratigraphic time determinants (see Paper 30).

In this case, we may use the initial time of peripheral destruction (Miocene) as an age designation of some significance. In this sense we cannot consider the peneplain as active today; although it may be getting smoother in spots, the smoothing is not determined by a base level corresponding to present sea level. Rather than increasing in area, the peneplain is diminishing in size by lateral encroachment.

We cannot leave this surface, inferred by the authors to be an old-age surface in the Davisian sense, without granting that it might have been a widespread pediplain, whose component pediments might have been softened into a more undulating topography at a later period. Our preference (following C. D. Holmes and others) would be to call planation surfaces "pediplains" only when the smoothness of the surface and other definite characteristics make it clear that the general surface is in fact made up of coalescing pediments, rather than the "valley sides" of Davis (Paper 14).

This is also the place to recall that in many cases, as in this one, we are obliged to deal with a double planation surface of sorts (Büdel, Paper 25): the planed rock surface and the genetically associated saprolite. In the present pedimented areas there is a tendency to incise the old peneplain margin down to the etched lower limit of the saprolites. The present paper conveys the impression that the exposed peneplain is almost entirely expressed by the saprolite. Residual bedrock knobs are rare (inselbergs, monadnocks). Is the peneplain defined by the soil cover or by the rock surface itself, which may be tens of meters below?

If we remove the deep weathered zone and expose the weathering front, is this a peneplain, or would it be more exact to call it an etchplain? Presumably, by the time the weathering front appears, many of the diagnostic characteristics associated with peneplains (as in this paper) would be removed, and the initial surface would become indeterminate unless preserved farther inland. We would have to call such rock planation surfaces, which are not made of coalescing pediments, etchplains instead of peneplains. This is consistent with the recommendation that all unetched regolithic planation surfaces be called peneplains unless they can be shown to consist of coalescing pediments.

The conclusions of Mulcahy and Bettenay have been sharply criticized by C. W. Finkle and H. J. Conacher in Paper 32, which follows without comment.

Mulcahy is given the last word in a short excerpt from a 1973 paper, "Landforms and Soils of Southwest Australia," in which he reviews the nature of the evidence for peneplains.

Landscape development

Preoccupation with questions of the age and development of laterite is a common and understandable characteristic of discussions of this topic in Western Australia. One or perhaps two periods of formation in the past are assumed (e.g., Prider, 1966), and thus the possibility of using a lateritized surface as a time marker is raised.

The evidence reviewed here shows that well-developed, deep laterite profiles may be found on surfaces as young as (early?) Pleistocene on the Swan Coastal Plain, where the process may still be continuing. They are most widespread on the older landscape elements of the Old Plateau, which Johnstone et al. (1973) considered to have been established in its present form by the mid-Cretaceous. The older laterites, except in the highest rainfall areas, now contain appreciable quantities of soluble salts in their pallid zones indicating that the necessary leaching conditions for their development are no longer operative. These facts, together with the detrital and transported nature of many lateritic materials, makes their use as a stratigraphic marker suspect, though they are, of course, likely to be preserved where erosional forces are least effective, i.e., on gentler slopes and in well-vegetated higher rainfall areas.

Age of landscape is much more likely to be indicated by a combination of landscape characteristics rather than the occurrence of a single characteristic such as laterite. These would include low relief, widespread deep weathering, ineffective drainage systems, and widespread retention within the landscape of weathering products and sediments derived from them. These are the conditions found most extensively inland of the Meckering Line.*

REFERENCE

Mulcahy, M. J., 1973. Landforms and soils of Southwest Australia. J. Roy. Soc. Western Australia, 56(1), p. 16–22.

*Reprinted from *Jour. Royal Soc. W. Australia*, 56(1), 16–22 (1973).

28

Reprinted from *J. Geol. Soc. Australia,* **10**, Pt. 1, 151–154, 156–163 (1963)

DEEP WEATHERING AND EROSION SURFACES IN THE DALY RIVER BASIN, NORTHERN TERRITORY

By R. L. WRIGHT

(WITH 4 PLATES AND 4 TEXT-FIGURES)

(*Received 17th July, 1962; read at Canberra, 27th November, 1962*)

ABSTRACT

Within the Daly River basin, Northern Territory, three erosion surfaces are described and their relationships to deep weathering are discussed. The Bradshaw surface is the highest and oldest surface recognized. It is of considerable perfection and forms main divides; it is associated with a deep lateritic profile with a strongly silicified horizon forming the lower part of the pallid zone and extending into rocks immediately beneath. The Maranboy surface now forms secondary divides, with related rock-cut terraces, below the level of the Bradshaw surface. In most areas it was produced by the stripping of the upper, less silicified parts of the Bradshaw pallid zone. The Maranboy surface is associated with a lateritic weathering profile less deep than the Bradshaw profile and mainly developed in the Bradshaw weathering mantle. A younger erosion surface, the Tipperary surface, advanced by the removal of the Maranboy re-weathered layer exposing the resistant Bradshaw silicified rock which commonly forms a base-level of denudation. The Tipperary surface consists of broad plains, gently undulating terrain, and dissected headwater valley floors. It is relatively unweathered and carries depositional mantles which are attributed to climatically induced slope instability.

The ages of the erosion surfaces and the possibility of climatic changes in the area are also briefly discussed.

I. INTRODUCTION

The stripping of deep weathered mantles is recognized as a major process in landscape evolution in many areas underlain by crystalline rocks. Linton (1955) ascribed the origin of British tors to this process and defined the stripped surface between them as the "basal platform", corresponding with the boundary between decomposed and solid rock. This boundary was designated the "basal surface" by Ruxton and Berry (1959), who emphasized that it need not demarcate the limit of weathering. Cotton (1961) has reviewed the work of Budel (1957) and others which emphasizes the importance of the stripping of regoliths in the evolution of savanna landscapes. More recently Mabbutt has proposed that the term "basal surface" be replaced by "weathering front" (Mabbutt, 1961a) and has shown that the level of an etchplain in Western Australia is controlled by the weathering front on interfluves but by the upper part of a "transition zone" in its lower parts (Mabbutt, 1961b).

This paper describes the relationships between erosion surfaces and deep weathering in an area which has a more complex denudation history and which is mainly underlain by sedimentary rocks.

II. PHYSICAL SETTING

The area of investigation (Fig. 1) was termed the "Daly River basin" by Christian and Stewart (1953). It mainly comprises the inland three-quarters of the Daly River catchment in the Northern Territory. The central parts of the area consist of gently undulating plains and low plateaux, with relief mainly less than 100 feet, which descend north-westwards from about 700 ft. in the south-east to about 100 feet above sea-level. The major divides consist of broad dissected plateaux and rugged hill lands which have a relief of up to 400 feet and which locally exceed 1,000 feet above sea-level. In their investigation of the Katherine-Darwin region, of which the present area forms the southern part, Christian and Stewart (1953) recognized a "Tertiary land surface" of low relief which was extensively lateritized by "Late Middle Tertiary" times. They stated that "the present cycle of erosion" was initiated by late Tertiary uplift which included warping and faulting, and from the present altitudes of residuals of their Tertiary land surface they deduced that the warping reached a maximum in the vicinity of Pine Creek (in the north of the present area) and fell gradually to the south. White (1954) subsequently recognized that denudation of a lateritized land surface in the Northern Territory resulted in the formation of a "detrital laterite" at a lower level.

Geologically, the area has been mapped and described by Noakes (1949) and, more recently, by Skwarko (1961), Malone (1962) and Randal (1962). It is a structural entity in which gently dipping Middle Cambrian limestone, sandstone, and siltstone overlie Lower Cambrian extrusive basalt and Proterozoic basement rocks which crop out on the margins of the basin. Remnants of a thin mantle of sub-horizontal Cretaceous sandstone and siltstone unconformably overlying the Cambrian and Proterozoic rocks survive in many areas, particularly in the south and south-east.

The climate (Slatyer, 1960; Bureau of Meteorology, 1961) is tropical, with a warm dry winter period from May to September, and a hot wet summer period from October to April. The mean annual rainfall varies from about 45 in. in the north-west to less than 30 in. in the south-east and has a moderate variability. Most of the rainfall is received between December and March inclusive and these months are characterized by intermittent showers interspersed with periods of dry weather which occasionally exceed a fortnight in length. The main rivers are perennial, but tributary streams flow only during the wet season and for a short time afterwards. High temperatures are characteristic throughout the year. At Katherine, mean monthly maxima range from 85° F. to 105° F. and minima from 50° F. to 80° F., and conditions throughout the remainder of the area will not differ greatly.

III. DEFINITIONS

In this paper, "deep weathered mantles" describes relatively thick profiles expressive of sub-surface rock rotting as well as sub-aerial weathering. The term has been applied to weathering forms ranging from laterites with pronounced profile zonation to undifferentiated kaolinized rock. In the Daly River area, the profiles are lateritic and exhibit the ferruginous, mottled, and bleached zones

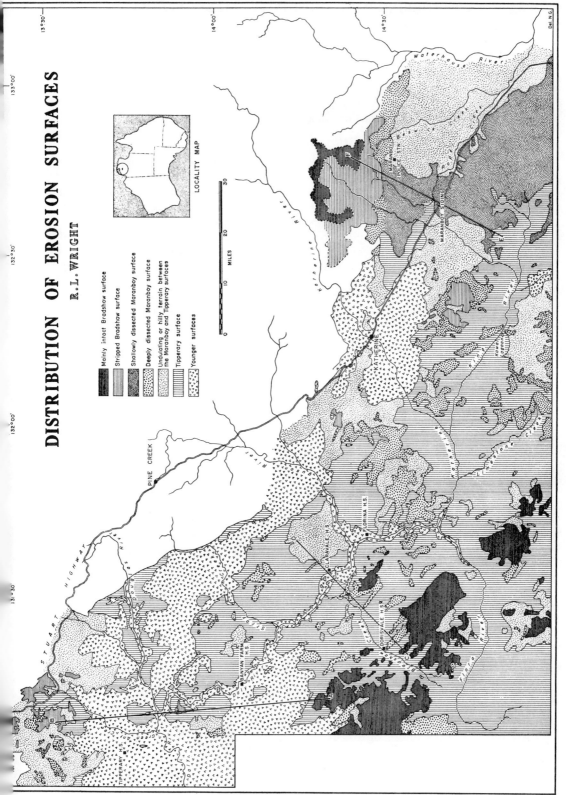

Fig. 1. Distribution of erosion surfaces.

first defined by Walther (1915). The bleached zone is here described as pallid zone after Whitehouse (1940), and the ferruginous zone rock is referred to as ironstone.

Such deep weathering occurs in vadose and phreatic ground water zones, particularly under warm, humid climates. It is a process of geologic duration, and such mantles, therefore, are commonly inherited from extensive ancient plains and possibly from earlier climates.

IV. THE EROSION SURFACES

Three main erosion surfaces have been recognized in the area and are mapped in Fig. 1 as the Bradshaw, Maranboy, and Tipperary surfaces. The mapping is based on field observations made during a survey in 1961 by a team from the Division of Land Research and Regional Survey, C.S.I.R.O., and on air photo interpretation. The erosion surfaces have been identified and mapped on the basis of relative altitude, continuity, weathering profiles, and escarpment relationships, and by the related rock-cut terraces and tributary valley floors which rise up-valley from the two younger surfaces and represent their headward extension into the oldest surface.

Figs. 2, 3 and 4 show the generalized topographic relationships of erosion surfaces along the section lines indicated in Fig. 1, together with their postulated relationships to weathering zones. These are based on field evidence at or near specific localities along the section lines.

[*Editor's Note:* Figure 2, Figure 4, and Plate 2 have been omitted.]

(a) Bradshaw Surface

(i) *Topographic Form and Distribution.* This is the highest and oldest surface recognized, and it survives as plateau summits more perfectly planed than younger erosion surfaces, forming main divides and also occurring in the central part of the basin. The type area of the Bradshaw surface is in the west of the basin where it is dissected up to 400 feet by tributaries of Bradshaw Creek. The surface is also extensively preserved in the east, north of the Maranboy Tin-Field, but remnants are more restricted along the northern margin of the area. In the areas referred to, the surface is mainly between 850 and 1,000 feet above sea-level, with marginal escarpments up to 400 feet high; it slopes gently towards the major valleys, and dissected outliers occur at about 400 feet above sea-level only a few miles from the Daly River near Claravale homestead.

The nature, distribution, and altitudinal relationships of remnants of the Bradshaw surface suggest that it originally consisted of an extensive, very gently sloping plain with broad undulations and with an internal relief of up to 200 feet, and that the Daly River and its major tributaries already occupied broad, shallow valleys within it.

In this area, the Bradshaw surface is everywhere preserved on the Cretaceous rocks, and the occurrence of these relatively unresistant, flat-lying strata appears to have facilitated its formation and perfection. However, the Bradshaw surface bevels these beds in the north of the area (Photo 1) where they are locally moderately to strongly dipping.

(ii) *The Bradshaw Weathering Profile.* Weathering of the Bradshaw surface is typically to a depth of 150 ft. The typical profile, 16 miles north-west of

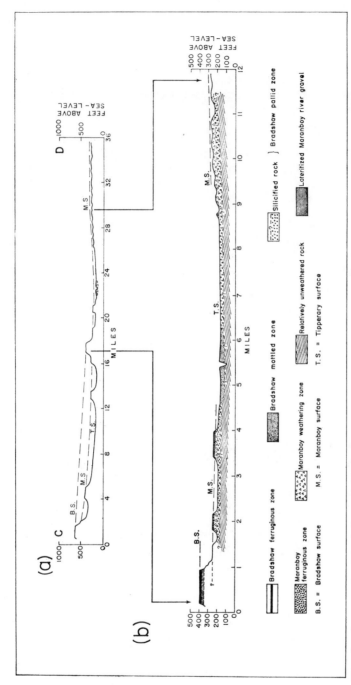

Fig. 3. Sections showing (a) generalized topographic relationships of the erosion surfaces along the section line CD shown in Fig. 1, and (b) postulated relationships between the surfaces and weathering zones at the indicated locality.

Dorisvale homestead (Fig. 2b and Photo 2), comprises lateritic red earths*
underlain by 20 feet of pisolitic ironstone, which becomes vermicular in the
lower part, overlying about 30 feet of mottled zone and 100 feet of pallid zone.
This profile is formed in the Cretaceous strata except for the lowest 30 feet,
which comprise pallid Cambrian sandstone and siltstone. In other localities, the
ironstone varies from vermicular to nodular or concretionary, but is generally
pisolitic in the upper parts. It ranges between 10 and 20 feet in thickness.
Mottled and pallid zones have been observed up to 40 feet and 100 feet thick
respectively.

Throughout the area, Bradshaw weathering has extended through the
Cretaceous rocks, which rarely exceed 100 feet in thickness (Skwarko, 1961),
into the underlying strata. In the north much of the pallid zone is in Lower
Proterozoic greywacke and siltstone which here form the sub-Cretaceous strata.
Elsewhere, the lower part of the pallid zone is formed in the Cambrian rocks.
For example, up to 100 feet of pallid zone is seen to be developed in dipping,
thinly-bedded Cambrian rocks along the entrenched Daly River between Florina
and Banyan Farm homesteads.

(iii) *Intact and Stripped Bradshaw Surfaces.* In Fig. 1, remnants of the
Bradshaw surface with mainly intact profiles are distinguished from stripped
remnants. In the areas of mainly intact Bradshaw surface, profiles are as described above, but with a shallowly stripped zone extending up to ¼ mile back
from escarpments. This stripped zone consists of stony slopes partly truncating
the ironstone and is interpreted as a waxing slope developed ahead of the
retreating escarpment. In the areas mapped as stripped Bradshaw surface,
however, the ironstone and much or all of the mottled zone has been removed,
and profiles comprise laterite, red or brown earths or skeletal soils overlying up to
20 feet of mottled zone or occurring directly on pallid zone.

Remnants of mainly intact Bradshaw surface are preserved in the west of the
area shown in Fig. 1, where the dissecting valleys are mainly more narrowly
entrenched, and also in the east of the area on the crest of the broad plateau
north of the Maranboy Tin-Field. Remnants of stripped Bradshaw surface occur
where dissection has been more vigorous, and more advanced, as in the north
of the area, and at the head of the King River in the east of the area where
shallow erosion appears to have regressed along a pre-existing valley on the
Bradshaw surface.

(iv) *Silicification in the Lower Part of the Bradshaw Profile.* Wherever
observed, the lower part of the Bradshaw pallid zone was strongly silicified.
Such exposures are more common in the central parts of the area, where up to 70
feet of silicified rock may be observed along the entrenched Daly River and its
tributaries. The silicified rocks take two forms — porcellanite and quartzite.

The porcellanite is a white to cream-coloured, fine-grained rock with characteristic conchoidal fracture. A thin section (Photo 3) shows mainly silica in the
form of a fine aggregate-polarizing mass of quartz, confirmed by X-ray identification. Floating in this quartz mass, and forming about 20 per cent. of the

---

* This and other references to soils are based on observations made by R. H. M. van de
Graaff of the Division of Land Research and Regional Survey, C.S.I.R.O.

entire section, are grains of feldspar and quartz sometimes together with remains identified by Dr. K. W. Crook as radiolaria and sponge spicules, now composed of fine quartz. There are also what appear to have been cavities which are now filled with quartz and rimmed with chalcedony.

These areas of intense silica replacement alternate with areas in which replacement appears to have been negligible and where the matrix between the radiolaria, sponge spicules, feldspar and quartz grains is more or less opaque. The junction between the areas of intense replacement and those of negligible replacement is mainly sharp and linear, but in some places it is gradational over a distance of up to 5 mm.

The porcellanite is interpreted as an argillite in which most of the clay material has been replaced by silica which is now in the form of extremely fine crystalline quartz. Macroscopically it resembles the porcellanite in pallid zones in many areas of Australia.

The porcellanite occurs in bedded sequences with thin, moderately silicified sandstone horizons and is characteristically developed in the Cretaceous beds but also in the Cambrian succession. Its upper limit against the unsilicified pallid zone is everywhere abrupt. It is underlain by the quartzite in many areas, but in the central parts it locally has an abrupt basal junction with calcareous Cambrian rocks, and on the margin of the basin it locally overlies the Proterozoic basement rocks with unconformity.

As shown in thin section (Photo 4), the quartzite consists of detrital quartz grains with optically continuous quartz overgrowths which weld the grains together forming an interlocking mass. The silica of the overgrowths has apparently been introduced, for there has been no appreciable solution of the detrital grains. The rock is interpreted as having formed by the cementation of sandstone by quartz.

The quartzite is a massive or thickly bedded, well-jointed rock, with a sugary texture on weathered surfaces. It occurs both in the basal part of the Cretaceous (e.g. at the head of Limestone Creek where it locally contains abundant fossil plant remains) and also in the Cambrian succession. It was always seen to be underlain by the Cambrian calcareous rocks.

Both porcellanite and quartzite occur in a sub-horizontal zone with extremely abrupt upper and lower limits. That this zone is not a stratigraphic unit is shown by the fact of its development in known Cretaceous rocks and in dipping Cambrian strata. It is stressed that these silicified rocks are quite different lithologically and in mode of occurrence from the chert lenses and nodules in the underlying Cambrian rocks.

From the geological setting it is evident that these strongly silicified rocks cannot be due to regional or thermal metamorphism. For the following reasons they are considered to be the result of silica movement associated with lateritic Bradshaw weathering and not due to diagenesis:

(1) Both rock types are always associated with the Bradshaw pallid zone; the porcellanite forming the lower part of the pallid zone and the quartzite occurring at its base. Sufficient exposures were observed to indicate that the sub-horizontal zone in which they occur is roughly accordant with the Bradshaw surface and its related lateritic weathering mantle.

(2) High grade diagenesis, such as would result in complete quartz cementation as exhibited by the quartzite, is usually associated with pronounced modification of quartz grain boundaries by solution (Dapples, 1959). This is lacking from the quartzite. Therefore, cementation probably took place under slight overburden — a conclusion supported by field evidence that the Cretaceous rocks covering the silicified zone probably never exceeded 200 feet in thickness. The lack of solution of detrital grains suggests that the cementing silica was introduced into the quartzite and it is difficult to envisage sufficient silica to produce total cementation being derived diagenetically from waters expelled during compaction of such a thin cover.

(3) The diagenetic development of secondary quartz probably requires a particular pressure-temperature regime (Coombs *et al.*, 1959). Therefore, diagenetic silicification would be expected to continue into sandstones beneath the silicified rocks described since the pressure-temperature regime would be exceeded there. This is not seen, for the silicified zone has an abrupt lower limit.

It is, therefore, suggested that the silicified rocks result from silicification in the lowest part of the Bradshaw pallid zone and in suitable rocks immediately beneath, and that the formation of porcellanite on the one hand and quartzite on the other is due to differences in the nature of the host rock. A somewhat similar differentiation of porcellanitic and quartzitic rocks in weathering zones has been recognized by Noakes (unpublished data).

(b) Maranboy Surface

(i) *Topographic Form and Distribution.* This surface is cut below the level of the Bradshaw surface and is separated from it by prominent escarpments. Two forms of the Maranboy surface are mapped in Fig. 1 — one of shallow dissection and one of deep dissection.

The shallowly dissected Maranboy surface comprises broad, very gently undulating secondary divides with extensive areas of relict lateritic soils. It occurs at about 700 ft. above sea-level at the head of the basin, forming the divide between drainage west to the Daly River and that which flows east to the Roper River. This divide includes the type area, immediately south of Maranboy Siding.

The deeply dissected Maranboy surface consists of scattered low plateaux and interfluve crests, commonly with stony surfaces, which decline from 700 feet in the east to about 250 feet above sea-level down-valley. It is preserved as secondary divides on both sides of the Katherine and Daly Rivers, and particularly between the Katherine and Fergusson Rivers. It locally includes a gravel terrace (Fig. 3b) west of the Daly River near Claravale homestead.

Related rock-cut terraces continue up the larger tributary valleys as valley-side benches, remnants of which also survive as accordant spur and hill crests (Photo 5).

(ii) *The Maranboy Profile.* In the type area the Maranboy profile consists of lateritic red and grey earths underlain by up to 6 feet of ironstone which is pisolitic in the upper parts, nodular and concretionary below, and characteristically has a sandy matrix. The ironstone is partly detrital and contains angular, broken fragments of recemented ironstone and inclusions of pallid zone rock (Photo 6).

The ironstone is underlain by at least 60 feet of pallid zone rock which is commonly powdery in argillaceous rocks, crumbly and honeycombed in coarser rocks, and which has secondary iron enrichment in the form of pisolitic ironstone or iron staining along joints and bedding planes in the upper parts (Fig. 4b).

Similar profiles characterize the deeply dissected remnants of the Maranboy surface to the west of the type area where the ironstone is locally up to 10 feet thick and is underlain by up to 100 feet of pallid zone rock.

Throughout the area there is a remarkable continuity or identity of level between the pallid rock underlying the Maranboy ferruginous zone and the Bradshaw pallid zone itself. The continuity of the two is best confirmed north of the type area, in the Bradshaw escarpments north of the Maranboy Tin-Field. Accordingly, the rock underlying the Maranboy ferruginous zone is attributed to the Bradshaw pallid zone.

At the head of Green Ant Creek in the extreme north of the area, where the Maranboy surface locally extends to the foot of Bradshaw escarpments (Fig. 2d), the weathered rock underlying the Maranboy ferruginous zone lies wholly below the level of the Bradshaw pallid zone. The weathered rock is between 30 and 40 feet thick and comprises Cambrian sandstone, siltstone and limestone.

(iii) *Weathering of the Maranboy Surface.* Although the pallid zone rock underlying the Maranboy ferruginous zone has been identified with the Bradshaw pallid zone, its powdery and honeycombed textures are in strong contrast with the resistant Bradshaw silicified rock which typically occurs beneath it, as exposed at the foot of Maranboy escarpments (Photo 7). This contrast, apparently resulting from desilicification of Bradshaw pallid zone rock, is attributed to reweathering associated with the Maranboy surface. This postulate of reweathering is further supported by secondary iron enrichment in the upper parts of the Maranboy profile, and by the occurrence of a distinct weathered layer where the Maranboy surface is cut below the base of the Bradshaw profile in the north of the area.

This characteristic leaching beneath the Maranboy ferruginous zone extends to a depth of 50 feet in the type area and to between 30 and 60 feet elsewhere, with quartzite or porcellanite typical of the Bradshaw profile locally exposed at its base. The basal junction with these rocks is irregular in that abrupt lateral changes from the silicified rock (usually quartzite) to the softer, leached pallid material are common. In part these lateral contrasts are strongly reminiscent of the observed pattern of silicification within and at the base of the Bradshaw pallid zone. Therefore, it appears probable that irregularity of original Bradshaw silicification, as controlled by lithological contrasts and joint patterns, may subsequently have induced irregularity of the Maranboy weathering front. This irregularity is greater in the quartzitic facies, which shows a more marked segregation into structural compartments.

(iv) *Factors Controlling the Level of the Maranboy Surface.* Except in the north of the area, the Maranboy surface is cut within the Bradshaw pallid zone, and the strongly silicified lower part of this zone appears generally to have set a lower limit to Maranboy erosion. However, at the head of tributary valleys, the level of the Maranboy surface is locally determined by resistant rocks forming the sub-Cretaceous surface, or by resistant sandstone layers within the weathered Cretaceous itself.

(c) Tipperary Surface

(i) *Topographic Form and Distribution.* This surface is cut below the level of the Maranboy surface and is separated from it by low, discontinuous escarpments and by dissected undulating or hilly terrain. As mapped in Fig. 1, the Tipperary surface comprises three main types of terrain. For the most part it consists of broad plains which extend downslope from the foot of Bradshaw and Maranboy escarpments and their flanking dissected tracts. Typically, the plains end at short, steep erosional slopes flanking the alluvial flats along the main rivers and their tributaries, but in the north-west of the basin and near Katherine a somewhat broader dissected zone intervenes. A second type of Tipperary terrain, occurring within these plains, consists of gently undulating terrain with relief mainly less than 30 feet. The third type comprises dissected headwater valley floors and occurs in the west and north of the area. These floors grade downvalley into the plains which extensively characterize the Tipperary surface.

(ii) *The Relationships of Forms of the Tipperary Surface to the Older Weathering Profiles.* Post-Maranboy deep weathering has nowhere been observed on the Tipperary surface, but older weathering mantles have exercised considerable control over the level and form of the Tipperary surface, just as the Maranboy surface was influenced by Bradshaw weathering features. Relationships to prior weathering vary in the three forms of the Tipperary surface described above.

The plains have formed mainly by the stripping of the Maranboy reweathered layer from the unaffected Bradshaw silicified rocks beneath (Figs. 2c, 3b, 4b; Photo 7). Tors up to 30 feet high (Photo 8) of rounded boulders occur locally at the foot of Maranboy escarpments, in the plains downslope, and along the dissected lower margins of the plains. Such tors consist mainly of the quartzite from the Bradshaw profile beneath the pallid zone, left by the removal of the surrounding weathered material. The separate tor mounds appear to reflect the structurally controlled, compartmented pattern of Bradshaw silicification and the resultant irregularity of the Maranboy weathering front typically observed from this zone.

In the gently undulating parts of the Tipperary surface, remnants of the Maranboy weathered layer survive on the low crests. These consist mainly of leached pallid zone rock with local ironstone detritus.

The headwater valley floors of the Tipperary surface are up to 200 feet below terrace remnants of the Maranboy surface and are hence cut below the level of prior weathering. Accordingly, residual soils developed from unweathered rock are moderately extensive upon them.

(iii) *Alluvial-colluvial Mantles on the Tipperary Surface.* Deep, relatively uniform mantles of locally derived material of alluvial-colluvial origin occur extensively on the Tipperary surface. In the higher parts of the gently undulating tracts, and in the dissected headwater valley floors, the material is mainly medium- to coarse-grained and contains broken, derived lateritic concretions; downslope on the plains it is characteristically fine- to medium-grained. On the plains the mantle has characteristically given rise to red earths; on the gently undulating tracts grey or brown sandy lateritic earths are common; red and yellow earths, which are locally lateritic, are typical of the mantle in the dissected headwater valley floors.

A number of deep holes were augered into the more extensive red earth mantle on the plains, but bedrock was only encountered in one locality, 21 miles east-south-east of Katherine, where Bradshaw silicified pallid zone rock occurred at a depth of 15 feet.

These extensive deposits may indicate conditions of surface instability such as could be induced by a relatively rapid change to a drier climate, resulting in a reduced vegetation cover. Under appropriate conditions of rainfall distribution this would lead to the movement of weathered material on slopes and to a reduction in the competence of tributary streams to carry away the derived material.

IV. DISCUSSION

1. Three main stages are recognized in the earlier evolution of the physical landscape in this area. The first of these resulted in the formation of an extensive, gently undulating plain termed the Bradshaw surface. Subsequently, the Bradshaw surface was dissected and a younger erosion surface, the Maranboy surface, was produced. The most likely cause of this dissection is uplift. Such uplift would have been greatest in the north of the area, since the Bradshaw surface has been most extensively destroyed in the north, and the Maranboy surface has been cut below the Bradshaw weathering front. Such uplift may have been a reactivation of the movements which locally tilted the Cretaceous strata in the north of the area, prior to their truncation and deep weathering during the Bradshaw stage of erosion.

The Maranboy surface was in turn dissected during a third stage of erosion, and an extensive younger erosion surface, the Tipperary surface, was produced. Little can be said concerning the cause of this dissection, although it implies rejuvenation of drainage throughout the basin.

The Tipperary surface has also been dissected, and still younger erosional and depositional surfaces have been produced.

There is little evidence for the ages of the surfaces described. Clearly the Bradshaw surface is post-Lower Cretaceous. The alluvial-colluvial mantles described as overlying the Tipperary surface may have resulted from climatic fluctuations similar to those dated as Pleistocene or later in other parts of Australia. Accordingly, the Tipperary surface is regarded as pre-Pleistocene. The shallowness of dissection of the Tipperary surface and the limited extent of the younger surfaces formed below, suggest that these latter are geologically extremely young.

2. The three erosion surfaces described above show successively shallower depths of weathering. The Bradshaw lateritic profile is the only very deep

weathering profile in the area; Maranboy profiles are generally less than 50 feet thick; whilst weathering on the Tipperary surface is shallow or negligible. This decrease in depth of weathering may be the result of continuing secular change towards a drier climate and/or of successively shorter periods of weathering. In the case of the Maranboy surface, the occurrence of a lateritic weathering profile, both within and locally below the Bradshaw weathering profile, suggests that the time factor may be more important. This is consistent with the lesser development and perfection of the Maranboy surface compared with the Bradshaw surface. In the case of the Tipperary surface, a drier climate seems a more acceptable explanation, for the extent of the surface suggests that there may otherwise have been sufficient time for at least some lateritic profile development in excess of the formation which has locally occurred. The climatic explanation is supported by evidence of slope instability in the form of the alluvial-colluvial mantles which may indicate drier or more strongly contrasted wet and dry seasonal conditions.

3. The stripping of the Bradshaw and Maranboy profiles has been a major factor in the evolution of the post-Bradshaw erosion surfaces. In most areas the Bradshaw profile was stripped of the upper, less silicified parts of the pallid zone to produce the Maranboy surface, and similarly the Tipperary surface advanced by the removal of the Maranboy reweathered layer near the base of the Bradshaw profile. In a sense, therefore, both the Maranboy and Tipperary surfaces are etchplains.

4. The zonation and relative thickness of the weathering profiles, together with the amount of dissection, also control the nature of the boundaries between successive erosion surfaces. The junction between the Bradshaw and Maranboy surfaces mainly consists of high escarpments capped by massive ironstone. In contrast, the junction between the Maranboy and Tipperary surfaces mainly consists of shallowly dissected undulating tracts, with remnants of more resistant ironstone cappings, cut in easily erodible, commonly desilicified rock.

V. ACKNOWLEDGMENTS

The author makes grateful acknowledgment of the helpful criticism and advice received from J. A. Mabbutt of the Division of Land Research and Regional Survey in the preparation of this paper. The Division's drawing office and K. Fitchett are thanked for preparing the figures. The author also thanks Dr. K. A. W. Crook of the Australian National University for his assistance with the section on silicification. Thin sections were kindly prepared by the Bureau of Mineral Resources and the Australian National University.

VI. REFERENCES

BUDEL, J., 1957: Die doppelten Einebnungsflächen in den feuchten Tropen. *Z. Geomorph.*, *1*, pp. 201-225.

BUREAU OF METEOROLOGY, 1961: Climatological Survey: Region I — Darwin-Katherine, Northern Territory.

CHRISTIAN, C. S., and STEWART, G. A., 1953: General Report on Survey of Katherine-Darwin Region, 1946. *Coun. sci. industr. Res. Aust. Land Res. Series, 1.*

COOMBS, D. S., ELLIS, A. J., FYFE, W. S., and TAYLOR, A. M., 1959: The Zeolite Facies, with Comments on the Interpretation of Hydrothermal Syntheses. *Geochim. et cosmoch. Acta, 17,* pp. 53-107.

COTTON, C. A., 1961: The Theory of Savanna Planation. *Geography, 46*, pp. 89-101.

DAPPLES, E. C., 1959: The Behaviour of Silica in Diagenesis. *Spec. Publ. Soc. econ. Paleont. Miner., 7*, pp. 36-54.

LINTON, D. L., 1955: The Problem of Tors. *Geogr. J., 121*, pp. 470-486.

MABBUTT, J. A., 1961a: "Basal Surface" or "Weathering Front". *Proc. Geol. Ass., Lond., 72*, pp. 357-358.

———, 1961b: A Stripped Land Surface in Western Australia. *Trans. Inst. Brit. Geogr., 29*, pp. 101-114.

MALONE, E. J., 1962: Explanatory Notes on The Pine Creek 1 : 250,000 Sheet Area, Northern Territory. *Bur. Min. Resour. Aust. Explan. Notes*, Sheet D 52/8.

NOAKES, L. C., 1949: Geological Reconnaissance of the Katherine-Darwin Region, Northern Territory, with Notes on the Mineral Deposits. *Bull. Bur. Min. Resour. Aust., 16.*

RANDAL, M. A., 1962: Explanatory Notes on the Fergusson River 1 : 250,000 Sheet Area, Northern Territory. *Bur. Min. Resour. Aust. Explan. Notes*, Sheet D 52/12.

———, 1962: Explanatory Notes on the Katherine 1 : 250,000 Sheet Area, Northern Territory. *Bur. Min. Resour. Aust. Explan. Notes*, Sheet D 53/9.

RUXTON, B. P., and BERRY, L., 1959: The Basal Rock Surface on Weathered Granite Rocks. *Proc. Geol. Ass., Lond., 70*, pp. 285-290.

SKWARKO, S. K., 1961: Progress Report on the Field Activities in the Northern Territory during 1961 Field Season. *Rec. Bur. Min. Resour. Aust., 1961/153* (unpubl.).

SLATYER, R. O., 1960: Agricultural Climatology of the Katherine Area, N.T. *Tech. Pap. Coun. sci. industr. Res. Aust. Div. Land Res. Reg. Surv., 13.*

WALTHER, J., 1915: Laterit in West Australien. *Z. dtsch. geol. Ges., 67*, pp. 113-132.

WHITE, D. A., 1954: Observations on Laterites in the Northern Territory. *Aust. J. Sci., 17*, pp. 14-18.

WHITEHOUSE, F. W., 1940: Studies in the Late Geological History of Queensland. *Pap. Dep. Geol., Univ. Qd., 2* (N.S.), 1.

R. L. Wright,
Division of Land Research and Regional Survey,
C.S.I.R.O.
Canberra, A.C.T.

Photo 1.   Stripped remnant of the Bradshaw surface bevelling the dipping
Cretaceous strata in the north of the area.

Photo 2.   Weathering of the Bradshaw surface north-west of Dorisvale
homestead, showing ironstone capping, about 30 feet of mottled zone, and
100 feet of white pallid zone.   The bench on the right marks the sub-
Cretaceous unconformity and is about 120 feet below the crest on the
left; it approximates to the level of adjacent remnants of the Maranboy
surface.   The Tipperary surface forms the foreground.

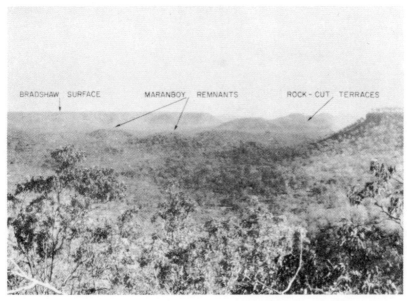

Photo 5.   Headwater remnants of the Maranboy surface comprising valley-side benches in the right middle distance and accordant hill crests in the centre middle distance.   The Bradshaw surface forms the skyline.

Photo 6.   Pisolitic Maranboy ferruginous zone, with inclusions of pallid zone rock, underlain by pallid zone rock attributed to the Bradshaw profile.

J.Geol.Soc.Aust.   Vol. 10, Pt. 1

425

Photo 7.   Maranboy escarpment with exposure of quartzitic silicified Bradshaw rock at base, overlain by Bradshaw pallid zone which has undergone Maranboy reweathering.

Photo 8.   Tors of quartzitic silicified Bradshaw rock on plains of Tipperary surface.

J.Geol.Soc.Aust.   Vol. 10, Pt. 1

426

29

Reprinted from *Zeit. Geomorph.*, **9**, 109–112, 114 (1965)

The Weathered Land Surface in Central Australia

by

J. A. MABBUTT, Canberra

[*Editor's Note:* In the original, material precedes this excerpt.]

7. Conclusions

Some general conclusions can be drawn from these regional studies:

1. In central Australia "weathered land surface" usefully designates an older landscape of more smoothly flowing contours and associated weathering which is being replaced by angular, rugged surfaces, only superficially weathered. The weathered profiles are in disaccord with present rainfall, and their heights above current watertables and base levels confirm their fossil character. They attest to the subsequent desiccation of central Australia.

2. "Weathered land surface" in this sense connotes the entire landscape which was subjected to weathering. It is a cyclically complex surface, consisting of pre-Cretaceous uplands, mainly on resistant sandstone, and younger duricrusted plains on crystalline and softer sedimentary rocks. The duricrusted plains are post-Lower Cretaceous and there is evidence from overlying beds that they were weathered before the late Tertiary.

3. Sub-aerial erosion of the uplands has been continuous from the Palaeozoic onwards, and earlier erosion or planation is reflected in subdued, rounded summit surfaces or bevels. The central folded ranges were reduced to moderate or subdued relief in pre-Cretaceous times, this extant relief increasing northwards towards the crystalline central ranges which were the watershed for a southwards transverse drainage. The Davenport-Murchison Ranges were planed in greater degree, and transverse drainage here is regarded as largely consequent on an uptilted summit plane. The younger duricrusted plains cycle is reflected in the uplands by the incision of weathered valleys, when the relief largely assumed its present form. Staged relief and monadnock survivals in the folded central ranges invalidate the concept of a single prior land surface linking the northern and southern plains and subsequently warped across the central ranges.

4. The weathered land surface was mainly erosional, although depositional intermont or piedmont "facies" indicate that cycles of weathering and slope instability reach back to the period of weathering. The contrast between this great extent of erosional surfaces, arguing for competent, integrated, out-going river systems, and the present conditions of interior and disorganized drainage

427

with extensive depositional lowlands is further argument for the desiccation of central Australia.

5. There is no geomorphic evidence for climatically determined periodicity of deep weathering in central Australia. Uplands and duricrusted plains represent differential landscape survivals on rocks of differing resistance and their weathering is not comparable. Staged duricrusts do not occur, and the weathering of sandstone on upland summit bevels is not to be distinguished from that on younger, weathered-valley slopes. There are no distinct "levels of weathering" corresponding to the two recognized stages of the weathered land surface; the evidence is that, just as the water-table in a humid area follows surface contours, so weathering has attacked the land surface in relief, with the weathering front correspondingly uneven. This is implicit in the application of the blanket term "weathered land surface" to a landscape which is seen to be cyclically and morphologically complex. Apart from the evidence of weathering, the shape of the complex weathered land surface in uplands indicates that the geomorphic régime which prevailed throughout its formation was that of a humid region; the only geomorphic evidence of a major change in morphogenesis is that which follows, brought on by the change to a drier climate.

Since land-surface stability and climate are both factors in the formation of a weathered profile, it is difficult to envisage how climatic oscillations could be established solely on the evidence of profile discontinuities. For instance, the disconformity between duricrust and deeper weathering zones in the high terrace of the western MacDonnells and the incorporation of previously weathered rock fragments in duricrusts point to instability of the land surface, but not necessarily to climatic fluctuation.

6. The flatness of the duricrusted plains must not be over-emphasized: slopes of 5 % are not unknown on the silcreted plain, whilst local relief of up to 50 ft [15 m] has commonly controlled the patterns of weathering on the piedmont northern plains. For this reason, interpretations based solely on the levels of duricrust cappings can be misleading (cf. STEPHENS, 1961) and use must be made of the stratigraphic evidence of the profile. Fig. 1 indicates that the duricrusted plains, although considered as a cyclic unit on the basis of continuity, range between 600 and 1700 ft [180 and 520 m] in the south and between 800 and 2300 ft [245 and 700 m] in the north of central Australia – adequate commentary on the value of altitudinal correlation on a continental scale.

7. The degree of planation of the duricrusted plains increased northwards and south-eastwards away from the central ranges. Only in the northernmost lateritized plains did it approach that of the old plateau of Western Australia, or that of the Bradshaw Surface of the Daly River Basin of northern Australia (WRIGHT, 1963). Hence, the formative physiographic setting of the duricrusted plains was that of today – continentally "central" with peripheral base levels. Relative to the older upland surface of the central ranges, the duricrusted plains lie 500 ft [150 m] lower in the south than in the north, a difference explained by the structural contrast between the stable Arunta Block in the north and the subsiding sedimentary basins in the south.

8. In northern Australia, a lateritized plain (the Maranboy Surface) is staged below the post-Lower Cretaceous, lateritized Bradshaw Surface (WRIGHT,

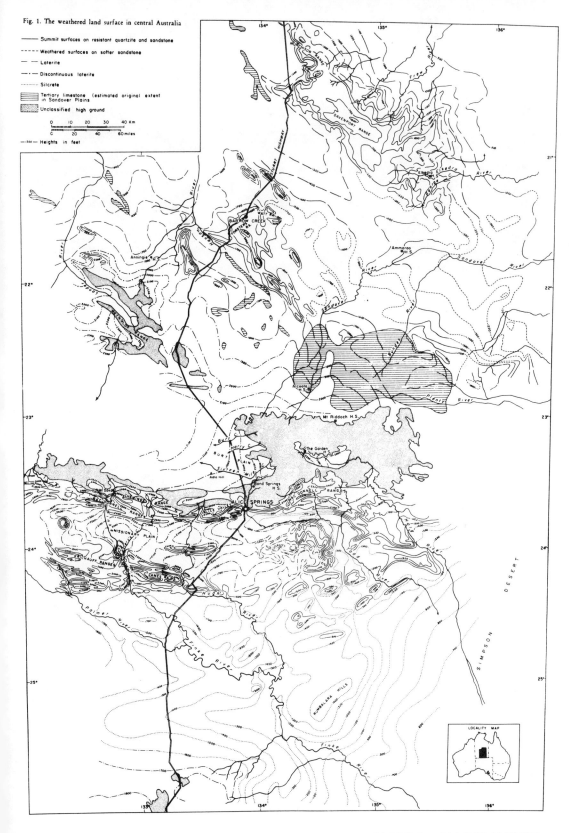

Fig. 1. The weathered land surface in central Australia

Summit surfaces on resistant quartzite and sandstone
Weathered surfaces on softer sandstone
Laterite
Discontinuous laterite
Silcrete
Tertiary limestone (estimated original extent in Sandover Plains)
Unclassified high ground

0 10 20 30 40 Km
0   20   40   60 miles

—1500— Heights in feet

Photo 1. Level sky-line of the silcreted plain in deeply-weathered Mesozoic claystone and sandstone, with prominent structural benches of underlying sandstone; Rumbalara Hills

Photo 2. Silcrete duricrust, showing crude laminar structure with botryoidal growth surfaces. Note the sharp contact with underlying soft rock of the pallid zone

1963); in central Australia all surfaces below the duricrusted plains (equated with the Bradshaw Surface) are unweathered or superficially weathered. On geomorphic evidence, therefore, lateritic weathering seems to have ceased at an earlier stage in central Australia than further north, foreshadowing the present continental climatic pattern of an arid "centre" and a wetter "north".

9. The controls of weathering in central Australia appear to have been climatic, topographic, and lithologic, the last two often acting in combination. One naturally looks to climate for the broadest control. In general, lateritic duricrusts are poorly developed here in comparison with those of northern Australia, although evidence has been brought that the laterites of the northern plains show increasing development northwards. It is suggested that, even at the time of weathering, the climate of central Australia may have been marginally dry for lateritization, and that rainfall increased northwards as today. Under such a marginal climate, relief and lithology would have maximum effect.

Further argument hinges on the relationship of the siliceous and lateritic crusts, a point on which landform evidence is not conclusive. The profile evidence, with minor lateritic elements in the siliceous profiles of the south-east and localized silicification in lateritic profiles of the northern plains, is consistent with silicification following lateritization. In this case, the contrast between the landscapes of central Australia and those described by JESSUP (1960) would be explained by the greater tectonic stability of the former area. Separation of the phases of lateritization and silicification allowing possible effects of interim change to a drier climate, might explain the association of siliceous layers in profiles regarded as evidence of desilicification in depth. Under this hypothesis also, the level of occurrence of silcrete (near-surface in the south-east and at moderate depth in the thinner lateritic profiles in the north) might be expressive of regional rainfall gradients parallel with those of today, in the manner suggested by STEPHENS (1961). Other factors favouring massive silicification near the surface in the south-east may have been more siliceous and impermeable bedrock, lesser relief, and lower gradients.

Under the hypothesis that lateritization and silicification *were* contemporary, such lithologic and topographic factors must be considered the overriding ones controlling the distribution of laterite and silcrete respectively.

Whatever the time-relationship of laterite and silcrete, however, lithologic control of weathering is clearly demonstrated in the sharp transition from silcrete with vestigial laterite on Mesozoic claystone to siliceous laterite on granite along the south margin of the silcreted plain. This may be read as further evidence for a drier climate in central Australia at the time of weathering than in northern Australia, where such lithologic control did not prevail and where siliceous Mesozoic rocks bear a thick laterite crust (WRIGHT, 1963).

Detailed topographic control of weathering is particularly apparent on crystalline rock, and is seen in the discontinuous lateritization of the piedmont northern plains, where lowlands were zones of thicker lateritic crusting, with thicker or localized pallid zones, and where the weathering front exaggerated the surface contours. Sloping and interfluvial surfaces, presumably with greater water-table fluctuation, have thicker mottled zones, even to the exclusion of a true pallid zone.

10. The control exercised by prior weathering over younger relief in central Australia has been very marked, for as noted earlier by BÜDEL (1957), change to a drier climate, bringing with it increased sensitivity to lithologic differences, provides optimum conditions for the exposure and exploitation of a prior weathering front. The importance of this control has, however, varied within the area, both with lithology and with tectonic setting. It is at its maximum on crystalline rocks, where the contrast in resistance offered by fresh and weathered rock is greatest; here the weathering front has been etched bare over large areas. On soft sedimentary rocks, as in the south-east, where neither weathered nor unweathered rock offers great resistance, the weathering front may exercise little control, but the contrast between duricrust and soft under-rock is here greatest and is well expressed in the mesa landforms. On the harder rocks of the uplands the weathered profiles undergo shallow stripping only, but differential weathering of softer beds may control the passage of later erosion to an important degree.

Control of landscape development by pre-weathering is at its strongest where rejuvenation has not exceeded the depth of weathering, as is generally the case in the northern plains and in the weathered valleys of the Davenport-Murchison Ranges. East of the Sandover plains, where downcutting by the headwaters of the Plenty River has exceeded this depth, the weathering front is not important in the younger landscape.

A combination of crystalline rock and a relatively stable tectonic setting has worked strongly for inheritance of relief in the northern plains, particularly where prior weathering was itself relief-controlled, as in the Burt Plain. Differential shallow etching of the weathered land surface is typical of such areas, and there has been little scope for landscape change in successive cycles of weathering and erosion.

11. Contrasts between current landforms and the older morphology are perhaps clearest in the weathered valleys and along the margins of uplands, where older slope facets on relatively resistant rocks survive. Two major changes are apparent. The first is that from a coarser-textured landscape with open, catenary valleys towards a more closely dissected younger landscape in which lithologic contrasts are fully expressed. In such areas, stream densities appear to have doubled with the onset of the arid régime. A second associated change is that from the concavo-convex slopes of the humid landscape to the more angularly contrasted system of pediments and rectilinear hill slopes of today. Such adjustments are typical of weathered valley slopes and of the transitional slopes between the uplands and duricrusted plains, where they commonly appear to have been facilitated by localized deeper weathering near the junction of hill and plain. In such areas, spectacular erosion of lower hill slopes may arise from profile adjustment alone, with little or no associated lowering of main base level. These processes – the proliferation of escarpments and their steepening to structurally-controlled angles, accompanied by regrading of pediments – have dominated wherever the weathered land surface had appreciable relief.

References

A<small>CKERMANN</small>, E. (1962): Büßersteine – Zeugen vorzeitlicher Grundwasserschwankungen. – Zeïtschr. f. Geom. NF 6: 148–182.

B<small>ÜDEL</small>, J. (1957): Die doppelten Einebnungsflächen in den feuchten Tropen. – Zeit. f. Geom. NF 1: 201–225.

C<small>LAYTON</small>, R. W. (1956): Linear depressions (Bergfüßniederungen) in savannah landscapes. – Geogr. Studies 3: 102–126.

C<small>RESPIN</small>, I. (unpublished data): Report of micropalaeontological examination of samples from the 16 mile Government Bore, west of Alice Springs, Northern Territory. – Aust. Bur. Min. Res. Unpub. Rec. No. 1950/48.

I<small>VANAC</small>, J. F. (1954): The geology and mineral deposits of the Tennant Creek Gold-Field, Northern Territory. – Aust. Bur. Min. Res. Bull. No. 22.

J<small>ACKSON</small>, E. A. (1962): Soil studies in central Australia. – CSIRO Aust. Soil Pub. No. 19.

J<small>ESSUP</small>, R. W. (1960): The lateritic soils of the south-eastern portion of the Australian arid zone. – Journ. Soil Sci., 11: 106–113.

J<small>UTSON</small>, J. T. (1934): The physiography (geomorphology) of Western Australia (2nd ed.). – Geol. Survey Western Aust. Bull. No. 95: 366 p.

M<small>ABBUTT</small>, J. A. (1961): A stripped land surface in Western Australia. – Inst. Brit. Geog. Trans. 29: 101–114.

– (1962): Geomorphology of the Alice Springs area. – CSIRO Aust. Land Res. Ser. No. 6: 163–184.

– (in press): Landforms of the western MacDonnell Ranges. – *In* Essays in Geomorphology, ed. G. H. D<small>URY</small>, London.

P<small>ATEN</small>, R. J. (1960): Lacustrine sandstones and limestones and spring sinters of far western Queensland. – *In* The Geology of Queensland. Journ. Geol. Soc. Aust. 7: 391–403.

P<small>RESCOTT</small>, J. A., & R. L. P<small>ENDLETON</small> (1952): Laterite and lateritic soils. – Commonwealth Bur. Soil Sci. Tech. Commn. No. 47.

P<small>RICHARD</small>, C. E., & T. Q<small>UINLAN</small> (1962): The geology of the southern half of the Hermanns-burg 1 : 250,000 Sheet. – Aust. Bur. Min. Res. Rept. No. 61.

Q<small>UINLAN</small>, T. (1962): An outline of the geology of the Alice Springs area. – CSIRO Aust. Land Res. Ser. No. 6: 129–146.

S<small>MITH</small>, K. G., J. R. S<small>TEWART</small> & J. W. S<small>MITH</small> (1961): The regional geology of the Davenport and Murchison Ranges, Northern Territory. – Aust. Bur. Min. Res. Rept. No. 58.

S<small>TEPHENS</small>, C. G. (1961): The soil landscapes of Australia. – CSIRO Aust. Soil Pub. No. 18.

S<small>TIRTON</small>, R. A., R. H. T<small>EDFORD</small> & A. H. M<small>ILLER</small> (1961): Cenozoic stratigraphy and vertebrate paleontology of the Tirari Desert, South Australia. – Rec. S. Aust. Mus. 14: 19–61.

S<small>ULLIVAN</small>, C. J., & A. A. Ö<small>PIK</small> (1951): Ochre deposits, Rumbalara, Northern Territory. – Aust. Bur. Min. Res. Bull. No. 8.

T<small>WIDALE</small>, C. R. (1962): Steepened margins of inselbergs from north-western Eyre Peninsula, South Australia. – Zeit. f. Geom. NF 6: 51–69.

W<small>HITEHOUSE</small>, F. W. (1940): The lateritic soils of western Queensland, *in* Studies in the late geological history of Queensland. – Univ. Queensland Papers, Dept. of Geol., 2 (n. s.), No. 1: 2–22.

W<small>OOLNOUGH</small>, W. G. (1927): The duricrust of Australia. – Proc. Roy. Soc. N. S. W. 61: 24–53.

W<small>OPFNER</small>, H. (1960): On some structural development in the central part of the Great Australian Artesian Basin. – Trans. Roy. Soc. S. Aust. 83: 179–193.

– (1961): The occurrence of a shallow groundwater horizon and its natural outlets in north-easternmost South Australia. – Trans. Roy. Soc. S. Aust. 85: 13–18.

W<small>RIGHT</small>, R. L. (1963): Deep weathering and erosion surfaces in the Daly River Basin, N. T. – Journ. Geol. Soc. Aust. 10: 151–164.

30

Reprinted from *J. Geol. Soc. Australia*, **17**, Pt. 1, 21–30 (1970)

THE AGE AND GEOMORPHIC CORRELATIONS OF DEEP-WEATHERING PROFILES, SILCRETE, AND BASALT IN THE ROMA-AMBY REGION, QUEENSLAND

By N. F. EXON, T. LANGFORD-SMITH & IAN McDOUGALL

(With 2 Text-Figures and 1 Table)

(*Received* 17 *January* 1969; *read at Canberra*, 27 *January* 1970)

ABSTRACT

A deep-weathering profile is developed in the Mesozoic sedimentary rocks of the Roma-Amby region, where it forms a capping to dissected tablelands. The profile consists of up to 25m of mottled and leached sediments, rarely in well-defined zones; crust in place with respect to the profile is seldom preserved. Tholeiitic basalts and olivine basalts with possible alkaline affinities were erupted along the axis of the Merivale Syncline in the Amby-Mt Hutton area; K-Ar dates indicate that they are about 23 m.y. old. The basalts were extruded after the period of deep-weathering, which occurred therefore in or before Early Miocene. The abundant silcrete in the area has formed at various times during the Cainozoic and is not necessarily related to the deep-weathering.

INTRODUCTION

The investigation reported in this paper arose out of a disagreement on the identity of an alleged duricrust pallid zone in a cutting on the Injune road 34 km north of Roma, Queensland. Langford-Smith, Dury & McDougall (1966, 1967) assigned a minimum age of 22.7 m.y. to the pallid zone by the K-Ar dating of a dolerite dyke which intruded it. Exon, Milligan & Day (1967) challenged the identification of the pallid zone as part of a duricrust profile. They considered the weathering of the profile to be post-duricrust, and drew attention to other sites in the area which, they claimed, provided unequivocal relationships between duricrust and basalt.

In an attempt to reconcile differences and examine new evidence, the present authors, representing both parties, arranged a joint investigation of the disputed site, followed by a survey of other relevant features in the adjacent area (Fig. 1). Unfortunately realignment of the road had obscured both the weathered profile and the dyke. However, examination of analogous stratigraphic units nearby showed that the profile at the site could be accounted for adequately in terms of lithology and normal weathering, the pallid zone being highly weathered mudstone and the overlying mottled zone a leached ferruginous sandstone. It was

agreed that the profile was a product of post-duricrust weathering; and that if a duricrust deep-weathering profile had once existed here, it had been stripped by landscape erosion of unknown but considerable vertical extent.

In this paper we present the results of our joint investigation, which covers the area shown in Figure 1. Also indicated in Figure 1 are the localities of the more significant sites examined. The survey established the main sequence of events in the geomorphic history of the area. It appears that only one period of deep-weathering, which produced a thick profile, can be identified. Although no site in itself was entirely free from ambiguity, evidence as a whole was overwhelming that the basalts of the area erupted subsequent to the development of the deep-weathering profiles; dating of the basalts has provided a minimum age for the time of deep-weathering.

DURICRUST, SILCRETE AND DEEP-WEATHERING: SOME GENERAL OBSERVATIONS

The field investigation brought to light important problems of definition in respect of the use of the terms duricrust and silcrete, and it is imperative to clarify these issues if errors in interpretation are to be avoided.

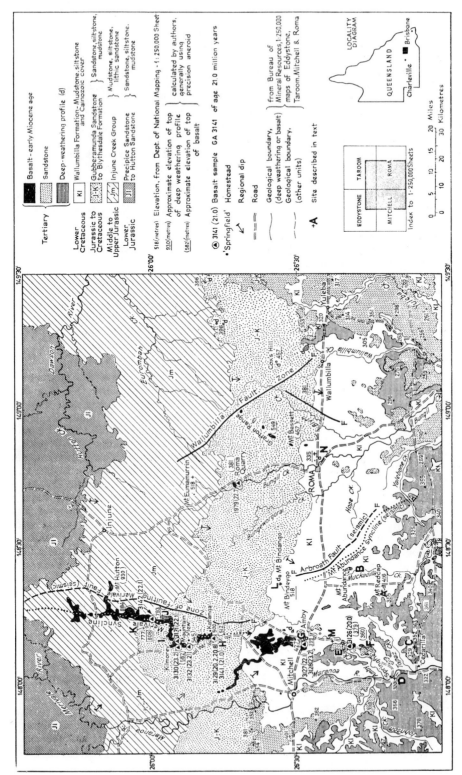

Fig. 1. Locality map and geological setting.

435

'Silcrete' was first used by Lamplugh (1902) rather loosely for 'masses . . . indurated by a siliceous cement'. He later (1907) described, as silcrete, extensive bands of 'hard sandstone or quartzite' in the Zambesi Basin 'knit together by a chalcedonic cement'. It is clear that Lamplugh intended silcrete as a very general term for any material indurated by silicification, without regard for size or shape of the constituent particles or for the presence or absence of associated weathering profiles. On the other hand Woolnough (1927), who coined the term duricrust, used it in reference to an indurated cap overlying a weathered profile. He emphasised the association of weathering with duricrust, concluding that 'in every instance there is a relatively hard crust, resting upon a decomposed or weakened substratum of rock material'. Woolnough further insisted that all duricrust, whatever its composition, was of the same age, being formed 'during an era of highly perfect peneplanation', that is he defined duricrust genetically as well as morphologically.

Since Woolnough, misunderstanding has arisen from time to time from failure to adhere to his rigid specifications, or from attempts to amend or add to them. Connah & Hubble (1961) have criticised a tendency to use the term laterite for 'the complete laterite profile, or any part of it, other than the ironstone material', and they are supported by Langford-Smith & Dury (1965) who point to confusion resulting from the application of laterite to duricrust in general. If it is evident that *laterite* and *duricrust* should not be regarded as interchangeable terms, it is just as clear from our own work at Roma that neither should *silcrete* and *duricrust* be used interchangeably. Similar conclusions have been reached from evidence in southwest Queensland, where multiple layers of silcrete without underlying profiles have been found (Senior, Galloway, Ingram & Senior, 1968).

The silcrete of the Roma-Amby area is closely similar to that of southwest Queensland which is described by Senior *et al.* (1968) as 'a predominantly pale grey, extremely indurated, highly siliceous rock with numerous angular quartz clasts distributed in an amorphous or cryptocrystalline matrix. Considerable variation exists in the ratio of quartz clasts to siliceous matrix, and in the size of the quartz fragments.'

In the Roma-Amby area, the formation of silcrete was not confined to any one period;

silcrete cobbles are present as inclusions in both silicified and unsilicified sandstones. A complication in field interpretation is the presence of silcrete boulder concentrations on the crests of hills and ridges of widely different altitudes, and it is apparent that much silcrete is no longer in its original place, but has been lowered by landscape stripping. In fact, only two or three localities could be found where silcrete appeared to be in place and overlying deep-weathering profiles.

In contrast to the various phases of silcreting in the area, there is evidence of only one period of deep-weathering, represented by distinctive profiles in a formerly extensive erosion surface, at some sites comprising both mottled and pallid zones. The crust of the profile is seldom preserved and hence the term 'deep-weathering profile' is preferred to 'duricrust profile' for this area.

Typically the profile consists of 12 to 25 m of weathered Cretaceous sediment (lithic sandstone, siltstone, and mudstone). Although some red mottling is usually present, most of the profile is leached white with common pink, yellow and violet beds. In this area a mottled zone is seldom distinctive, although both mottled and pallid zones are commonly recognized farther south (Thomas & Reiser, 1968). Usually the upper part of the profile, at least, is tougher than the fresh material, although the location and degree of induration in the profile may vary from site to site.

The grainsize, distribution of rock types, and bedding characteristics are unchanged by the deep-weathering. Within the profile calcite has been completely leached, while feldspar has broken down to kaolin, some of which forms an interstitial matrix. Kaolin is also derived from labile rock fragments, and from alteration of the original clays of the matrix. Iron oxide has come from breakdown of grains of iron ore and ferromagnesian minerals. Quartz grains and resistant lithic grains have been unaffected, while white mica has broken down only high in the profile.

GEOLOGICAL SETTING

The general geological setting of the area, which forms part of the northern margin of the Surat Basin, is shown in Figure 1. The geology of various parts of the area has been discussed in detail by Exon, Milligan, Casey & Galloway (1967); Day (1964); Mollan,

Dickins, Exon & Kirkegaard [in press]; and Mollan, Forbes, Jensen, Exon & Gregory [in preparation]. The exposed strata are mainly of Jurassic and Cretaceous sedimentary rocks which dip gently to the south ($\frac{1}{4}°$ average). These rocks are little disturbed in the shelf areas in the east and west which are separated by the gentle fault-controlled downwarps of the Merivale and Mt Abundance Synclines. The north-trending Merivale-Arbroath fault zone consists of *en échelon* faults with total displacement, down to the west, of about 1,370 m in pre-Permian rocks (Exon, Milligan, Casey & Galloway, 1967). The trough so developed in the west was completely filled with Permian and Triassic sediments; but compaction, further faulting, and warping has resulted in a broad syncline, with flank dips of less than 2°, in the overlying Jurassic and Cretaceous sediments.

A major line of weakness extends from the Arbroath and Merivale Faults northward for about 370 km as far as Emerald, where it is represented by the Springsure-Serocold Anticline. Extrusion of Cainozoic basalts in the region was mainly along this line; Oligocene basalt occurs in the Springsure region as a large sheet extending over 160 km south of Emerald (Webb & McDougall, 1967). Extrusion of these basalts presumably was from fissures in the anticline represented today by basalt dykes (Mollan *et al.* [in press]). The basalts of the Amby area are younger than those to the north and are separated from them by a distance of 53 km south of Carnarvon Gorge.

GEOMORPHOLOGY

The geomorphology of the area has been strongly influenced by stratigraphy and structure. After deposition of the Jurassic and Lower Cretaceous sediments there was differential compaction, warping and faulting along dominantly meridional axes, and the whole area was tilted to give a slight declination to the south. There was then a long period of bevelling and truncation of beds, probably by pediplanation, during which a remarkably planar surface was developed. This surface, which we have termed the upper pediplain, was deeply weathered and in places developed a duricrust which varied from ferruginous to siliceous. As already noted, the character of the original crust in the Roma-Amby area (if in fact there was a distinctive crust here) is obscure.

Morphological evidence suggests that pedi-planation and deep-weathering occurred over a very long interval of time, during which stable climatic conditions favourable to deep weathering prevailed. These conditions were probably terminated by epeirogenic uplift accompanied by a climate favouring appreciable run-off, for the upper pediplain was eroded and the deep-weathering profile largely removed, valleys were incised and widened, and sandstones and conglomerates deposited on the new valley floors.

The pattern of the drainage carved in the deeply weathered surface was largely determined by the lithology and structure of the bedrock, and by the southern regional tilt. Thus stream valleys conformed to synclines and eroded anticlines, and the streams themselves flowed south. Remnants of the upper pediplain remain as isolated hills and ridges to the north; more extensive mesas, cuestas, and plateaus lie to the south, where the surface is lower.

As will be discussed later, volcanic activity took place about 23 m.y. ago, and over a short period of time. Basalts were extruded into the valleys along the Merivale Syncline forming a sequence of flows in places more than 80m thick. It is probable that basalt over-topped some valley interfluves and thinly capped the deep-weathering profile, but has since been stripped.

Erosion continued after the period of volcanism. Streams were laterally displaced by the lava flows, and new channels incised between the flows and the old valley walls. Some interfluves, consisting of sediments relatively more susceptible to erosion than the basalt, were completely removed, leaving the flows as residual ridges and giving rise to topographic inversion. The Merivale/Maranoa River probably moved from an old course down the Merivale Syncline as the topography became reversed. The Dawson and Maranoa River systems stripped the deep-weathering profile from most of the north of the area.

The latest significant geomorphic event was the backwearing of valley slopes to produce a newer erosion surface which we have designated the 'lower pediplain'. This lower surface is usually quite readily distinguished in the field from the upper pediplain, even when extremely narrow. It is perhaps most extensive where it comprises the broad 'rolling downs' country typical of much of the good grazing land of the Roma area.

SIGNIFICANT SITES

The sites considered significant are described in Appendix I. They illustrate typical deep-weathering profiles, the relationship between the profiles and the younger basalts, and the relationship of the silcretes to the profiles and the basalts. A general discussion of the sites, in their natural groupings, follows.

The dissected tablelands area in the south: typical profiles

The upper pediplain is best preserved in the south, where it forms a dissected tableland with a gently undulating surface; it persists well south of this area. It is deeply weathered with a maximum preserved profile thickness of about 25m. At Mt Abundance (Site B) a siliceous and ferruginous crust is preserved, but nowhere else has a crust been seen which is indisputably in place. In most places the mottled and pallid zones cannot be separated. Silcrete boulders and ferruginous gravel are common on the surface but these do not necessarily represent a duricrust. To the north and northeast the upper pediplain is replaced by the

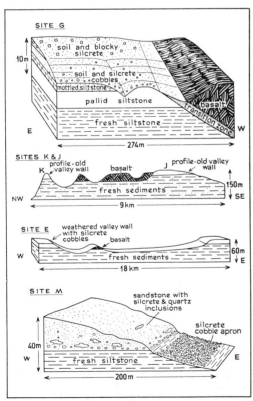

Fig. 2. Silcrete, deep weathering, and basalt relationships.

lower pediplain, the upper pediplain being represented merely by isolated residuals which increase in elevation northwards.

The Merivale Syncline—Amby area: basalt and the profile

In this area there is a clear connection between the deeply weathered upper pediplain, which falls steadily southwards, the stream valleys that incised the pediplain following the post Lower Cretaceous regional tilt, and the basalt that flowed into those valleys. The sites discussed below are located in Figure 1: four of them are illustrated in Figure 2.

Site G, north of the Amby quarry (Fig. 2), is probably the most significant. Here an unweathered basalt flow sits in a valley cut in a weathering profile. Its top is at the same level as the top of the profile, and the two are separated laterally by only a few metres of transitional material. The obvious interpretation of this site is that the basalt is younger than the deep-weathering.

At Site J in the north (Fig. 2), a fresh basalt flow lies below a deeply weathered rise of Cretaceous sediment, which is interpreted as the old valley wall. The basalt was deposited in a valley cut in the upper pediplain. The other side of the valley is probably represented by the deeply-weathered pinnacle at Site K (Fig. 2). An apparently similar situation is represented by Site E in the south (Fig. 2). At these sites also the obvious interpretation is that the basalt is younger than the deep-weathering.

Silcretes

The relationship of silcretes to the deep-weathering profile undoubtedly varies. In some places it may represent the original duricrust capping formed during deep-weathering, while in others it certainly does not. Sites B, D, G, K, L and N fall into the first category, and Site M into the latter. In many cases where silcrete overlies the profile it is possible that it merely represents later silicification of younger soil or sand cover. When the silcrete sits as a solid mass on the profile, a connection is more likely than when the profile is overlain by silcrete rubble which may have been lowered hundreds of metres. In general, however, there is little evidence about the nature of the duricrust, or indeed whether there ever was a true duricrust capping all the deep-weathering profiles of the area.

Site M (Fig. 2) shows that there was more than one period of post-Cretaceous silcreting.

Silcrete pebbles and cobbles are included in a partially silicified Tertiary* sandstone and the original silcrete crust of the sandstone is now represented merely by an apron of cobbles: the silcrete pebbles are older than the crust developed on the sandstone. Site H shows that there was pre-basalt silcreting, while silcrete is still forming in stream valleys at the present day.

BASALTS OF THE AMBY AREA

Basalts crop out discontinuously along a line striking essentially due north, just to the west of Amby (Fig. 1). They extend for 90 km north of Amby, but south of Amby there are only isolated outcrops. The outcrop of the

sedimentary rocks. Hence by dating the basalts a minimum age for the deep-weathering in this region can be obtained. In addition it is of interest to date the basalts in connection with studies on the age distribution of the Cainozoic volcanic activity in eastern Australia (McDougall & Wilkinson, 1967; Webb & McDougall, 1967; Webb, Stevens & McDougall, 1967).

The basalts occur as massive to somewhat vesicular flows; individual lavas range from about 6m to 30m in thickness. All the lavas examined contain some olivine (up to 10%), which is fresh. Plagioclase, clinopyroxene and glass are the other major constituents, with iron oxide and a yellow or green mineraloid as

TABLE

Potassium-argon dates on whole-rock basalt samples from the Amby area, near Mitchell, Queensland

Sample No. GA—	Rock Type	%K	Radiogenic Ar (10^{-11} moles/g)	Radiogenic ^{40}Ar (%)	Calculated Age (m.y.)	Locality
3126	Tholeiitic basalt	0·459, 0·459	1·642 1·653	54·3 56·6	20·0±0·3 20·1±0·4 }	16 km south of Amby
3127	Tholeiitic basalt	0·420, 0·420	1·683	74·7	22·4±0·4 }	Quarry 4 km NW of
3128	Tholeiitic basalt	0·391, 0·392	1·626	62·2	23·2±0·4 }	Amby
3129	Olivine basalt	1·098, 1·101	3·989 4·080	75·3 72·1	20·3±0·3 20·8±0·4 }	Spur, near Katanga Homestead turnoff,
3141	Olivine basalt	1·008, 1·010	3·779	73·5	21·0±0·4 }	36 km north of Amby
3132	Olivine basalt	0·944, 0·945	3·921	86·9	23·2±0·4	Spur. Mt View Homestead road 45 km north of Amby
3130	Olivine basalt	0·951, 0·952	3·923	92·3	23·1±0·4	1·6 km SW of Kilmorey Homestead 52 km N of Amby
3131	Olivine basalt	1·001, 1·001	3·953	81·9	22·1±0·4	Falls near Kilmorey Homestead
3142	Tholeiitic basalt	0·577, 0·579	2·372	88·2	22·9±0·4	Falls 12 km NE of Kilmorey Homestead

$\lambda_e = 0.584 \times 10^{-10}\text{yr}^{-1}$ $\lambda_\beta = 4.72 \times 10^{-10}\text{yr}^{-1}$ $^{40}\text{K} = 1.19 \times 10^{-2}$ atom %

basalts is rarely more than 2.4 km in width, except locally north of Amby, where the width reaches 8 km. The basalts may exceed 80m in thickness in the north but in general total 30m or less; in some sections up to three flows have been identified. The base of the basalt rises from about 370m altitude 16 km south of Amby, to over 600m north of Kilmorey Homestead, 64 km north of Amby. Structurally this valley follows quite closely the axis of the gently southward plunging Merivale Syncline in the underlying Mesozoic sedimentary rocks.

The valley into which the basalts were extruded was eroded after deep-weathering of the extensive surface developed on the Mesozoic

minor additional phases. The mineraloid generally occurs as vesicle fillings. The grainsize of the main minerals lies in the range 0.1 to 0.5 mm, and rarely up to 1 mm. The proportion of glass is always relatively large, but variable (10 to 30%). In most rocks the glass is unaltered and essentially isotropic, but in a few cases it is weakly anisotropic suggesting that devitrification may have commenced.

From petrographic examination two basalt types are recognised. Samples from three of the flows (GA 3126, GA 3127, GA 3128, GA 3142) contain the clinopyroxene pigeonite, in addition to the augite, and are clearly tholeiitic olivine basalts. The low potassium contents

---

* The generally quartzose, well consolidated, post-Rolling Downs Group sediments of Queensland are conventionally regarded as of Tertiary age, although some may be Late Cretaceous in age.

(0.4 to 0.6%) of these basalts (see Table) are in keeping with this observation. The other basalts examined differ in that no pigeonite was identified, and the potassium contents are considerably higher (0.95 to 1.1%). Whether this second group comprises alkali olivine basalts or transitional types is not certain in the absence of major element chemical analyses. However, there is no doubt that tholeiitic olivine basalts and olivine basalts with alkaline affinities were erupted in this area.

Of the samples collected a final choice for dating was made on the basis of stratigraphic importance and after examination of thin sections. Only fresh, essentially unaltered rocks were regarded as acceptable. The high proportion of glass in all the samples was of some concern as potassium is likely to be concentrated in this material, and it is known that volcanic glass may lose radiogenic argon, particularly if alteration or devitrification has occurred. Brief petrographic descriptions of the dated samples are given in Appendix II together with details of locality. The results are given in the Table.

The techniques employed in the potassium-argon dating have been described in detail previously (McDougall, 1964; 1966; Cooper, 1963). Whole rock samples were used throughout; 5 to 10g of rock was employed for each argon extraction. Argon was determined by isotope dilution and potassium by flame photometry. The uncertainty in the calculated ages in the Table is given at the 68% level of confidence, and is based upon the errors observed in the physical measurements, including the error in tracer calibrations.

Discussion of results

The measured ages range from 20.0 to 23.2 m.y. (Table). The spread is about 15%, much greater than that due to experimental error, and may indicate that eruption occurred over a period of the order of 3 m.y. Alternatively, the outpouring of the basalts may have been restricted to a short interval of time with the spread in indicated age the result of variable loss of radiogenic argon from samples.

The average indicated age is 22.1 ± 0.5 m.y. (standard deviation of the mean), giving unit weight to each locality dated. If the results from the two localities that yield the younger dates of 20.0 and 20.7 m.y. (samples 3126, 3128, 3129) are excluded, then the mean age for the remaining five localities is 22.8 ± 0.2

m.y. This is less than the uncertainty expected from experimental error alone. Hence the data from these five sites may be regarded as concordant, providing confidence that the mean age closely approaches the true age of eruption. Both tholeiitic basalts and basalts with alkaline affinities occur in the group from which concordant results were obtained. Of the two flows that yield younger dates one is tholeiitic and the other belongs to the alkali group. The glass in these samples is slightly anisotropic, indicating that some devitrification has taken place, and suggesting that partial loss of radiogenic argon from the glass has occurred, resulting in apparent younger ages.

On this basis the preferred interpretation is that eruption of the basalts took place over a short period of time at 23 ± 1 m.y. ago. As the age of the Oligocene-Miocene boundary is at about 26 m.y. (Funnell, 1964) the basalts are regarded as Early Miocene.

The minimum date of 22.7 m.y. obtained previously on a basalt cutting the Westbourne Formation to the east of the Amby area (Langford-Smith et al., 1966) is identical to the ages reported in the present paper. There seems little doubt that this basalt belongs to the same period of volcanic activity.

The dates obtained on basalts from the Amby area of about 23 m.y. are significantly younger than the 26 to 27 m.y. ages found for the basalts and related rocks of the Springsure region 250 km to the north (Webb & McDougall, 1967). The relationship of the Springsure basalts to the deep weathering needs investigation. The basalts extend some 130 km south of Springsure but none of these have yet been dated; there is a gap of 53 km between these basalts and the basalts of the Amby area. Dating on rocks from other volcanic provinces in southern Queensland and northern New South Wales shows that during the period from Late Oligocene to Middle Miocene volcanic activity was very widespread (McDougall & Wilkinson, 1967; Webb & McDougall, 1967; Webb et al., 1967).

DISCUSSION

The evidence from the Roma-Amby area of southern Queensland suggests that a single period of deep-weathering has affected the Mesozoic sedimentary rocks. The basalts of this area, dated at about 23 m.y., were erupted subsequently to the deep-weathering, which therefore must have occurred in Early Miocene times or before. As there was a considerable period of erosion after the deep-weathering and prior to the extrusion of the basalts, it is suggested that the deep-weathering took place during the Oligocene or possibly even earlier. One point we wish to stress is that in this area development of silcrete has continued from pre-Miocene times until the present, and is not necessarily related to the deep-weathering profile.

Similar deep-weathering profiles are found over wide areas of Queensland and New South Wales, and have already been dated as older than mid-Miocene (Dury, Langford-Smith, & McDougall, 1969). As many workers have suggested that deep-weathering pertained to the same general period of geological time, our conclusions may well apply to a much larger region.

In the Roma-Amby area the youngest rocks that are deep-weathered are the early Albian sediments of the Coreena Member of the Wallumbilla Formation. As there was a period of bevelling before deep-weathering, we suggest a maximum age of Late Cretaceous for the deep-weathering. In the Eromanga Basin it has been reported that the youngest rocks to be deep-weathered are the sediments of the Winton Formation, which probably range into the Late Cretaceous (Senior et al., 1968). As it is highly likely that deep weathering occurred in both basins at the same time, the age of the deep-weathering probably lies between Late Cretaceous and late Oligocene.

The basalts of the Roma-Amby area belong to the same period of volcanism as those dated in the Main Range near Toowoomba (Webb et al., 1967), but are younger than the Oligocene basalts of the Springsure area to the north (Webb & McDougall, 1967), from the southern continuation of which they are separated by a gap of only about 50 kilometres. Both tholeiitic basalts and olivine basalts with alkaline affinities were extruded in the Amby area at the same time.

ACKNOWLEDGMENTS

Technical assistance in the K-Ar dating was ably provided by I. H. Ingram and Miss S. Dinter. The field work of one of us (T. L-S) was assisted by a grant from the Australian Research Grants Committee, while N.F.E. acknowledges permission to publish from the Director, Bureau of Mineral Resources. The figures were drawn by A. Bartlett.

REFERENCES

CONNAH, T. H., & HUBBLE, G. D., 1960: Laterites, in The Geology of Queensland, chapt. XII. J. geol. Soc. Aust., 7, pp. 373-386.

COOPER, J. A., 1963: The flame photometric determination of potassium in geological materials used for potassium-argon dating. Geochim. cosmochim Acta, 27, pp. 525-546.

DAY, R. W., 1964: Stratigraphy of the Roma-Wallumbilla area. Publs geol. Surv. Qd, 318, pp. 1-23.

DURY, G. H., LANGFORD-SMITH, T., & McDOUGALL, I., 1969: A minimum age for the duricrust. Aust. J. Sci., 31, pp. 362-363.

EXON, N. F., MILLIGAN, E. N., CASEY, D. J., & GALLOWAY, M. C., 1967: The geology of the Roma and Mitchell 1:250,000 Sheet areas, Queensland. Rec. Bur. Miner. Resour. Geol. Geophys. Aust., 1967/63 [unpublished].

EXON, N. F., MILLIGAN, E. N., & DAY, R. W., 1967: Age of the duricrust in southern Queensland. Aust. J. Sci., 30, p. 110.

FUNNELL, B. M., 1964: The Tertiary Period. In The Phanerozoic Time-scale. Q. Jl geol. Soc. Lond., 120S, pp. 179-191.

LAMPLUGH, G. W., 1902: 'Calcrete', Geol. Mag., 9, p. 575.

———, 1907: The geology of the Zambesi Basin around the Batoka Gorge (Rhodesia), Q. Jl geol. Soc. Lond., 63, pp. 162-216.

LANGFORD-SMITH, T. & DURY, G. H., 1965: Distribution, character, and attitude of the duricrust in the northwest of New South Wales and the adjacent areas of Queensland. Am. J. Sci., 263, pp. 170-190.

———, DURY, G. H., & McDOUGALL, I., 1966: Dating the duricrust in southern Queensland. Aust. J. Sci., 29, pp. 79-80.

———, DURY, G. H., & McDOUGALL, I., 1967: Reply to Exon, N. F. et al., Aust. J. Sci., 30, p. 111.

McDOUGALL, I., 1964: Potassium-argon ages from lavas of the Hawiian Islands. Bull. geol. Soc. Am., 75, pp. 107-128.

———, 1966: Precision methods of potassium-argon isotopic age determination on young rocks, in Methods and Techniques in Geophysics, 2, pp. 279-304. Interscience, London.

———, & WILKINSON, J. F. G., 1967: Potassium-argon dates on some Cainozoic volcanic rocks from northeastern New South Wales. J. geol. Soc. Aust., 14, pp. 225-234.

MOLLAN, R. G., DICKINS, J. M., EXON, N. F., & KIRKEGAARD, A. G., in press: The geology of the Springsure 1:250,000 Sheet area, Queensland, Rep. Bur. Miner. Resour. Geol. Geophys. Aust., 123.

———, FORBES, V. R., JENSEN, A. R., EXON, N. F., & GREGORY, C. M.: The geology of the

Eddystone and Taroom 1:250,000 sheet areas and the western part of the Mundubbera 1:250,000 sheet area. *Rep. Bur. Miner. Resour. Geol. Gophys. Aust.* [in preparation].

SENIOR, B. R., GALLOWAY, M. C., INGRAM, J. A., & SENIOR, DANIELLE A., 1968: The geology of the Barrolka, Eromanga, Durham Downs, Thargomindah, Tickalara and Bulloo 1:250,000 Sheet areas, southwest Queensland. *Rec. Bur. Miner. Resour. Geol. Geophys. Aust.* 1968/35 [unpublished].

THOMAS, B. M., and REISER, R. F., 1968: The geology of the Surat 1:250,000 Sheet area, Queensland. *Rec. Bur. Miner. Resour. Geol. Geophys. Aust.*, 1968/56 [unpublished].

WEBB, A. W., & McDOUGALL, I., 1967: A comparison of mineral and whole rock potassium-argon ages of Tertiary volcanics from Central Queensland, Australia. *Earth Planet. Sci. Lett. 3*, pp. 41-47.

———, STEVENS, N. C., & McDOUGALL, I., 1967: Isotopic age determination on Tertiary volcanic rocks and intrusives of south-east Queensland, *Proc. R.Soc. Qd, 79*, pp. 79-92.

WOOLNOUGH, W. G., 1927: The duricrust of Australia. *J. Proc. R.Soc. N.S.W., 61*, pp. 24-53.

N. F. Exon,
Bureau of Mineral Resources,
P.O. Box 378,
Canberra City, A.C.T. 2601.

T. Langford-Smith,
Dept of Geography,
University of Sydney,
Sydney, N.S.W. 2006.

Ian McDougall,
Dept of Geophysics & Geochemistry,
Australian National University,
Canberra, A.C.T. 2600.

APPENDIX I

DESCRIPTION OF SITES

Dissected tablelands area south of Amby

Site A: Mt Redcap, elevation 416m. Typical exposure of mottled and leached siltstone, mudstone and very fine sandstone in 25m capping. No crust but profile tough. Cannot differentiate mottled and leached zones.

Site B: Mt Abundance, elevation 470m. Capping is regolith up to 1m thick of red soil mixed with silcrete and ferricrete pebbles. Underlain by ferruginous and siliceous fine sandstone and silcrete in place. In quarry on northeast side up to 2m of bedded silcrete overlies 6m of poorly defined mottled pallid zone, which overlies pallid zone. Capping probably represents original crust on deep-weathering profile.

Site C: Edge of plateau near Cedarilla Homestead, elevation 352m. 5m of leached and mottled sandstone and siltstone containing small ferruginous concretions. Silcrete cobbles and large ferruginous concretions rest on plateau surface. Neither show definite relationship to profile.

Site D: Opposite Springfield Homestead. Boulders of silcrete at 320m represent crust, which may have been slightly lowered. 6m below in gully, is 6m of white and red heavily weathered Cretaceous sediment. Maranoa River valley floor (lower pediplain) is only 15m below silcrete (upper pediplain).

Merivale Syncline—Amby area

Site E: (Fig. 2). 16 km south of Amby. Basalt lava flow (GA 3126) forms ridge within valley, with its top at 379m. Silcrete pebbles on flow, but no deep-weathering profiles nearby. Probable that lava was erupted into stream valley cut through deeply weathered surface.

Site F: 3km south of Site E. Bole exposed beneath same basalt.

Site G: (Fig. 2). 2.4km north of Amby basalt quarry. Large silcrete boulders rest on 10m scarp of mottled and pallid Cretaceous siltstone, which rests on fresh Cretaceous siltstone. Silcrete may represent crust of profile, possibly lowered. Immediately west, and at level of profile is basalt flow more than 10m thick (almost certainly same flow as exposed in quarry) (GA 3127, 3128) resting on fresh siltstone. Between basalt and profile is transition zone several metres wide. As basalt is fresh, obvious interpretation is that it is post-deep-weathering and was poured into valley cut in upper pediplain. Basalt boulders on deeply-weathered surface suggest over-topping of valley.

Site H: Hill 3km southwest of Katanga Homestead, elevation 512m. 12m of basalt (GA 3129, GA 3141) caps hill. About 15m below is 10m of poorly exposed Tertiary sandstone—thickly bedded, quartzose, poorly sorted. Contains rounded pebbles of quartz, quartzite and silcrete, and clasts of white siltstone. Cross-bedding suggests derivation from west. Silcrete cobbles in sandstone, and boulders to 1.2m out of place above sandstone. Overlies normal Cretaceous sandstone. Tertiary sandstone was deposited in valley in Cretaceous sediments, and later overlain by basalt. Silcrete pebbles and boulders suggest nearby high silcrete source, possibly duricrust.

Site I: Waterfall 1.5km southeast of Kilmorey Homestead. Basalt flow (GA 3131) in present day stream valley with vertical gorge. Below basalt is Mesozoic sandstone with surface sloping steeply down to southwest; basalt has flowed down this slope. Thus pre-basalt topography preserved at this point.

Site J: (Fig. 2). Cliff 13km northeast of Kilmorey Homestead. Consists of leached white Mesozoic siltstone and sandstone, with some red mottling. Overlain by slope of weathered material. Total weathered thickness 25m or more. Weathered sediment rises above basalt and represents old deeply-weathered valley wall. Top is 792m.

Site K: (Fig. 2). Pinnacle west of basalt, north of Kilmorey Homestead, elevation 777m. Capped with 3m of coarse silicified Cretaceous sandstone. Below is 12m scree slope of somewhat altered and silicified fine sandstone, overlying fresh sandstone. Probably an instance of deep-weathering giving silcrete capping, rather than more recent silcrete formation. Porous quartzose sandstone would have been ideal for silcrete development.

Other Sites

Site L: Mt Bindeygo 22km northeast of Amby, elevation 518m. 6m silcrete capping is breaking up. Silcrete is silicified breccia, may represent old soil horizon. Underlain by 12m of white leached and slightly silicified claystone, which overlies fresh Cretaceous siltstone. Probably another instance of deep-weathering profile with silcrete

capping in place, although silcrete could be younger than profile.

Site M: (Fig. 2). 8km south of Amby. Thick apron of silcrete cobbles and boulders around foot of hill of fine- to medium-grained Tertiary sandstone. Sandstone is more than 15m thick, base at 344m. Contains rounded pebbles of silcrete and quartz, and is itself silicified in part. Overlies fresh Cretaceous siltstone. Sandstone was probably deposited in a valley cut in the deep-weathering profile, and later silicified, particularly at top. Silcrete cobbles of apron are larger than those seen in sandstone, and are probably remnants of a capping. Site suggests that at least two periods of silcrete formation occurred in area. In streams in the south silcretes are forming at the present day, and these frequently contain older silcrete cobbles.

Site N: Hill 6km east of Roma, elevation 335m, about 30m above surrounding flats of lower pediplain. 3m of white weathered siltstone (leached zone) is overlain by a layer of silcrete cobbles which may, or may not, represent original crust. Neighbouring hills and ridges at approximately same level support view that site represents upper pediplain.

APPENDIX II

PETROGRAPHY AND LOCATION OF DATED SAMPLES

GA 3126: Tholeiitic olivine basalt. Hill at site E, south of Amby. Mitchell 1:250,000 map sheet (SG 55-11), grid reference 634684. Altitude about 380m. Both pigeonite and augite present in the basalt; glass (15%) is isotropic to partly anisotropic, but unaltered. A few per cent of a yellow mineraloid occurs in vesicles.

GA 3127, 3128: Tholeiitic olivine basalt. Quarry 8m deep in a thick basalt flow, 4 km northwest of Amby. Mitchell 1:250,000, 636704. Altitude about 400m. GA 3127 from north wall and GA 3128 from west wall of quarry, 90m from GA 3127. Petrographically similar to GA 3126, with about 5% of green mineraloid in vesicles.

GA 3129, 3141: Olivine basalt. Spur at Site H, 0.8 km east of Mitchell-Tooloombilla road, 3.2 km north of turnoff to Katonga Homestead, north of Amby. Mitchell 1:250,000, 637741. Altitude about 500m. A single flow at least 20m thick overlying Tertiary sandstone. No pigeonite identified in groundmass. Glass (30%) is somewhat altered and anisotropic. Green mineraloid present in minor amount.

GA 1332: Olivine basalt. Spur 45 km north of Amby. Mitchell 1:250,000, 637750. Altitude about 550m. At least two basalt flows occur

overlying Blythesdale Formation (Lower Cretaceous). Glass (10-15%) is essentially isotropic.

GA 3130: Olivine basalt. Hill about 1.6 km east of Mitchell-Tooloombilla road, about 1.6 km southwest of Kilmorey Homestead, 52 km north of Amby. Mitchell 1:250,000, 635758. Altitude about 580m. Two or three lava flows occur in this section, and are physically continuous with those exposed at the locality of GA 3132. Well developed ophitic texture. No pigeonite found. Pale brown glass (15%) is isotropic. Minor yellow mineraloid present.

GA 3131: Olivine basalt. Kilmorey Falls (Site I) near Kilmorey Homestead. Mitchell 1:250,000, 637758. Altitude about 560m. Thick, hackly jointed basalt filling a steep-sided valley cut in Cretaceous sediments. Probably same lava occurs in GA 3130. Contains about 30% brown glass which may be somewhat altered, but is isotropic. Minor yellow mineraloid in vesicles.

GA 3142: Tholeiitic olivine basalt. From waterfall, 13.8 km north of Kilmorey Homestead, 60 km north of Amby. Mitchell 1:250,000, 646764. Altitude about 800 m. Pigeonite and augite present. About 10% of pale brown, isotropic glass occurs in groundmass.

31

Reprinted from *Search,* **2**(11–12), 433–434 (1971)

The Nature of Old Landscapes

M.J. MULCAHY, E. BETTENAY *

Present day concepts of the nature of old landscapes are derived very largely from W.M. Davis' ideas on landscape development as expressed in his classical essay on the geographical cycle (Davis, 1899). His genetic classification of landscapes, and the concepts of base levelling and peneplanation, or the reduction of a landmass to low relief by long continued subaerial weathering and erosion have had far reaching effects on the thinking of several generations of geologists and geographers.

His views are, of course, open to criticism, and have received it on various grounds. Some, such as King (1962), influenced by Penck (1924), object to the concept of 'normality' in slope development, resulting in the smooth concave-convex slopes of the temperate climates where most of Davis' experience was gained. Dury (1963) points to the possibly misleading use of metaphor when considering Davis' work as technical writing, and Chorley (1962) argues on the basis of general systems theory that the geographical cycle, as a closed system model, inherently imposes certain restrictions upon the interpretation of landforms. Kirk Bryan (1940) points out that Davis failed to submit his inferences to the test of experimental verification or further field observation. On this last point Davis' lapse may well be

* CSIRO Division of Soils,
Western Australian Laboratories,
Floreat Park.

excusable, since experimental simulation of geomorphological process is notoriously difficult, and old landscapes, in his sense, are extremely rare (Mulcahy, 1961). The observations summarized here, however, suggest that such an old landscape exists in the south-western corner of the Australian continent. Further, it corresponds to a remarkable degree with Davis' predictions as to its nature, and thereby establishes criteria for recognition of old landscapes elsewhere.

Davis considered that there are certain regions whose lack of structural control of drainage, small relief, and deep 'soils' cannot be explained without supposing them to have, in effect, passed through a number of stages characteristic of his 'geographical cycle', and that composite topography belonging to two or more cycles of erosion may be present in the one landmass. Such appears to be the case in south-western Australia, regarded by Hills

(1961) as being stable since Mesozoic times, where the older, interior, landscape is separated along the line – the Meckering line (Mulcahy, 1967) — from a downstream area in which streams are more sharply incised and drainage patterns more intense due to rejuvenation following epeirogenic uplift (Fig. 1).

It is proposed here that the older, interior landscape, unaffected as it is by the glaciations of the Pleistocene, or by tectonic disturbance, offers an opportunity of putting Davis' theories to the test of further observations. Earlier workers such as Woolnough (1927) and Jutson (1934) believed that the drainage of the whole of this region, not only that of the marginal zone defined by the Meckering line, had been rejuvenated, following epeirogenic uplift. This was on the grounds that the main trunk valleys of the interior appeared to be cut down to the base of a widespread, ancient profile of lateritic

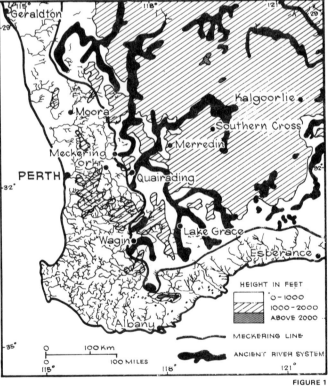

HEIGHT IN FEET

0 – 1000
1000 – 2000
ABOVE 2000

MECKERING LINE

ANCIENT RIVER SYSTEM

FIGURE 1

Drainage systems of south-western Australia

weathering, which was widely preserved on the divides. However recent work by Bettenay and Hingston (1964), and Bettenay *et al.* (1964) has shown that the *in situ* weathered profiles are also preserved beneath the inland valleys, where they are extremely deep, though buried by thin, relatively unweathered, alluvial and colluvial deposits derived from zones of more active erosion on the valley sides, where shallow soils and fresh rock outcrops are frequent. It therefore seems more reasonable to regard the Meckering line as the limit of drainage rejuvenation subsequent to uplift, and hence the country inland of it should display to some extent the features which Davis inferred to be characteristic of old landscapes.

It is now possible, drawing on data from the *Atlas of Australian Soils,* 1960-68 (Northcote, 1960) and the work of Bettennay and Hingston (1964) and Bettenay, *et al.* (1964) to say what these features are. They may be summarized as follows:

(1) low relief rising to only about 2000 feet on the divide some 300 miles inland from the coast, with local relief amounting to some 400 feet or less;

(2) braided drainage channels of an ancient river system (Fig. 1), which lie in valleys with gradients of only about 1 foot per mile. These are now characterized by saline formations and playa lakes, and are choked by the debris which they are incompetent to remove;

(3) deep profiles are widespread, consisting largely of rock saprolites, in which the original rock fabric is largely preserved in the form of resistant quartz grains in a matrix of kaolinized pseudomorphs of feldspars and ferromagnesian minerals. These 'pallid zones' (Walther, 1915) are associated with extensive lateritization, and may frequently exceed 150 feet in thickness, particularly under the valley floors;

(4) the extensive divides or interfluves are almost entirely cloaked by a mantle of lateritic 'sandplain' deposits which may be 20 feet or more in thickness (Mulcahy, 1967). These deposits are largely colluvial, and consist of the resistant residues from the original granitic country rock, mainly in the form of iron oxide stained quartz grains of sand size. Boulders and rock fragments are absent, and sand size grains ($<$2 mm dia.) form 90% or more of the deposit, with small amounts of kaolinitic clay;

(5) in the valleys, particularly near the saline lake chains there are thin alluvial and colluvial deposits, normally less than 6 feet thick, in which some bases are retained and clays including variously hydrated illites are developed. Evaporation of the saline groundwaters from the lake surfaces, followed by deflation, provides a mechanism whereby the less soluble salts such as gypsum are retained in dunes, while the more soluble salts are removed by the occasional sheet flow along the regional drainage lines;

(6) present day erosional modification is of only limited extent, and is confined to areas of local instability. Characteristically the only active modification is by sheet flow on the valley sides, where fresh rock may be exposed, and by wind deflation of the floors of the salt lakes. The latter is the source of 'lake parna' (Bettenay, 1962), aeolian deposits of limited extent flanking the lake shores.

It is evident that this part of Western Australia has the characteristics of low relief and sluggish drainage which Davis has attributed to old age associated with long continued stability and subaerial weathering. It has also, in the form of the sandplains, his 'waste sheet . . . of great thickness', as well as widespread deep and intensive weathering, which in itself must be evidence of long continued landscape stability with functional drainage under pluvial conditions (Woolnough, 1927). Today the climate is seasonally arid, with average annual rainfall below 20 inches, which, with low gradients and sluggish drainage, has resulted in the formation of playa lakes. The surfaces of the lakes when dry are wind eroded and thus lowered; that is to say, by the process that Davis (after Passarge, 1904) called 'levelling without base levelling', and which was recognized as such many years ago by Jutson (1919).

The old landscape is thus characterized by a set of superficial deposits, retained within the system because of the incompetence of the drainage to remove them. Noteworthy among these are the extensive 'waste sheets' of the sandplain deposits, perhaps a more useful indicator of old landscapes than deep weathered profiles or the crude topographic criteria frequently employed by geomorphologists. In this respect, a comparison with the plain lands of southern Africa or South America would be interesting, but unfortunately beyond the scope of this paper.

Some of the evidence on which our argument is based is already published, and reviewed by Mulcahy (1967). Other material on river valley form, drainage pattern and associated superficial deposits and weather profiles has been accepted for publication elsewhere (Bettenay and Mulcahy, in press; Mulcahy and Bettenay, in press).

Received 30 July 1971

References

BETTENAY, E. (1962) *J. Soil Sci.* **13** 10.
BETTENAY, E. and HINGSTON, F.J. (1964) *Aust. J. Soil Res.,* **2** 173.
BETTENAY, E., BLACKMORE, A.V. and HINGSTON, F.J. (1964) *Aust. J. Soil Res.,* **2** 187.
BETTENAY, E. and MULCAHY, M.J. (in press) *J. Geol. Soc. Aust.*
BRYAN, KIRK (1940) *Assoc. Amer. Geog. Ann.,* **30** 254.
CHORLEY, R.J. (1962) U.S. Geol. Surv. Prof. paper 500-B.
DAVIS, W.M. (1899) *Geogr. J.,* **14** 481.
DURY, G.H. (1963) *Aust. Geographer,* **9** 67.
HILLS, E.S. (1961) *Q. J. geol. Soc. Lond.,* **117** 77.
JUTSON, J.T. (1919) *Amer. J. Sci.,* **48** 435.
JUTSON, J.T. (1934) Bull. Geol. Surv. Western Australia.
MULCAHY, M.J. (1961) *Z. Geomorph.* **5** 211.
MULCAHY, M.J. (1967) In *Landform Studies from Australia and New Guinea,* p.211. Canberra, Australian National University Press.
MULCAHY, M.J. and BETTENAY, E. (in press) *J. Geol. Soc. Aust.*
NORTHCOTE, K.H. (comp.) (1960-) *Atlas of Australian Soils,* 1960-68. CSIRO in assoc. with Melbourne Univ. Press.
PENCK, W. (1924) Die morphologische Analyse. Ein Kapitel der physikalischen Geologie. Geographische Abhandlungungen. 2. Reihe, Heft 2, Engelhorn, Stuttgart.
PASSARGE, S. (1904) Rumpffläche und Inselberge. *Zeitschrift der Deut. Geol. Gesellschaft,* LVI, 193.
WALTHER, J. (1915) *Z. Dtsch. geol. Ges.,* **67B** 113.
WOOLNOUGH, W.G. (1927) *J. Proc. Roy. Soc. New South Wales,* **61** 17.

Reprinted from *Search,* 3(3), 54 (1972)

HOW OLD IS OLD?

Sir — In a recent research communication, Mulcahy and Bettenay [*Search* 2 (11—12) 433] have attempted to interpret the landsurface of south-western Australia in terms of 'features which Davis inferred to be characteristic of old landscapes' and to establish 'criteria for recognition of old landscapes elsewhere'. Since Davis's inferences, in the words of Mulcahy and Bettenay, have not been submitted 'to the test of experimental verification', the addition of further purely morphological features not based on any such tests can only perpetuate the Davisian 'lapse'.

Although their argument is prefaced with references to Davis's 'geographical cycle' and to 'peneplanation', and whilst Davis receives frequent mention through-out the paper, nowhere do they explicitly state that this landsurface is a peneplain, and some of their diagnostic criteria (for example, braided drainage channels and playas) are more pertinent to the hypothetical end result of Davis's (1905) arid cycle. Thus it is not clear whether Mulcahy and Bettenay consider the south-western Australian landsurface east of the so-called 'Meckering Line' to be a peneplain.

If peneplanation is not inferred, then their use of the term 'old' is unfortunate. As Ollier (1967) has pointed out, use of the stage names 'youth', 'maturity' and 'old age' should be avoided in landform description because these terms were devised to describe stages in the erosion cycle; they 'invariably colour ideas about landscape genesis, and their implications are best avoided by using purely descriptive terms'. Moreover, it is not clear on what basis this landsurface is considered to be 'old' if it is not considered to be' the end result of the erosion cycle, since different criteria, based on several concepts, are used. In this respect Dury and Langford-Smith (1966) have emphasized some of the perils which arise when two or more incompatible concepts develop around the same term.

In what sense is the term old used in 'old landscape'? How old is 'old'? Older than what? In addition to these absolute and relative connotations of 'old', Mulcahy and Bettenay also imply age of con-figuration of the landsurface (their criterion 1), age of soils (criterion 3), age of surficial materials (criteria 4 & 5), largely ineffectual processes (criterion 6) and the length of time for which certain processes have operated (for example, their penultimate paragraph).

Their argument is based on constantly shifting premises and has no crux. For example, it is evident that the sequence of events reconstructed by Mulcahy and Bettenay bears no relation to either of Davis's erosion cycles because, as they point out, climates and their associated geomorphic processes have changed from pluvial to arid to seasonally arid. If, however, the focus of the argument shifts from peneplanation, or the Davisian arid cycle, to other connotations of 'old', then the various components of the landsurface under consideration are related to different combinations of processes which have operated at varying magnitudes and frequencies over different periods of time, including the present. To label this landsurface an 'old landscape' begs the questions. Indeed, we would suggest that the only valid use of the term 'old' is on the basis of an accurately determined absolute age in relation to an agreed, but arbitrary limit. In this respect what may be considered 'old' to a pedologist may not be regarded as such by a sedimentologist; whilst to a geomorphologist, concerned with surface configuration, perhaps the only meaningful use of the term is in relation to the length of time for which specified processes have operated, since shape as such cannot be dated.

CHARLES W. FINKLE, JNR
International Nickel Australia Limited
1205 Hay Street
West Perth, Western Australia

ARTHUR J. CONACHER
Department of Geography
University of Western Australia
Nedlands, Western Australia

References

DAVIS, W.M. (1905) *J. Geol.,* **13** 381.
DURY, G.H. and LANGFORD-SMITH, T. (1966) *Aust. J. Sci.,* **28** 291.
OLLIER, C.D. (1967) *Aust. Geog. Studies,* **5** 73.

SUMMARY

The preceding array of benchmark papers indicates that there are more to come, since many questions are still to be resolved, ranging from the precise meaning of terms to the interpretation of past climates.

In spite of this situation, the following appear to be areas of general agreement concerning planation surfaces:

1. Planation surfaces do exist, and are not merely steady-state topographic products of random dissection.

2. The most distinctive unit surface is the pediment.

3. In attempting to estimate the emergence of a land mass from present altitudes of planation surfaces, their origins, base levels, and probable regional slopes must be considered.

4. Most planation surfaces are or have been graded with respect to some base elevation, usually sea level, but often to an endorheic basin or to a climatic control above sea level.

5. The most important time aspect of a planation surface is from the latest possible time of initiation of the cycle that produced it to the earliest possible time that it ceased being shaped (i.e., its terminal date) because of either burial or uplift.

6. A planation surface need not be presently under the same climatic regime as when it was initially developed.

7. Planation surfaces should be considered in the general framework of plate tectonics, sea-floor spreading, and the resulting change of continental areas with respect to the position of climatic zones.

Summary

I have selected these statements with some trepidation but believe that they express a consensus. The idea of stipulating areas of agreement stems from Lester King's *Canons of Landscape Evolution* (Paper 18). As suggested by King, the reader should examine the seven statements listed and decide how many are personally acceptable. Regardless of the number, I believe that it is necessary to determine what can be stipulated before a problem can be further pursued.

ADDITIONAL BIBLIOGRAPHY

Ahnert, F., 1970. Functional relation between denudation, relief and up-lift in large mid-latitude drainage basins. Amer. J. Sci., 268, p. 243–263.

Ambrose, J. W., 1964. Exhumed peneplains of the Precambrian shield of North America. Amer. J. Sci., 262, p. 817–857.

Atwood, W. W., Sr., and Atwood, W. W. Jr., 1938. A working hypothesis for the physiographic history of the Rocky Mountain region. Bull. Geol. Soc. America, 49, p. 957–958.

———, 1948. Tertiary-Pleistocene transition at the east margin of the Rocky Mountains. Bull. Geol. Soc. America, 59, p. 605–608.

Awad, H., 1955. Un problème de gémorphologie aride, les pédiments. Soc. Géog. Egypte, Le Caire, 28, p. 5–19.

Bakker, J. P., 1957. Quelques aspects du problème des sédiments cor-rélatif en climat tropical humide. Zeit. Geomorph., 1(1), no. 1.

———, and Levelt, T. W., 1964. An enquiry into the probability of a polyclimatic development of peneplains and pediments (etchplains) in Europe during the Senonian and Tertiary periods. Publ. Serv. Luxembourg, 14, p. 27–75.

———, and Strahler, A. N., 1956. Report on a quantitative treatment of slope-recession problems. Congr. Intern. Geogr. Rio de Janeiro, p. 1–12.

Barrell, J., 1915. Central Connecticut in the geologic past. Conn. Natural Hist. Surv. Bull. 23, p. 1–44.

Bascom, F., et al., 1909. Philadelphia Folio. U.S. Geol. Surv.

Baulig, H., 1957. Peneplains and pediplains. Bull. Geol. Soc. America. 68, p. 913–930.

Behrmann, W., 1959. Die Flächenbildung in den feuchten Tropen und die Rolle solcher fossiler Flächen in anderen Klimazonen. Deut. Geographentag, Würzburg, 1959.

Berkey, C. P., and Morris, F. K., 1927. The geology of Mongolia. 2, p. 475.

Bethune, P. de, 1948. Geomorphic studies in the Appalachians of Pennsylvania. Amer. J. Sci., 246, p. 1–22.

Bettenay, E., and Hingston, F. J., 1964. Development and distribution of soils in Western Australia. Australian J. Soil Res., 2, p. 173–186.

——, and Mulcahy, M. J., 1972. Soil and landscape studies in Western Australia. J. Geol. Soc. Australia, 18, p. 349–357.

Bigarella, J. J., 1965. Paleogeographische and Paleoclimatishe Aspecte des Kainozoikuns in Süd-Brasil. Zeit. Geomorph., 8, p. 286–312.

Birot, P., 1950. Sur le problème des pédiments. Congr. Intern. Geogr., Lisbon, Compte Rendu, Tome II, Sect. II, Geogr. Phys., p. 9–18.

——, 1968. The cycle of erosion in different climates. Batsford, London (translated by C. I. Jackson and K. M. Clayton).

——, and Joly, F., 1952. Observations sur les glacis d'érosion et les reliefs granitiques du Maroc. Centre de Documentation Cartographique et Géographique, Paris, No. 3, p. 7–56.

Blackwelder, E., 1928. Mudflow as a geologic agent in semi-arid mountains. Bull. Geol. Soc. America, 39, p. 465–484.

——, 1954. Geomorphic processes in the desert. In Geology of Southern California, R. H. Jahns (ed.), Calif. Dept. Natural Resources, Div. of Mines, Bull. 170, p. 11–20.

Bradley, W. H., 1936. Geomorphology of the north flank of the Uinta Mountains. U.S. Geol. Surv. Prof. Paper 185-I, p. 163–199.

——, 1940. Pediments and pedestals in miniature. J. Geomorph., 3, p. 244–255.

Bremer, H., 1973. Der Formungsmechanismus in tropischen Regen wäldern. Zeit. Geomorph., Suppl. Band 17, p. 195–222.

Bretz, J. H. Geomorphic history of the Ozarks of Missouri. Missouri Geol. Surv. Water Resources Bull., 41, 147 p.

Bridge, J., 1950. Bauxite deposits in the southeastern United States. In F. G. Snider (ed.), Symposium on Mineral Resources in the Southeastern United States, University of Tennessee Press, Knoxville, Tenn., p. 170–201.

Brown, E. H., 1961. Britain and Appalachia: a study in the correlation and dating of planation surfaces. Inst. Brit. Geogr. Trans., 29, p. 91–100.

Bruckner, W. D., 1957. Laterite and bauxite profiles of West Africa as an index of rhythmical climatic variations in the tropical belt. Ecologae Geologicae Helvetiae, 50, p. 239–256.

Bryan, K., 1932. Pediments developed in basins with through drainage. Bull. Geol. Soc. America, 43, p. 128–129.

——, 1934. Geomorphic processes at high altitudes. Geogr. Rev., 24, p. 655–656.

——, 1935. The formation of pediments. Rept. 16th Intern. Geol. Congr., Pt. 2, p. 765–775.

——, 1935–1936. Processes of formation of pediments at Granite Gap, N. Mex. Zeit. Geomorph., 9(4), p. 125–135.

——, 1940. The retreat of slopes. Assoc. Amer. Geogr. Ann., 30, p. 254–268.

——, and McCann, F. T., 1936. Successive pediments and terraces of the upper Rio Puerco in New Mexico. J. Geol., 44(2), p. 145–172.

Bucher, W. H., 1932. "Strath" as a geomorphic term. Science, 75, p. 130–131.

Büdel, J., 1955. Relief Generation und plio-pleistcener Klimawandel im Hoggar-Gebirge (Central Sahara). Erdkunde, 9(2), p. 100–115.

———, 1963. Klima-genetische Geomorphologie. Geogr. Rundshau, 7, p. 269–286.

Bulla, B., 1965. Tertiary levelled surfaces (peneplains) in Hungary: geomorphological problems of the Carpathians. I: Evolution of relief in the Tertiary. Slovensk-Acad. Vied., Bratislava 18, Hungary.

Buxton, B. P., 1958. Weathering and sub-surface erosion in granite at the piedmont angle, Bolas, Sudan. Geol. Mag., 95, p. 353–377.

———, 1961. Weathering profiles and geomorphic position on granite in two tropical regions. Rev. Géomorph. Dynamique, 12, p. 16–31.

Cady, J. G., 1950. Rock weathering and soil formation in the North Carolina piedmont region. Soil Sci. Soc. America Proc., p. 337–342.

Cahen, L. and Lepersonne, J., 1952. Equivalence entre le système du Kalahari de Congo Belge et les Kalahari beds d'Afrique Australe. Mém. Soc. Belge Géol., 8, p. 3–64.

Cailleux, A., 1950. Ecoulement liquide en nappes et applanissements. Rev. Géomorph. Dynamique, 1, p. 243–270.

———, and Tricart, J., 1950. La Surface infra-tertiaire dans le bassin de Paris. Congr. Intern. Géogr., Lisbon, 1949, 2, p. 651–658.

Campbell, M. R., 1903. Geologic development of northern Pennsylvania and southern New York. Bull. Geol. Soc. America, 14, p. 277–296.

———, 1933. Chambersburg (Harrisburg) peneplain in the Piedmont of Maryland and Pennsylvania. Bull. Geol. Soc. America, 44, p. 553–573.

Chadwick, G. H., 1948. Peneplains in Maine. Bull. Geol. Soc. America, 59, p. 1315–1316.

Chapman, R. W., 1949. Spheroidal weathering of igneous rocks. Amer. J. Sci., 247, p. 407–429.

Chemakov, Y. F., 1964. Origin and evolution of denudational-planated surfaces in folded areas. In Problems of planated surfaces. Hayka, Moscow, p. 151–164.

Chorley, R. J., 1965. A re-evaluation of the geomorphic system of William Morris Davis. In Frontiers in geographical teaching, Methuen, London, p. 21–38.

———, Dumont, A. J., and Bechindale, R. P., 1964. History of the study of landforms, Vol. 1, Geomorphology before Davis. Methuen, London.

Churchward, H. M., and Bettenay, E., 1973. Physical significance of conglomerate sediments and associated laterites in the valleys of the Darling Plateau, New Haney, Western Australia. J. Geol. Soc. Australia, 20(3), p. 309–318.

Cleland, H. F., 1920. A Pleistocene peneplain in the Coastal Plain. J. Geol., 28, p. 702–706.

Cole, W. S., 1935. Rock resistance and peneplain expression. J. Geol., 43, p. 1049–1062.

———, 1941. Nomenclature and correlation of Appalachian surfaces. J. Geol., 49, p. 129–148.

Cooke, R. U., and Warren, A., 1974. Geomorphology in deserts. University of California Press, Berkeley, Calif., 374 p.

Corbel, J., 1967. Pédiments d'Arizona. Centre Documents Géogr. Mém. Doc., Paris, 9(3), p. 552–570.

Cotton, C. A., 1942. Climatic accidents, Chapter 7 and 8. Whitcombe and Tombs, Wellington, New Zealand.

———, 1955. Peneplanation and pediplanation. Bull. Geol. Soc. America, 66, p. 1213–1214.

———, 1962. Plains and inselbergs of the humid tropics. Trans. Roy. Soc. New Zealand, 88, p. 269–277.

———, 1965. The theory of savannah planation. Geography, 46, p. 89–101.

Crowl, G. H., 1952. Erosion surfaces of the Adirondacks. Bull. Geol. Soc. America, 61, p. 1565.

Dake, C. L., and Bridge, Josiah, 1932. Buried and resurrected hills of the southern Ozarks. Amer. Assoc. Petrol. Geologists Bull. 16, p. 629–652.

Dapples, E. C., 1959. The behavior of silica in diagenesis. Spec. Pub., Soc. Econ. Paleontologists Mineralogists, 7, p. 36–54.

Darton, N. H., 1950. Configuration of bedrock surfaces of the District of Columbia and vicinity. U.S. Geol. Surv. Prof. Paper 217, 41 p.

Davies, J. L., 1971. Tasmanian landforms and Quaternary climates. In Landform studies from Australia and New Guinea, J. N. Jennings and J. A. Mabbutt (eds.), Australian National University Press, Canberra.

Davis, W. M., 1890. The rivers of Northern New Jersey with notes on the classification of rivers in general. Natl. Geogr. Mag., 2, p. 81–110.

———, 1892. The convex profile of badland divides. Science, 20, p. 245.

———, 1896b. The physical geography of southern New England. Natl. Geogr. Soc. Monograph 1, p. 269–304.

———, 1898 (rev. 1909). The peneplain: geographic essays. Ginn & Co., Boston, p. 350–380. (See Paper 3.)

———, 1902. Base level, grade, and peneplain: geographic essays. Ginn and Co., Boston, p. 381–410. (Reprinted by Dover, 1954.)

———, 1903. Stream contest along the Blue Ridge. Geogr. Soc. Philadelphia, Bull. 3, p. 213–244.

———, 1905. Complications of the cycle. 8th Intern. Geogr. Congr., p. 150–163.

———, 1905. The geographic cycle in an arid climate. J. Geol., 13, p. 381–407.

———, 1923. The cycle of erosion and the summit level of the Alps. J. Geol., 31, p. 1–41.

———, 1923. The scheme of the erosion cycle. J. Geol., 31, p. 10–25.

———, 1933. Granite domes of the Mohave desert, California. Trans. San Diego Soc. Natural Hist., 7, p. 211–258.

———, 1936. Geomorphology of mountainous deserts. Rept. 16th Intern. Geol. Congr., Pt. 2, p. 703–714.

———, 1938. Sheetfloods and streamfloods. Bull. Geol. Soc. America, 49, p. 1337–1416.

Denny, C. S., 1967. Fans and pediments. Amer. J. Sci., 265, p. 81–105.

Derruau, M., 1956. Les Formes périglaciaires du Labrador–Ungava central comparé à celle d'Island centrale. Rev. Géogr. Dynamique, 7, p. 12–16.

———, 1956. Précis de geomorphologie, Chap. 8: Les Glacis, la pédiplaine, l'inselberg. Masson et Cie, Paris.

D'Hoore, J., 1954. L'Accumulation des sesquioxides libres dans les sols tropicaux. Publ. I.N.E.A.C., Sér. Sci., 62.

Dietrich, R. V., 1958. Origin of the Blue Ridge Escarpment directly southwest of Roanoke, Virginia. Virginia Acad. Sci. J., 9, p. 233–246.

Dixey, F., 1945. Main peneplains of Central Africa. Geol. Soc. London Quart. J., 403–404, 101 (3–4), p. 243–253.

———, 1955. Erosion surfaces in Africa. Geol. Soc. South Africa Trans. 56, p. 265–280.

Douglas, I. D., 1969. The efficiency of humid tropical denudation systems. Inst. Brit. Geogr. Trans., 46, p. 1–16.

Drake, C. L., Worzel, J. L., and Beckman, W. C., 1954. Geophysical investigation on the emerged and submerged Atlantic Coastal Plain: Part IX, Gulf of Maine. Bull. Geol. Soc. America, 65, p. 957–970.

Dresch, J., 1947. Pénéplaines africaines. Ann. Géogr., 56(302), p. 125–137.

———, 1949. Pénéplaines en Afrique noire française. Rapport Comm. Cartogr., Surface d'applanissement. Preparé pour le Congr. Intern. Géogr., Lisbon, p. 140–148.

———, 1950. Sur les pédiments en Afrique mediterranienne et tropical. Congr. Intern. Geogr., Lisbon, 1949, 2, p. 19–28.

———, 1957. Pédiments et glacis d'érosion, pédiplaines et inselbergs. L'Inform. Géogr., 21, p. 183–196.

———, 1962. Remarques sur une division géomorphologique des régions arides et les caractères originaux des régions arides méditerranées: I.G.U. Arid Zone Colloquium, Greece.

Dumas, B., 1966. Les Glacis, formes de convergence. Bull. Assoc. Géogr. Fran., 345, p. 34–47.

———, 1966. Les Mécanismes d'élaboration des glacis d'après l'exemple du Levant espanol. Compt. Rend. Acad. Sci. Ser. D21.

———, 1967. Place et signification des glacis dans le Quaternaire. Assoc. Géogr. Fran. E. Quat., fasc. 3, p. 223–244.

———, 1970. The origin of glacis. In Fundamental problems of relief planation, M. Pesci (ed.), Studies in Geography in Hungary, 8, Geog. Res. Instit., Hung. Acad. Sci., Budapest, p. 113–118.

Dury, G. H., and Langford-Smith, T., 1964. The use of the term peneplain in descriptions of Australian landscapes. Australian J. Sci., 27(6), p. 171–175.

———, 1965. Pediment, peneplain: Australian landform examples. Australian Geographer, Sidney, p. 386–387.

———, Langford-Smith, T., and McDougall, I., 1969. A minimum age for the duricrust. Australian J. Sci., 31, p. 362–363.

Eggler, D. H., Larson, E. E., and Bradley, W. C., 1969. Granite, grusses and the Sherman erosion surface, southern Laramie range, Colorado, Wyoming. Amer. J. Sci., 267, p. 510–522.

Fair, T. J., 1947. Slope form and development in the interior of Natal South Africa. Geol. Soc. South Africa Trans., 50, p. 110.

———, 1948. Hillslopes and pediments of the semi-arid Karoo. South African Geog. J., 30, p. 71–79.

Fairbridge, R. W., 1968. Fall line, zone. In Encyclopedia of geomorphology, R. W. Fairbridge (ed.), Van Nostrand Reinhold, New York, p. 344–345.

Fermer, L., 1911. What is laterite? Geol. Mag., 5, p. 454–462, 507–516, 559–566.

Field, R., 1935. Stream carved slopes and plains in desert mountains. Amer. J. Sci., 5 ser., 29, p. 313–322.

Finkle, C. W., Jr., and Churchward, H. M., 1973. Etched land surfaces of southwestern Australia. J. Geol. Soc. Australia, 20(3), p. 259–307.

Flemal, R. C., 1971. An attack on the Davisian system of geomorphology: a synopsis. J. Geol. Educ., 19(1), p. 3–13.

Flint, R. F., 1963. Altitude, lithology and the Fall Zone in Connecticut. J. Geol., 71, p. 683–697.

Folster, H., 1964. Morphogenese der Südsudanische Pediplane. Zeit. Geomorph., 8, p. 393–423.

Fridley, H. M., and Nolting, J. P., 1931. Peneplains of the Appalachian Plateau. J. Geol., 39, p. 749–755.

Frye, J. C., 1959. Climate and Lester King's uniformitarian nature of hillslopes. J. Geol., 67, p. 111–113.

———, and Smith, H. T. U., 1942. Preliminary observations on pediment-like slopes in the central High Plains. J. Geomorph., 5, p. 215–221.

Garner, H. K., 1968. Tropical weathering and relief. In Encyclopedia of geomorphology, R. W. Fairbridge (ed.). Van Nostrand Reinhold, New York, p. 1161–1172.

Gavrilovic, D., 1972. Experiment in climatic geomorphology: stream terraces and pediments. Zeit. Geomorph., 16(3), p. 315–331.

Gellert, J. F., 1968. The climatic-morphologic and paleoclimatic relations of the Inselberg relief in Southwest Africa. 21 Intern. Geol. Congr., New Delhi.

Geyl, W., 1960. Geophysical speculations in the origin of stepped erosion surfaces. J. Geol., 68(2), p. 154–176.

Gilbert, G. K., 1909. The convexity of hillslopes. J. Geol., 17, p. 344–350.

Gilluly, J., 1937. Physiography of the Ajo region. Bull. Geol. Soc. America, 48, p. 323–348.

Glinka, K. D., 1927. The great soil groups of the world and their development. English translation by C. F. Marbut. Edwards Bros., Ann Arbor, Mich., 235 p.

Glock, W. S., 1932. Premonitory planations in Western Colorado. Pan-Amer. Geol., 57, p. 29–37.

———, 1936. Desert cliff recession and lateral regional planation. Pan-Amer. Geol., 66, pp. 81–86.

Goldthwait, J. W., 1914. Remnants of an old graded surface in the Presidential Range of the White Mountains. Amer. J. Sci., 187, p. 451–463.

Goudie, A. C., 1972. The chemistry of world calcrete deposits. J. Geol., 80, p. 449–463.

——, 1973. Duricrusts in tropical and sub-tropical landscapes. Oxford University Press, New York, 174 p.

Groot, J. J., 1955. Sedimentary petrology of the Cretaceous sediments of Northern Delaware in relation to paleogeographic problems. Delaware Geol. Surv. Bull. 5, 157 p.

Guilcher, A. 1949. La Surface post-hercynniene dans l'Europe occidentale. Ann. Géogr., 58(310), p. 97–112.

Hack, J. T., 1961. Geomorphology of the Shenandoah Valley, Va. and W. Va., and the origin of residual ore deposits. U.S. Geol. Surv. Prof. Paper 484, 84 pp.

——, 1966. Interpretation of the Cumberland escarpment and the Highland Rim, south-central Tennessee and northeastern Alabama. U.S. Geol. Surv. Prof. Paper 524C, 16 pp.

Hadley, R. F., 1967. Pediments and pediment-forming processes. J. Geol. Educ., 15, p. 83–89.

Hayes, C. W., and Campbell, M. R., 1894. Geomorphology of the southern Appalachians. Natl. Geogr. Mag., 6, May.

Hernandez-Pacheca, F., 1950. Las Rañas de las Sierras Centrales de Estremadura. Congr. 1898, Intern. Geogr., Lisbon, 1949, 2, p. 87–105.

Hettner, H., 1972. Surface features of the land: problems and methods in geomorphology. Translated by P. Tilley, Macmillan, New York.

Hill, R. T., and Vaughan, T. W., 1896. Geology of the Edwards Plateau and the Rio Grande Plain, etc. U.S. Geol. Surv. Ann. Rept. 18, p. 193–321.

Hobbs, W. H., 1901. The river systems of Connecticut. J. Geol., 9, p. 469–485.

Holmes, C. D., 1955. Equilibrium and humid climate physiographic processes. Amer. J. Sci., 260, p. 427–438.

Howard, Alan D., 1965. Geomorphological systems—equilibrium dynamics. Amer. J. Sci., 263, p. 302–312.

Howard, A. D., 1942. Pediment passes and the pediment problem. J. Geomorph., 5, p. 2–31, 95–136.

——, and Smith, D. D., 1956. Studies bearing on erosion surfaces in the United States and Canada—1952–1955—A review. 9th gen. Assn. and 18th Intern. Geogr. Congr. Rio de Janeiro, 8th Report. Comm. Study Correlation of Erosion Surfaces Around the Atlantic (North America).

Hunt, C. B., Averill, P., and Miller, R. L., 1953. Geology and geography of the Henry Mtns., Utah. U.S. Geol. Surv. Prof. Paper 228, p. 190.

James, P. E., 1959. The geomorphology of eastern Brazil as interpreted by Lester King. Geog. Rev., 49, p. 240–246.

Johnson, D. W., 1916. Plains, planes, and peneplanes. Geogr. Rev., 8, p. 443–447.

——, 1932. Miniature fans and pediments. Science, 76, p. 546.

——, 1932. Rock fans in arid regions. Amer. J. Sci., 223, p. 389–416.

——, 1932. Rock planes in arid regions. Geogr. Rev., 22, p. 656–665.

——, 1940. Contributions to the symposium "Walther Penck's contributions to geomorphology. Assoc. Amer. Geogr., 4, p. 219–284.

Joly, F., 1950. Pédiments et glacis d'érosion dans le sud-est du Maroc. Congr. Intern. Géogr., Lisbon, 1949, 2, p. 110–125.

Jutson, J. T., 1934. The physiography of western Australia. Bull. Geol.

Surv. Western Australia, 95, 366 p. Also Amer. J. Sci., 1919, 48, p. 435.

Kartashov, I. P., 1970. Valley pediments or denuded terraces. In Fundamental problems of relief planation, M. Pesci (ed.), Studies in Geography in Hungary, 8, Geogr. Res. Inst., Hung. Acad. Sci., Budapest, p. 127–132.

King, L. C., 1947. Landscape studies in South Africa. Proc. Geol. Soc. South Africa, 50, p. 32–52.

———, 1948. A theory of bornhardts. Geogr. J., 112, p. 83–87.

———, 1949. The pediment landform: some current problems. Geol. Mag., 86, p. 245–250.

———, 1950. The study of the world's plainlands: a new approach to geomorphology. Quart. J. Geol. Soc. London, 106, p. 101–127.

———, 1951. The cyclic landsurfaces of Australia. Proc. Roy. Soc. Victoria, 62, p. 79–95.

———, 1951. South African scenery. Oliver & Boyd, Edinburgh, 379 p.; 3rd ed., 1963, Hafner, New York, 308 p.

———, 1953. The uniformitarian nature of hillslopes. Trans. Edinburgh Geol. Soc., 17, p. 81–102.

———, 1956. Pediplanation and isostasy: an example in South Africa. Quart. J. Geol. Soc. London, 111, p. 353–359.

———, 1966. The origin of bornhardts. Zeit. Geomorph., 10, p. 97–98.

———, 1967. The morphology of the earth, 2nd ed. Hafner, New York, 726 p.

———, 1972. The coastal plain of Southeast Africa: its form, deposits and development. Zeit. Geomorph., 16, p. 239.

———, 1972. The natal monocline. Geol. Dept. Univ. Natal, Durban, South Africa, 113 p.

King, L. H., 1972. Relation of plate tectonics to the geomorphic evolution of the Canadian Atlantic provinces. Bull. Geol. Soc. Amer., 83, p. 3083–3090.

Langford-Smith, T., and Dury, G. H., 1965. Distribution, character and attitude of the duricrust in the northwest of New South Wales and adjacent areas of Queensland. Amer. J. Sci., 263, p. 170–190.

Lawson, A. C., 1890. Notes on the pre-Paleozoic surface of Archean terranes of Canada. Bull. Geol. Soc. America, 1, p. 163–173.

———, 1932. Rainwash erosion in humid regions. Bull. Geol. Soc. Amer., 43, p. 703–724.

Lee, W. T., 1923. Peneplains of the Front Range and Rocky Mountain National Park, U.S. Geol. Surv. Bull. 730, p. 1–17.

Linton, D. L., 1955. The problem of tors. Geogr. J., 121, p. 471–487.

Lobeck, A. K., 1917. The position of the New England peneplain in the White Mountain region. Geogr. Rev., 3, p. 53–60.

Louis, H., 1965. Completion and destruction of peneplains in Tanganyika. 20th Intern. Geogr. Congr., London, Albs., p. 95.

Lovasz, G., 1970. Surfaces of planation in the Mecsek mountains. In Fundamental problems of relief planation, M. Pesci (ed.), Studies in Geography in Hungary, 8, Geogr. Res. Inst., Hungar. Acad. Sci., Budapest, p. 65–72.

Mabbutt, J. A., 1952. A study of granite relief from Southwest Africa. Geol. Mag., 89, p. 82–96.

————, 1961. A stripped land surface in Western Australia. Inst. Brit. Geogr. Trans., 29, p. 101–114.

————, 1966. Mantle-controlled planation of pediments. Amer. J. Sci., 264, p. 78–91.

Macar, P., ed., 1967. L'Évolution des versants. Congr. Colloq. Univ. Liège, 40, p. 23–41.

Mackin, J. H., 1937. Erosional history of the Big Horn Basin, Wyoming. Bull. Geol. Soc. America, 48, p. 813–894.

————, 1947. Altitude and local relief of the Big Horn area during the Cenozoic. Wyoming Geol. Assoc. Field Conf. in Big Horn Basin, Guidebook, p. 103–120.

Malott, C. A., 1921. Planation stream piracy. Indiana Acad. Sci. Proc., 30, p. 249–260.

Mammerick, J., 1964. Quantitative observations on pediments in the Mohave and Sonoran deserts. Amer. J. Sci., 262, p. 429–430.

Martin, E. A., 1940. On peneplains and related landforms: a discussion. Quart. J. Geol. Soc. London, 382, v. 96, Pt. 2, p. 279–312.

Martonne, E. de, 1940. Problèmes morphologiques du Brésil Tropique Atlantique. Ann. Géogr., 49(277), p. 1–27.

Maxson, J. H., and Anderson, G. H., 1935. Terminology of surface forms of the erosion cycle. J. Geol., 43, p. 88–96.

McGee, W. J., 1888. Three formations of the Middle Atlantic Slope. Amer. J. Sci., 3rd ser., 35, p. 120–143.

Mensching, H., 1958. Glacis–Füssfläche–Pediment. Zeit. Geomorph., 3, p. 165–186.

————, 1970. Planation in arid sub-tropic and tropic regions. In Fundamental problems of relief planation, M. Pesci (ed.), Studies in Geography in Hungary, 8, Geogr. Res. Inst., Hung. Acad. Sci., Budapest, p. 73–84.

————, 1973. Pediment und Glacis, ihre Morphogenese und Einordnung im System der klimatischen Geomorphologie auf Grund von Beobachten in Trockgebiet Nord-Americas, U.S.A. and Nordmexico. In Contributions to Climatic Geomorphology, Zeit. Geomorph., Suppl. Band 17, p. 133–155.

Meyerhoff, H. A., and Hubble, M., 1928. The erosional landforms of central Vermont. Vermont State Geol. Surv., 16th Ann. Rept., p. 315–381.

Miller, V. C. Pediments and pediment forming processes near House Rock, Arizona. J. Geol., 58, p. 634–644.

Mistardis, G., 1950. Essai d'une classification des vestiges des surfaces pédimentaires et de leur couverture de depôts grossiers conservée dans les pays méditerranéennes. Congr. Intern. Géogr., Lisbon, 1949, 2, p. 126–131.

————, 1950. Les pédiments arides et sémi-arides de l'Atlantique centrale. Congr. Intern. Géogr., Lisbon, 1949, 2, p. 137–147.

Morris, L., 1954. Origin of pediments. Intern. Geol. Congr. (19th), Algeria, 1953, C.R. Sec. 7, f. 7, p. 131–133.

Mulcahy, M. J., 1966. Peneplains and pediments in Australia. Australian J. Sci., p. 290–291.

————,1967. Landscapes, laterites, and soils in Southwest Australia. In Landform studies from Australia and New Guinea, J. H. Jennings and J. A. Mabbutt (eds.), p. 211–230, Australian National University Press, Canberra.

Murray, H. W., 1947. Topography of the Gulf of Maine. Bull. Geol. Soc. America, 58, p. 153–196.

Nagell, R. H., 1962. Geology of the Serra do Novio Manganese District, Brazil. Econ. Geol., 57, p. 481.

Nikiforoff, C. C., 1943. Introduction to paleopedology. Amer. J. Sci., 241, p. 194–200.

Ogilvie, I. G., 1905. The high altitude conoplain. Amer. Geol., 36, p. 27–34.

Ollier, C. D., 1959. A two cycle theory of tropical pedology. J. Soil Sci., 10(2), p. 138–147.

———, 1965. Some features of granite weathering in Australia. Zeit. Geomorph., 9, p. 285–304.

———, 1969. Weathering. American Elsevier, New York, 304 p.

———, and Tuddenham, W. G., 1962. Slope development at Coober Pedy, South Australia. J. Geol. Soc. Australia, 9, p. 91–105.

Olmstead, D. W., and Little, L. S., 1946. Marine planation in southern New England. Bull. Geol. Soc. America, 57, p. 1271.

Paige, S., 1912. Rock-cut surfaces in the desert ranges. J. Geol., 20, p. 442–450.

Pallister, J. W., 1954. Erosion levels and laterite in Buganda Province, Uganda. Intern. Geol. Congr., 19th, Algiers, Pt. 2, f. 21, p. 193–199.

Palmer, J., and Neilson, R. A., 1962. The origin of granite tors on Dartmoor, Devonshire. Proc. Yorkshire Geol. Soc., 33, p. 315–340.

Pennekoek, A. J., 1967. Generalized contour maps. Zeit. Geomorph., 11(2), p. 169–182.

Passarge, S., 1904. Rumpfflächen und Inselberge. Zeit. Deut. Geol. Ges., 56.

———, 1929. Das Problem der Inselberglandschaften. Zeit. Geomorph., 4.

Pécsi, M. (ed.), 1970. Fundamental problems of relief planation. Studies in Geography in Hungary, 8, p. 11, 12, Geogr. Res. Inst., Hung. Acad. Sci., Budapest.

———, 1970. Surfaces of planation in the Hungarian mountains and their relevance to pedimentation. In Fundamental problems of relief planation, M. Pesci (ed.), Studies in Geography in Hungary, 8, p. 29–40, Geog. Res. Inst., Hung. Acad. Sci., Budapest.

Peel, R. F., 1941. Denudational landforms in the central Libyan desert. J. Geomorph., 4, p. 3–23.

———, 1960. Some aspects of desert morphology. Geography, 45, p. 241–262.

Peel, R. G., 1966. The landscape in aridity. Inst. Brit. Geogr. Trans., 38, p. 1023.

Penck, A., 1919. Die Gipfelflur der Alpen. Sitzber. Preuss. Acad. Wiss. (Math. Phys.), 17, p. 256–268.

Pinces, Z., 1970. Planated surfaces and pediments of the Bukk mountains. In Fundamental problems of relief planation, M. Pesci (ed.), Studies in Geography in Hungary, 8, Geo. Res. Inst., Hung. Acad. Sci., Budapest, p. 55–64.

Piotrovsky, M. V., 1970. The problem of pediments and morpho-tectonics. In Fundamental problems of relief planation, M. Pesci (ed.), Studies in Geography in Hungary, 8, Geog. Res. Inst., Hung. Acad. Sci., Budapest, p. 119–127.

Pough, J. C., 1966. The landforms of low latitudes. In Essays in geomorphology, G. Dury (ed.), American Elsevier, New York, p. 121–138.

Powell, J. W., 1875. Exploration of the Colorado River of the West. Smithsonian Institution, Washington, D.C., 291 p.

Price, W. A., 1933. Reynosa problem of South Texas and the origin of caliche. Amer. Assoc. Petrol. Geol. Bull., 17(5), p. 488–522.

Prouty, W. F., 1946. Atlantic Coastal Plain floor and continental slope of North Carolina. Amer. Assoc. Petrol. Geol. Bull., 30, p. 1919–1920.

Pugh, J. C., 1950. Isostatic adjustment in the theory of pediplanation. Quart. J. Geol. Soc. London, 111, p. 361–369.

Pumpelly, R., 1891. Relation of secular rock disintegration to certain transitional crystalline schists. Bull. Geol. Soc. America, 2, p. 209–224.

Radhadkrishnan, B. P., 1952. The Mysore Plateau; its structural and physiographic evolution. Bull. Mysore Geol. Assoc., 57.

Rahn, P. H., 1959. Inselbergs and nickpoints in southwestern Arizona. Zeit. Geomorphol., 3(10), p. 217–225.

Ramsey, A. C., 1873. The pre-Miocene Alps and their subsequent waste and degradation. Proc. Roy. Inst. (Gt. Brit.), 7, p. 455.

Renner, G. T., Jr., 1927. Physiographic interpretation of the Fall Line. Geogr. Rev., 17, p. 278–286.

Ribeiro, O., and Frio, M., 1950. Les Depôts de type "rañas" au Portugal. Congr. Intern. Géogr., Lisbon, 1949, 2, p. 152–159.

Richter, G. D., and Kamanine, L. G., 1956. Caractéristiques comparatives morphologiques des boucliers de la partie européenne de l'URSS. Essais Géogr. Moscow Acad. Sci., p. 82–92.

Rosenkranz, E., 1970. Pediments on the northeastern border of the Thuringian forest. In Fundamental problems of relief planation, M. Pesci (ed.), Studies in Geography in Hungary, 8, p. 133–136, Geogr. Res. Inst., Hung. Acad. Sci., Budapest.

Rudberg, S., 1970. The Sub-Cambrian peneplain in Sweden and its slope gradient. Zeit. Geomorph., Suppl. Vol. 9, New contributions to slope evolution, p. 157–167.

Ruellan, F., 1950. Les Surfaces d'érosion de la région sub-orientale du Plateau Central Brésilien. Congr. Intern. Géogr., Lisbon, 1949, 2, p. 659–673.

Ruhe, R. V., 1956. Geomorphic surfaces and the nature of soils. Soil Sci., 82, p. 441–455.

Ruxton, B. P., and Berry, L., 1957. The weathering of granite and associated erosional features in Hong Kong. Bull. Geol. Soc. America, 68, p. 1263–1292.

———, 1959. The basal rock surface on weathered granite rocks. Proc. Geol. Assoc. London, 70, p. 285–290.

Scheidegger, A. E., 1961. Mathematical models of slope development. Bull. Geol. Soc. America, 72, p. 37–50.

Schumm, S. A., 1956. Evolution of drainage basins and slopes in badlands at Perth Amboy, New Jersey. Bull. Geol. Soc. America, 67, p. 597–646 and 1962, 73, p. 719–724.

———, 1956. The role of creep and rainwash in retreat of badland slopes. Amer. J. Sci., 254, p. 693–706.

———, 1963. Disparity between present rates of denudation and orogeny. U.S. Geol. Surv. Prof. Paper 454-H, 13 pp.

Shaler, N. S., 1899. Spacing of rivers with reference to the hypothesis of base-leveling. Bull. Geol. Soc. America, 10, p. 263–276.

Shaw, E. W., 1918. Ages of peneplains of the Appalachian Province. Bull. Geol. Soc. America, 29, p. 575–586.

Shaw, W. S., 1918. Pliocene history of northern and central Mississippi. U.S. Geol. Surv. Prof. Paper 108, p. 125–162.

Simons, M., 1962. The morphological analysis of landforms: a new review of the work of Walther Penck. Inst. Brit. Geogr., Trans., 31, p. 1–14.

Smalley, I. J., and Vita-Finzi, C., 1969. The concept of systems in earth science. Bull. Geol. Soc. America, 80, p. 1591–1594.

Smith, K. G., 1958. Erosional processes and landforms in Badlands National Monument, South Dakota. Bull. Geol. Soc. America, 69, p. 975–1008.

Smith, W. T. S., 1899. Some aspects of erosion in relation to the theory of peneplains. Univ. Calif. Dept. Geol. Bull. 2, p. 155–178.

Spreitzer, H., 1951. Die Piedmonttreppen in der regionalen Geomorphologie. Erdkunde, Bonn, 5(4), p. 294–305.

Steven, T. A., 1968. Critical review of the San Juan peneplain. U.S. Geol. Surv. Prof. Paper 594I, p. 196.

Strahler, A. N., 1950. Davis' concepts of slope development viewed in the light of recent quantitative investigations. Assoc. Amer. Geogr. Ann., 40, p. 209–213.

———, 1950. Equilibrium theory of erosional slopes approached by frequency distribution. Amer. J. Sci., 248, p. 673–696.

Su Ting, 1950. Geomorphology of the Tarim Basin and the formation of pediments. Congr. Intern. Géogr., Lisbon, 1949, 2, p. 166.

Swan, S. B. St. C., 1972. Land surface evolution and related problems with reference to a humid tropical region: Johor, Malaya. Zeit. Geomorph., 16(2), p. 160–181.

Swart, A. M. J. de, 1964. Laterization and landscape development in parts of Equatorial Africa. Zeit. Geomorph., 8, p. 313–333.

Symposium, 1964. Problems of planated surfaces. Hayka, Moscow.

Symposium, 1965. Geomorphic problems of the Carpathians: Slovensk. Acad. Vied, Bratislava, Hungary.

Szekely, A., 1970. Landforms of the Matron mountains and their evolution with special reference to surfaces of planation. In Fundamental problems of relief planation, M. Pesci (ed.), Studies in Geography in Hungary, 8, p. 41–54, Geogr. Res. Inst., Hung. Acad. Sci., Budapest.

Tator, B. A., 1952. The climatic factor and pedimentation. Comp. Rend. Intern. Geogr. Congr., Algeria.

———, 1952. Pediment characteristics and terminology. Assoc. Amer. Geogr. Ann., 42, p. 293–317.

Teale, E. O., 1950. The river system in Tanganyika in relation to tectonic movements. Congr. Intern. Géogr., Lisbon, 1949, 2, p. 233–241.

Te Punga, M., 1956. Altiplanation terraces in southern England. Biol. Periglac., 4, p. 331–338.

Thomas, M. F., 1965. An approach to some problems of landform analysis in tropical environments. In Essays in geography, J. B. Whitten and A. D. Wood (eds.), Austin Miller, Reading, England, p. 118–144.

———, 1965. Some aspects of domes and tors in Nigeria. Zeit. Geomorph., 9, p. 63–81.

Thompson, W. F., 1960–1961. The shape of New England mountains: Pt. 1, Appalachia, 33, p. 145–159; Pt. 2, p. 316–335.

Trendall, A. F., 1960. The formation of apparent peneplains by a process of combined laterization and surface wash. Zeit. Geomorph., 6, p. 183–197.

Tricart, J., 1961. Les Caractéristiques fondamentales due système morphogenetique des pays tropicaux humides. Inform. Géogr., 25, p. 455–469.

———, 1968. À propos de la genèse des glacis. Bull. Assoc. Géogr. Fran. l'Étude quaternaire, Paris, 5(17), p. 316–318.

———, 1960. Le Modèle des régions sèches: Fasc. I, Le milieu morphoclimatique, les mécanismes morphogénétiques des régions sèches, Société d'édition d'enseignement supérieur, Paris.

Tuan, Y.-Fu, 1959. Pediments of southeastern Arizona. Univ. Calif. Publ. Geog., 13, 140 p.

Twidale, C. R., 1956. Chronology of denudation in northwest Queensland. Bull. Geol. Soc. America, 67, p. 867–882.

———, 1959. Some problems of slope development. J. Geol. Soc. Australia, 6, p. 131–147.

———, 1962. Steepened margins of inselbergs from northwestern Eyre Peninsula, South Australia. Zeit. Geomorph., 6, p. 51–69.

———, 1964. A contribution to the general theory of domed inselbergs. Inst. Brit. Geogr. Trans., 34, p. 91–113.

———, 1967. Origin of the piedmont angle as evidenced in South Australia. J. Geol., 75, p. 393–411.

———, 1972. The neglected third dimension. Zeit. Geomorph., 16, p. 283–300.

———, et al., 1964. General report on lands of the Leichhardt–Gilbert area, Queensland. Land Res. Ser., 11, p. 113–128.

Van der Boeck, J. M. M., and Van der Vaals, L., 1967. The late Tertiary peneplain of South Limburg (the Netherlands). Soil Surv. Papers, 3, Neth. Soil Inst., 24 p.

Van Tuyl, F. M., and Lovering, T. S. Physiographic development of the Front Range. Bull. Geol. Soc. America, 46, p. 1291–1300.

Ver Steeg, Karl, 1930. Windgaps and watergaps of the northern Appalachians; their characteristics and significance. N.Y. Acad. Sci. Ann., 32, p. 87–220.

———, 1940. Correlation of Applachian peneplains. Pan-Amer. Geol. 73, p. 203–210.

Wahrhaftig, C. A., 1965. A stepped topography of the southern Sierra Nevada (California). Bull. Geol. Soc. America, 76, p. 1165–1190.

Walther, J., 1915. Laterit in West Australien. Zeit. Deut. Geol. Ges., 67, p. 113–132.

————, 1924. Das Gesetz der Wüstenbildung, 4th ed. Quelle & Meyer, Leipzig, 421 p.

Ward, Freeman, 1930. The role of solution in peneplanation. J. Geol., 38, p. 262–270.

Warnke, D. A., 1969. Pediment evolution in the Halloran Hills, Central Mohave Desert, California. Zeit. Geomorph., 13(4), p. 357–387.

White, W. A., 1950. Blue Ridge Front: a fault scarp. Bull. Geol. Soc. Amer., 61, p. 1309–1346.

Willis, B., 1936. East African plateaus and rift valleys. Carnegie Institution of Washington, Publ. 470.

Wood, A., 1942. The development of hillside slopes. Geol. Assoc. London Proc., 53, p. 128–140.

Wooldridge, S. W. 1935. Erosion surfaces. Nature, 136, p. 897–898.

————, 1939. Peneplains and related landforms. Nature, 143(3622), p. 569–570.

Woolnough, W. G., 1918. The physiographic significance of laterite in Western Australia. Geol. Mag., 55, p. 385–393.

————, 1928. Origin of white clays and bauxite and chemical criteria of weathering. Econ. Geol., 23, p. 887–894.

Wopfner, H., and Twidale, C. R., 1967. Geomorphological history of the Lake Eyre Basin. In Landform studies from Australia and New Guinea, J. H. Jennings and J. A. Mabbutt (eds.), no. 7, p. 119–143, Australian National University Press, Canberra.

Wright, F. J., 1928. The erosional history of the Blue Ridge. Dennison Univ. J. Sci. Lab., 23, p. 321–344.

Zeissink, H. E., 1969. Mineralogy and geochemistry of nickeliferous laterite profile. Dept. Geol. Univ. Queensland, Australia, no. 10, p. 132–151.

AUTHOR CITATION INDEX

SUBJECT INDEX